AF581680

Heat Transfer in Energy Conversion Systems

Heat Transfer in Energy Conversion Systems

Editors

Alessandro Mauro
Nicola Massarotti
Laura Vanoli

MDPI • Basel • Beijing • Wuhan • Barcelona • Belgrade • Manchester • Tokyo • Cluj • Tianjin

Editors
Alessandro Mauro
Università degli Studi di Napoli "Parthenope"
Italy

Nicola Massarotti
Università degli Studi di Napoli "Parthenope"
Italy

Laura Vanoli
Università degli Studi di Napoli "Parthenope"
Italy

Editorial Office
MDPI
St. Alban-Anlage 66
4052 Basel, Switzerland

This is a reprint of articles from the Special Issue published online in the open access journal *Energies* (ISSN 1996-1073) (available at: https://www.mdpi.com/journal/energies/special_issues/heat).

For citation purposes, cite each article independently as indicated on the article page online and as indicated below:

LastName, A.A.; LastName, B.B.; LastName, C.C. Article Title. *Journal Name* **Year**, *Volume Number*, Page Range.

ISBN 978-3-0365-0750-7 (Hbk)
ISBN 978-3-0365-0751-4 (PDF)

Contents

About the Editors

Alessandro Mauro (Prof.) is Associate Professor of Applied Thermodynamics and Thermal Science and technical manager of LaTEC lab at the Department of Engineering of the University of Naples "Parthenope". His Ph.D. thesis was awarded with the Emerald Engineering Outstanding Doctoral Research Award in the category Numerical Heat Transfer & CFD. He is the author of more than 100 papers in the fields of thermo-fluid dynamics and energy systems.

Nicola Massarotti (Prof.) is a Full Professor of Fluid and Thermal Sciences. He received his Ph.D. at the University of Wales Swansea. He is the author of more than 150 papers in the fields of thermo-fluid dynamics and energy systems, and he is the head of the Laboratory of Thermo-fluid dynamics, Energy, and HVAC systems (LaTEC) at the Department of Engineering of the University of Naples "Parthenope".

Laura Vanoli (Prof.) is a Full Professor of Fluid and Thermal Sciences. She is the author of more than 150 papers in the fields of thermodynamic and thermo-economic analysis of advanced energy systems, energy saving, renewable energy sources, dynamic modeling of energy conversion systems, innovative energy conversion systems, and energy planning.

Preface to "Heat Transfer in Energy Conversion Systems"

In recent years, the scientific community's interest towards efficient energy conversion systems has significantly increased. One of the reasons is certainly related to the change in the temperature of the planet, which appears to have increased by 0.76 °C with respect to pre-industrial levels, according to the Intergovernmental Panel on Climate Change (IPCC), and this trend has not yet been stopped. The European Union considers it vital to prevent global warming from exceeding 2 °C with respect to pre-industrial levels, since this phenomenon has been proven to result in irreversible and potentially catastrophic changes. These climate changes are mainly caused by the emissions of greenhouse gasses related to human activities, and can be drastically reduced by employing energy systems, for both heating and cooling of buildings and for power production, characterized by high efficiency levels and/or based on renewable energy sources.

This Special Issue, published in the journal *Energies*, includes 12 contributions from across the world, including a wide range of applications, such as HT-PEMFC, district heating systems, a thermoelectric generator for industrial waste, artificial ground freezing, nanofluids, and others.

Finally, we wish to express our deep gratitude to all the authors and reviewers who have significantly contributed to this Special Issue. Sincere thanks also go to the editorial team of MDPI and *Energies* for providing the opportunity to publish this book and helping in all possible ways, especially Ms. Julyn Li for her support and availability.

Alessandro Mauro, Nicola Massarotti, Laura Vanoli
Editors

Article

Thermodynamic Modeling and Performance Analysis of a Combined Power Generation System Based on HT-PEMFC and ORC

Hyun Sung Kang [1,2], Myong-Hwan Kim [3] and Yoon Hyuk Shin [1,*]

[1] Eco-Friendly Vehicle R & D Division, Korea Automotive Technology Institute, 303 Pungse-Ro, Pungse-Myeon, Cheonan-Si 330-912, Korea; hskang@katech.re.kr

[2] Department of Mechanical Engineering, Korea University, 409 Innovation Hall Building, Anam-Dong, Sungbuk-Gu, Seoul 02841, Korea

[3] Hydrogen Fuel Cell Mobility R&D Center, Korea Automotive Technology Institute, 303 Pungse-Ro, Pungse-Myeon, Cheonan-Si 330-912, Korea; kimmh@katech.re.kr

* Correspondence: yhshin@katech.re.kr; Tel.: +82-41-559-3284; Fax: +82-41-559-3235

Received: 12 October 2020; Accepted: 23 November 2020; Published: 24 November 2020

Abstract: Recently, the need for energy-saving and eco-friendly energy systems is increasing as problems such as rapid climate change and air pollution are getting more serious. While research on a power generation system using hydrogen energy-based fuel cells, which rarely generates harmful substances unlike fossil fuels, is being done, a power generation system that combines fuel cells and Organic Rankine Cycle (ORC) is being recognized. In the case of High Temperature Proton Exchange Membrane Fuel Cell (HT-PEMFC) with an operating temperature of approximately 150 to 200 °C, the importance of a thermal management system increases. It also produces the waste heat energy at a relatively high temperature, so it can be used as a heat source for ORC system. In order to achieve this outcome, waste heat must be used on a limited scale within a certain range of the temperature of the stack coolant. Therefore, it is necessary to utilize the waste heat of ORC system reflecting the stack thermal management and to establish and predict an appropriate operating range. By constructing an analytical model of a combined power generation system of HT-PEMFC and ORC systems, this study compares the stack load and power generation performance and efficiency of the system by operating temperature. In the integrated lumped thermal capacity model, the effects of stack operating temperature and current density, which are important factors affecting the performance change of HT-PEMFC and ORC combined cycle power generation, were compared according to operating conditions. In the comparison of the change in power and waste heat generation of the HT-PEMFC stack, it was shown that the rate of change in power and waste heat generation by the stack operating temperature was clearly changed according to the current density. In the case of the ORC system, changes in the thermal efficiency of the ORC system according to the operating temperature of the stack and the environmental temperature (cooling temperature) of the object to which this system is applied were characteristic. This study is expected to contribute to the establishment of an optimal operation strategy and efficient system configuration according to the subjects of the HT-PEMFC and ORC combined power generation system in the future.

Keywords: high temperature proton exchange membrane fuel cell; thermal management; organic rankine cycle; plate heat exchanger; waste heat recovery; cooling system; thermodynamic modeling

---

1. Introduction

Today, the need to expand power generation systems utilizing eco-friendly and waste heat energy to tackle climate changes is increasing, and active research on hydrogen fuel cell generation (electricity generation) and a cogeneration system capable of utilizing waste heat is being done. Unlike engine or boiler-based power generation systems that generate power through a combustion process that uses fossil fuels to produce thermal energy and emission, hydrogen fuel cell systems is an eco-friendly power generator of electricity, thermal energy, and water through the chemical bonding process of hydrogen and oxygen.

The Solid Oxide Fuel cell (SOFC) based micro-cogenerative power system is being actively researched, and modeling research for predicting appropriate operating conditions is being considered for important research project purposes. Arpino et al. investigated the factors that influence the measurement uncertainty for combined heat and power design using SOFC [1]. In addition, they studied the correlation between the 0D model of those SOFC-based systems and the collected data. Additionally, an effective thermal management strategy through fuel utilization adjustments was presented for optimizing cogenerative power system operation [2]. Duhn et al. conducted an analytical study of the cooling plate design to improve operational efficiency by ensuring the pressure drop uniformity of the gas distributor of the SOFC system [3]. As described above, in a fuel cell-based power generation system having a high operating temperature, optimum control of the working fluid is important in addition to proper operating temperature and pressure drop formation in order to improve the efficiency and performance of the system.

In the case of High Temperature Proton Exchange Membrane Fuel Cell (HT-PEMFC), there is an advantage in that it can utilize waste heat at a relatively high temperature (150 °C or higher) with highly efficient power generation. In order to secure the power efficiency and reliability of such HT-PEMFC, a thermal management system is essential to maintain a high operating temperature [4,5]. As the operating temperature of the stack must be kept constant, the stack coolant must be used within a controlled temperature. Consequently, a thorough examination of the appropriate operating range of waste heat utilization (heat exchange) reflecting the respective stack thermal management and system control thereto, should be performed for the optimal design of cogeneration using HT-PEMFC and waste heat recovery [6,7].

Among the fuel cells, the PEMFC exhibits a relatively high power density and power efficiency, and it can minimize noise and residual emissions. It is divided into Low Temperature Proton Exchange Membrane Fuel Cell (LT-PEMFC) with an operating temperature of 60 to 80 °C and HT-PEMFC with an operating temperature of 100 °C or more. The power efficiency of HT-PEMFC appears to be approximately 45 to 60% [8]. Currently, research is being carried out on the cogeneration system suitable for each operation characteristic of each PEMFC type [9]. The advantage of HT-PEMFC is the simplification of the water management device configuration due to the high operating temperature as well as the generation of highly useful waste heat. Specifically, if the liquid cooling system is applied to HT-PEMFC thermal management with an operating temperature of 100 °C or higher, waste heat exchange with higher utilization is possible, which is advantageous for cogeneration and trigeneration systems [10]. Najafi et al. compared the performance and efficiency of the HT-PEMFC trigeneration system according to operation strategies during a certain operating period while the research team carried out a study on a trigeneration system to which both LT-PEMFC and HT-PEMFC were applied. Furthermore, research on warm-up strategies to quickly increase the stack operating temperature when the HT-PEMFC starts up was conducted [11]. Thus, based on the previous studies, it seems that the HT-PEMFC-based cooling and heating system can be selectively applied according to the operation strategy and subject.

It is important to secure the performance and efficiency of the waste heat recovery system in order to expand the subjects for application and functionality of the combined power generation system using such HT-PEMFC. This is one of the most important factors in selecting a target building and power system to secure electric energy with high utilization at a certain level depending on the operating environment [12,13]. Therefore, today, active research on Organic Rankine Cycle (ORC) system using stack waste heat energy in addition to fuel cell systems is being done [14]. Dickes' research team experimentally examined the temperature distribution of the working fluid for power generation in the evaporator heat exchanger in the ORC power system, and Jang's research team conducted a study on the performance of the compact ORC system at the 1 kW-level using a heat source in the range of 100 to 140 °C [15,16]. In addition, Jeong's research team conducted a study on the heat exchange performance and characteristics of the plate heat exchanger for each working fluid operating condition applied to the ORC system [17]. S.C. Yang et al. conducted a pilot study on an ORC power generation system at the 3 kW-level, capable of utilizing waste heat at 100 °C [18]. In the case of HT-PEMFC, the inlet temperature and mass flow rate of the coolant must be kept relatively constant in order to secure appropriate power efficiency at a relatively high operating temperature [19]. This is a limiting factor that must be reflected in the operational design of waste heat utilization systems such as ORC power generation and is the reason for the need to optimize the integrated system linked to the stack thermal management system. To this end, it may be useful to introduce a HT-PEMFC and ORC power generation integrated system modeling, as well as a confirmation and verification process of power generation performance and efficiency range according to the application subject and operating conditions.

In this study, through an analytical method based on the existing HT-PEMFC and ORC power generation system model, the power generation performance and efficiency range according to the operating conditions of the HT-PEMFC and ORC combined power generation system considering stack thermal management were confirmed, and the rate of change of power generation performance (effect on power generation performance change) for each control factor for the combined power generation system was presented. For this, a system analysis was conducted to predict system performance and efficiency according to changes in operating conditions such as stack operating temperature, current density, and ORC cooling temperature for a combined power generation system composed of a lumped thermal capacity model.

2. System Description Based on HT-PEMFC and ORC

Figure 1 shows the model composition for performance prediction and comparative analysis of the combined system consisting of the HT-PEMFC subsystem for cooling of the HT-PEMFC and the ORC subsystem for waste heat recovery power generation. The HT-PEMFC subsystem and the ORC subsystem, each with a fluid flow diagram, share an evaporator. The coolant of the HT-PEMFC subsystem was selected as Tri-ethylene glycol (TEG) since its phase does not change at the operating temperature of HT-PEMFC, which is 423~463 K. It follows the black solid line in Figure 1. The HT-PEMFC subsystem consists of an auxiliary heater/cooler, thermal storage, 3-way valve, cooling pump, and evaporator heat exchanger. The auxiliary heater/cooler is configured to keep the inlet temperature of PEMFC constant regardless of the influence of the ORC subsystem when the thermal power of the HT-PEMFC and the heat supplied to the ORC is not the same during the initial system startup. The 3-way valve is configured to meet the same flow conditions as the thermal power of HT-PEMFC and the heat exchange amount of the ORC evaporator. The cooling pump was controlled so that it would meet the flow condition in which the temperature difference between the inlet and outlet of HT-PEMFC satisfies 5 K.

To simplify the analysis of this system, the study followed subsequent assumptions:

1. All equipment of the system follows the lumped model and ignores heat loss.
2. The pressure loss in the pipe through which the stack coolant and working fluid travel is ignored.
3. Temperature and cell voltage are evenly distributed over the entire electrode of HT-PEMFC, and the reaction gas mixture is an ideal gas fluid.
4. The cathode charge transfer coefficient and the anode charge transfer coefficient are the same.
5. The isentropic efficiency of the expander is 60%, and the overall efficiency of the refrigerant pump is 50%.
6. In the condenser of the ORC system, the working fluid is sufficiently subcooled, and the existing superheat of the evaporator is 5 K.

7. The pressure loss of the evaporator reflected only the pressure loss on the heat source side, and only the loss due to friction was considered.

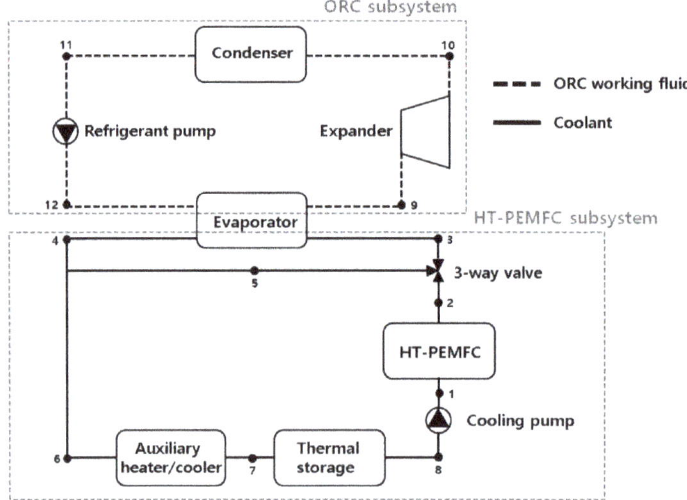

Figure 1. Schematic diagram of combined system.

3. Combined HT-PEMFC and ORC System Modeling

3.1. Thermodynamic Model of HT-PEMFC

The HT-PEMFC model applied in this analysis can be used by setting the operating temperature range of 120 to 200 °C, the cathode stoichiometric ratio range of 2 to 5, and the CO concentration range of 0.1 to 10%. Using a previously studied model as a reference [20,21], it is a one-dimensional isothermal model, HT-PEMFC based on PBI, and the electrochemical reaction in the fuel cell is as follows [22]:

$$Anode\text{: } H_2 \rightarrow 2H^+ + 2e^- \quad (1)$$

$$Cathode\ 2H^+ + \frac{1}{2}O_2 + 2e^- \rightarrow H_2O \quad (2)$$

$$Overall\ reaction\ H_2(g) + \frac{1}{2}O_2(g) \rightarrow H_2(g) \quad (3)$$

The overall cell voltage of the stack is calculated through the overpotential loss acting on the cathode and the overpotential loss acting on the anode at the ideal standard potential as shown in the Equation (4). The overpotential is divided into the activation overpotential (η_{act}), ohmic overpotential (η_{ohmic}), and concentration overpotential (η_{conc}) [23]. The activation overpotential is affected by the Tafel equation and the charge transfer coefficient while the ohmic overpotential is affected by the thickness of the membrane and the catalyst. The last term, concentration overpotential, represents the effect of cathode stoichiometry.

The ideal standard potential (V_{ideal}) is calculated by the variation of Gibbs free energy (Δgf) through electrochemical reaction and Faraday Constant (F) as shown in Equations (5) and (6), but in this model, the open circuit voltage of a reference [20] is used.

$$V_{cell} = V_{ocv} - \eta_{act} - \eta_{ohmic} - \eta_{conc} \quad (4)$$

$$\Delta gf = gf_{H_2O} - gf_{H_2} - gf_{O_2} \quad (5)$$

$$V_{ideal} = -\frac{\Delta gf}{2F} \quad (6)$$

Activation overpotential acting on cathode and anode is obtained from the following equations:

$$\eta_{act} = \frac{RT_{cell}}{4\alpha_c F}\ln\left(\frac{I+I_0}{I_0}\right) + \frac{RT_{cell}}{\alpha_a F}sinh^{-1}\left(\frac{I}{2k_{eh}\theta_{h2}}\right) \quad (7)$$

$$\alpha_c = a_0 T_{cell} + b_0 \quad (8)$$

$$I_0 = a_1 e^{-b_1 T_{cell}} \quad (9)$$

where R is the universal gas constant, T_{cell} is the operating temperature, α_c is the cathode charge transfer coefficient, F is the faraday constant, I is the current density, k_{eh} is the hydrogen electro-oxidation rate constant, θ_{h2} is the hydrogen surface coverage, I_0 is the exchange current density, λ_{air} is the cathode stoichiometry ratio, and α_a is the anode charge transfer coefficient, and it is assumed to be the same as the cathode charge transfer coefficient.

Ohmic overpotential and concentration overpotential acting on the cathode is given by:

$$\eta_{ohmic} = R_{ohmic} I \quad (10)$$

$$R_{ohmic} = a_2 T_{cell} + b_2 \quad (11)$$

$$\eta_{conc} = \frac{R_{conc}}{\lambda_{air} - 1} I \quad (12)$$

$$R_{conc} = a_3 T_{cell} + b_3 \quad (13)$$

Linear regression was used for the cathode charge transfer coefficient, ohmic resistance (R_{ohmic}), and concentration resistance (R_{conc}), and the exchange current density was expressed as an exponential function type. The values of the regressions used are shown in Table 1.

Table 1. Numerical value for regressions used in the High Temperature Proton Exchange Membrane Fuel Cell (HT-PEMFC) model.

Parameters	Values	Unit
Charge transfer constant, a_0	2.761×10^{-3}	[K^{-1}]
Charge transfer constant, b_0	−0.9453	-
Limiting current constant, a_1	3.3×10^{3}	[A]
Limiting current constant, b_1	−0.04368	-
Ohmic loss constant, a_2	-1.667×10^{-4}	[Ω/K]
Ohmic loss constant, b_2	0.2289	[Ω]
Diffusion limitation constant, a_3	-8.203×10^{-4}	[Ω/K]
Diffusion limitation constant, b_3	0.4306	[Ω]

It was assumed that all cell unit performances of the HT-PEMFC were the same, and the electric power (W_{FC}) and thermal power (Q_{FC}) of HT-PEMFC were calculated in proportion to the number of cells (N_{cell}) and single cell active area (A_{cell}) as in Equations (14) and (15). Moreover, the power efficiency (η_{FC}) of HT-PEMFC can be obtained as in Equation (16) based on the lower heating value (LHV) of hydrogen. Table 2 shows the parameters used in the HT-PEMFC model

$$W_{FC} = N_{cell} V_{cell} I A_{cell} \quad (14)$$

$$Q_{FC} = N_{cell}\left(\frac{LHV}{2F} - V_{cell}\right) I A_{cell} \quad (15)$$

$$\eta_{FC} = \frac{W_{FC}}{N_{cell}\frac{LHV}{2F} I A_{cell}} \quad (16)$$

Table 2. Operating parameters and empirical parameters used in the HT-PEMFC model.

Parameters	Values	Unit
Open circuit Voltage, V_{ocv}	0.95	[V]
Number of cells, N_{cell}	880	-
Single cell active area, A_{cell}	300	[cm^2]
Operating temperature, T_{cell}	433	[K]
Current density, I	0.4	[A/cm^2]
Universal gas constant, R	8.314	[J/mol·K]
Faraday constant, F	96485.3	[C/mol]
Cathode stoichiometry ratio, λ_{air} [24]	3	-
Hydrogen electro-oxidation rate constant, k_{eh} [24]	1.63818	[A/cm^2]
Hydrogen surface coverage, θ_{h2} [24]	0.14212	-
Low heating Value of hydrogen, LHV	239.92	[kJ/mol]

The thermal power generated by the HT-PEMFC was assumed to be heat-exchanged with the coolant by the Dittus-Boelter Equation (17), and the heat generated by auxiliary devices such as the cooling pump was ignored.

$$Nu = 0.023Re^{0.8}Pr^{0.4} \tag{17}$$

Furthermore, the pressure drop on the coolant side of the HT-PEMFC was reflected by the curve fitting the pressure drop according to the flow rate based on the experimental value. In general, the pressure drop on the coolant side of the stack is dependent on the flow path design of the cooling plate, and in this study, the pressure drop test value of the most widely commercialized vehicle stack with a level similar to the reaction area was applied to the model.

$$\Delta P_{FC} = 4074 + 1.86 \times 10^6 Q_{FC} + 3.184 \times 10^9 {Q_{FC}}^2 \tag{18}$$

As a cooling pump model used to transport Triethylene glycol (TEG), which is a stack coolant, a commercial pump for cooling of a maximum 100 kW stack that has a performance curve as shown in Figure 2 was applied [25].

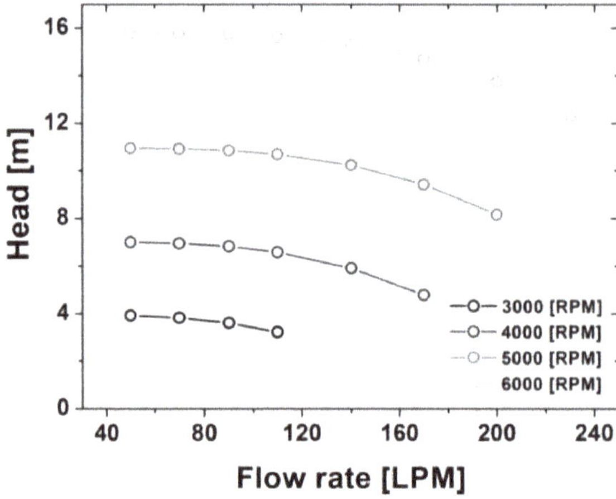

Figure 2. The performance curve of the cooling pump.

3.2. Thermodynamic Model of ORC

Figure 3 shows the T-s diagram of the ideal ORC cycle and conceptually shows each state and system. In Figure 3, the movement from point 12 to point 9 refers to the section in which the liquid working fluid changes to the gaseous state through the evaporator and is the section to recover waste heat from HT-PEMFC in the combined system. The vaporized working fluid generates power through the expander, which is the travel section from point 9 to 10, reducing the pressure. The working fluid with reduced pressure and temperature is liquefied in the travel section from point 10 to 11 through the condenser and maintains the pressure difference while transporting the liquefied working fluid through the pump.

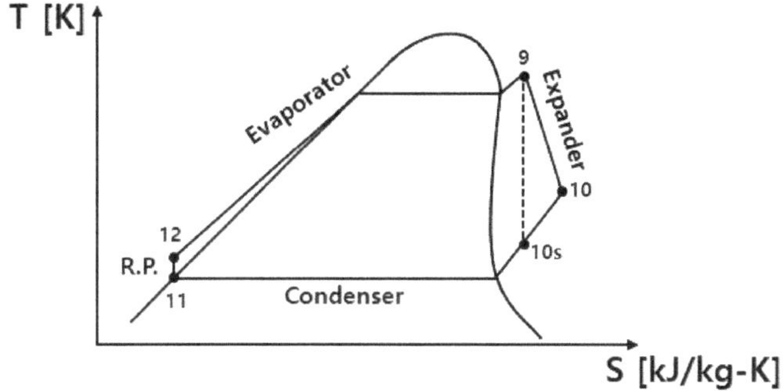

Figure 3. Thermodynamic T-s diagram for ideal Organic Rankine Cycle (ORC) cycle (R.P. is the meaning of refrigerant pump).

R245fa was selected as the working fluid used in this study considering the operating temperature range of the HT-PEMFC. The pressure on the evaporator side of the ORC power generation of the combined power generation system was set to 12 bar, and by setting the temperature range at the outlet side of the condenser to 293~308 K, the saturation pressure appropriate for the respective temperature was used.

All waste heat generated from the HT-PEMFC is heat-exchanged through the evaporator and is calculated as shown in Equation (19). $\dot{Q}_{eva}$ is the amount of heat exchange of the evaporator, and $\dot{m}_{ORC}$ is the mass flow rate of the working fluid of the ORC system. h means the enthalpy according to the temperature and pressure for each location indicated by each number on the T-s diagram.

$$\dot{Q}_{eva} = \dot{m}_{ORC}(h_9 - h_{12}) \tag{19}$$

The amount of power generated through the expander is W_{exp} and is calculated as in the Equation (20).

$$W_{exp} = \dot{m}_{ORC}(h_9 - h_{10}) \tag{20}$$

The amount of heat dissipated through the condenser is $\dot{Q}_{con}$ and is calculated as in the Equation (21).

$$\dot{Q}_{con} = \dot{m}_{ORC}(h_{10} - h_{11}) \tag{21}$$

The power consumption of the refrigerant pump is W_{rp} and is calculated as in the Equation (22), taking into account the overall efficiency η_{rp}.

$$W_{rp} = \frac{\dot{m}_{ORC}(h_{12} - h_{11})}{\eta_{rp}} \tag{22}$$

The net power generated through the ORC system is calculated by the power generated by the expander and the power consumed by the refrigerant pump as shown in the Equation (23).

$$W_{ORC} = W_{exp} - W_{rp} \tag{23}$$

The thermal efficiency of the ORC system is η_{ORC} and is calculated by the net power of the ORC and the endothermic reaction through the evaporator as in the Equation (24).

$$\eta_{ORC} = \frac{W_{ORC}}{\dot{Q}_{eva}} \tag{24}$$

3.3. Analytical Model of the Evaporator Heat Exchanger

A heat exchanger model for the evaporator was constructed to calculate the mass flow rate required by the heat source according to the mass flow rate of the working fluid of the ORC power generation system and the inlet temperature of the heat source (stack coolant) side. The plate heat exchanger used in the experiment in Jeong's study was used as a reference for the shape information of the respective evaporator and is shown in Table 3. [17]. The basic geometric characteristics of the chevron plate heat exchanger are shown in Figure 4.

Table 3. Operating parameters of the chevron plate heat exchanger.

Parameters	Values	Unit
Effective width of plate, L_h	0.111	[m]
Vertical distance between ports, L_v	0.466	[m]
Plate thickness, t	0.0004	[m]
Chevron configuration pitch, λ_p	0.007	[m]
Plate channel gap, g	0.002	[m]
Flow channel hydraulic diameter, D_h	0.003389	[m]
Effective heat transfer area, A_{hx}	0.06105	[m^2]
Plate chevron angle, β	35	[°]
Plate thermal conductivity, k	15	[w/m-K]
Surface enlargement factor, φ	1.18	-
Working fluid channel number, N_{wf}	21	-
Heat source channel number, N_{hs}	22	-

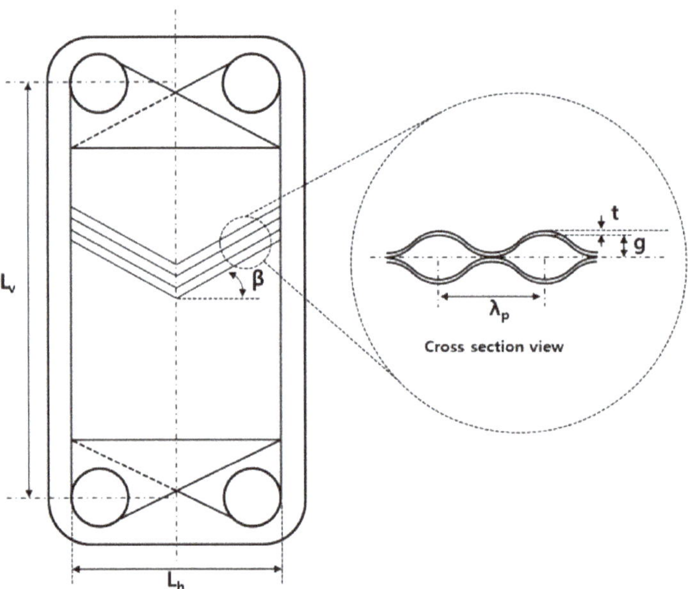

Figure 4. Basic geometric characteristics of chevron plate heat exchanger.

The overall heat transfer coefficient (U) of the Evaporator Heat Exchanger is calculated by the convective heat transfer coefficient of the working fluid (h_{wf}), the convective heat transfer coefficient of the heat source (h_{hs}), and the conduction heat transfer coefficient of the heat exchanger (k_p) as shown in the Equation (25).

$$\frac{1}{U} = \frac{1}{h_{wf}} + \frac{t}{k_p} + \frac{1}{h_{hs}} \tag{25}$$

The heat source is in a single-phase state in all operating areas, and the convective heat transfer coefficient in the single-phase state followed the Muley and Manglik correlation. Based on the Reynolds number, it followed the Equation (26) at 400 or below and followed the Equation (27) at 800 or more [26]. Moreover, the Nusselt number was interpolated using the transitional algorithm in the transition zone. Re_L is the Reynolds number in the liquid state, whereas Pr_L is the Prantle number in the liquid state. In addition, k_L is the heat transfer coefficient in the liquid state, while D_h is the hydraulic diameter of the plate heat exchanger.

$$h_{hs} = 0.44\left(\frac{\beta}{30}\right)^{0.38} {Re_L}^{0.5} {Pr_L}^{0.33}\left(\frac{k_L}{D_h}\right) \tag{26}$$

$$D_0 = 0.2688 - 0.006967\beta + 7.244 \times 10^{-5}\beta^2$$

$$D_1 = 20.78 - 50.94\varphi + 41.1\varphi^2 - 10.51\varphi^3$$

$$D_2 = 0.728 + 0.0543sin\left(\frac{\pi\beta}{45} + 3.7\right)$$

$$h_{hs} = D_0 D_1 {Re_L}^{D_2} {Pr_L}^{0.33}\left(\frac{k_L}{D_h}\right) \tag{27}$$

The convective heat transfer coefficient in the two phase region of the evaporator's working fluid follows the Yan and Lin correlation and is as shown in Equation (28) [27]. Re_{eq} is the equivalent Reynolds number, Bo is the boiling number, G_{eq} is the equivalent mass flux. G is the channel mass flux. q is the heat flux. x is the vapor quality. μ_L is the dynamic viscosity of the liquid. ρ_L is the density of the liquid, and ρ_v is the density of the vapor.

$$h_{wf} = 1.926{Re_{eq}}^{0.5}{Pr_L}^{0.33}Bo^{0.3}\left[1 - x + x\left(\frac{\rho_L}{\rho_v}\right)^{0.5}\right]\left(\frac{k_L}{D_h}\right) \tag{28}$$

$$Bo = \frac{q}{GA_{hx}} Bo = \frac{q}{GA_{hx}} \tag{29}$$

$$G_{eq} = G\left[1 - x + x\left(\frac{\rho_L}{\rho_v}\right)^{0.5}\right] \tag{30}$$

$$Re_{eq} = \frac{G_{eq}D_h}{\mu_L} \tag{31}$$

To compare and verify the analysis results based on the evaporator heat exchanger model constructed as described above and the previous experimental results, The validation analysis was conducted using the same conditions as in Jeong's study using R245fa as the working fluid and water as the heat source. As shown in Figure 5, the difference between Jeong's heat exchanger performance data that was previously studied and the analysis results in this study were within 3% approximately, and the analysis model constructed accordingly was confirmed to have a certain level of reliability.

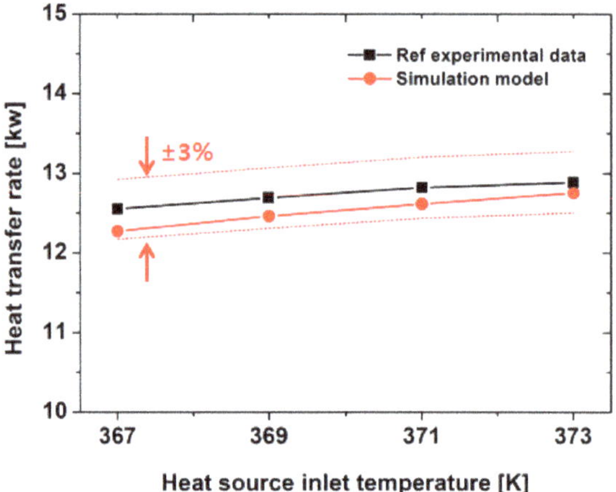

Figure 5. Validation of heat exchanger performance.

The pressure loss on the heat source side of the evaporator heat exchanger is ΔP_{eva} and is defined as the Equation (32) considering only the loss due to friction.

$$\Delta P_{eva} = f_{hs}\frac{L_v N_{hs} G^2}{2D_h \rho_L} \tag{32}$$

f_{hs} follows the Darcy friction factor and is expressed as in the Equation (33) [17].

$$f_{hs} = 72.5{Re_L}^{-0.045} \tag{33}$$

3.4. Performance of the Combined System (HT-PEMFC and ORC)

The total electric power produced by the combined system consisting of the HT-PEMFC and ORC is expressed as the sum of the power generated by the HT-PEMFC (W_{FC}), the power consumed by the cooling pump for the HT-PEMFC (W_{cp}), the power generated through ORC (W_{exp}), and the power consumed by the refrigerant pump (W_{rp}) as shown in the Equation (34).

$$W_{total} = W_{FC} - W_{cp} + W_{exp} - W_{rp} \tag{34}$$

In addition, the power efficiency of the combined system is expressed as the ratio of the total power of the combined system according to the *LHV* of HT-PEMFC, as shown in the Equation (35).

$$\eta_{system} = \frac{W_{total}}{N_{cell}\frac{LHV}{2F}IA_{cell}} \tag{35}$$

The whole system is analyzed using the commercial program Flomaster based on the law of conservation of energy (36), the law of conservation of mass (37), and the law of conservation of species (38).

$$\sum\left(\dot{m}h\right)_{in} + \sum\dot{Q}_{in} = \sum\left(\dot{m}h\right)_{out} + \sum\dot{Q}_{out} \tag{36}$$

$$\sum\left(\dot{m}\right)_{in} = \sum\left(\dot{m}\right)_{out} \tag{37}$$

$$\sum\left(\dot{m}x\right)_{in} = \sum\left(\dot{m}x\right)_{out} \tag{38}$$

4. Results and Discussion

4.1. Effect of Stack Temperature

In order to check the performance change according to the operating temperature and current density of the stack, the performance curves for each operating temperature (433 K, 443 K, 453 K, 463 K) and current density (0~0.5 A/cm^2) were verified. As shown in Figure 6a, as the temperature increased, the stack's single cell voltage and efficiency increased as well because of the decrease in the cell activation overpotential. In addition, it tended to decrease when the current density increased. Furthermore, the stack electric power and thermal power showed a tendency to increase as the current density increased, but the percentage of increase in the electric power decreased although the percentage of increase in the thermal power increased. As the temperature of the stack increased, the stack electric power increased thanks to the increase in power efficiency, whereas the stack thermal power decreased. When the current density was 0.1 A/cm^2 and 0.4 A/cm^2 at a stack temperature of 433 K, the single cell voltage was 0.66 V and 0.47 V, the stack power efficiency was 53.7% and 38.1%, the stack electric power was 17.6 kW and 50 kW, and the stack thermal power was 15.2 kW and 81.3 kW.

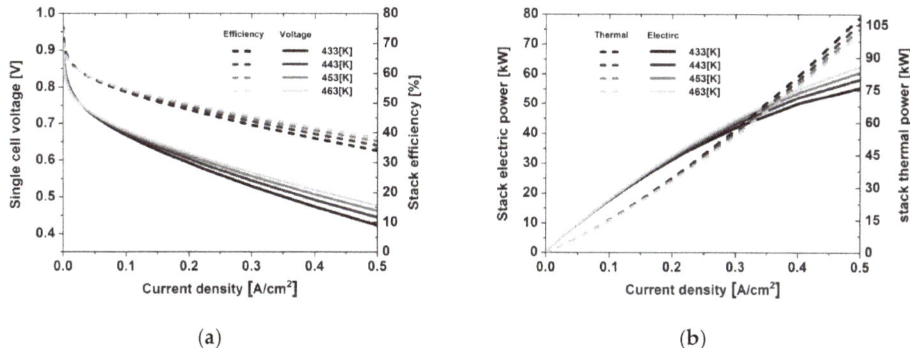

Figure 6. HT-PEMFC performance curve according to stack temperature and current density. (**a**) Single cell voltage and stack power efficiency; (**b**) Stack electric power and thermal power.

4.2. Effect of Working Fluid Mass Flow Rate in the ORC System

The performance change was analyzed by applying the evaporator model configured to calculate the performance according to the mass flow rate of R245fa, the working fluid of the ORC system, and the heat exchange amount of the evaporator. The evaporator pressure was selected as 12 bar considering the temperature level of the waste heat of the stack, and the condenser pressure was selected as 2.2 bar considering the extreme summer outdoor temperature. Moreover, the flow rate of the heat source (stack coolant, TEG) in which the superheat of the evaporator satisfies 5 K was calculated according to the corresponding inlet temperatures of 428 K, 448 K, and 468 K.

As shown in Figure 7a, as the mass flow rate of R245fa increased, the ORC net power increased linearly by the power consumption of the expander and the power consumption of the refrigerant pump. Although there was a change in performance according to the mass flow rate of the working fluid, the efficiency of the ORC system was relatively constant at about 7.69%, because all the conditions satisfied 5 K of superheat. In addition, as shown in Figure 7c, since the mass flow rate of the heat source (TEG) side where the superheat of the evaporator satisfies 5 K required a higher heat transfer coefficient as the inlet temperature of the heat source decreased, the mass flow rate increased. For the R245fa mass flow rate of 0.3 kg/s, the TEG-required mass flow rate was a maximum of 0.83 kg/s.

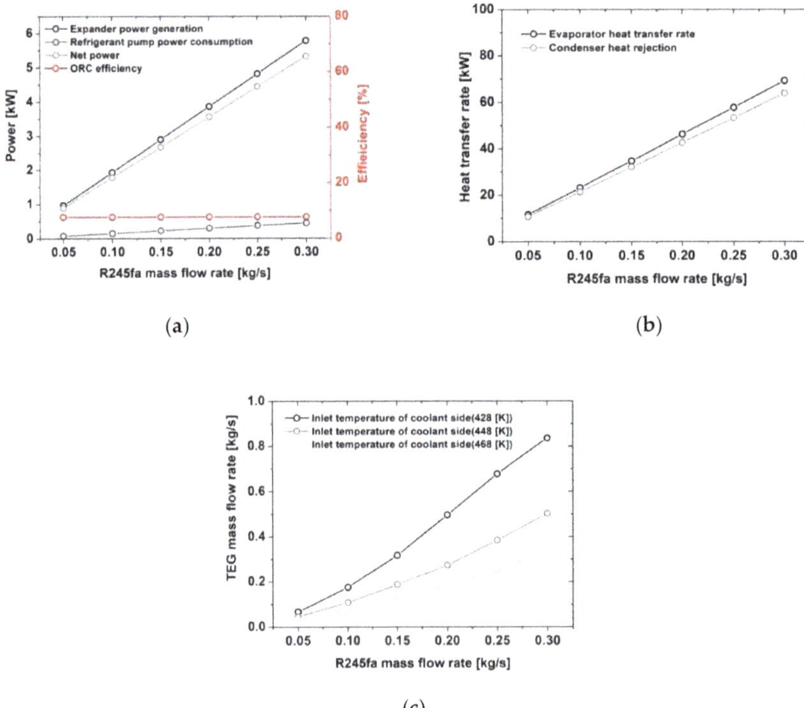

Figure 7. ORC system performance curve according to R245fa mass flow rate. (**a**) Electric power; (**b**) Evaporator and condenser heat transfer rate; (**c**) Mass flow rate of Tri-ethylene glycol (TEG) that satisfies superheat 5 K in the evaporator.

4.3. Effect of Stack Inlet Temperature in the Combined System

In order to analyze the combined system that merged the HT-PEMFC subsystem and the ORC subsystem, the transport pump controlled the mass flow rate so that the temperature difference at the inlet and outlet of the stack was 5 K. The mass flow rate was controlled through a 3-way valve so that all thermal power generated from the stack was exchanged with the evaporator of the ORC subsystem. The system performance was compared and analyzed after the inlet temperature conditions of the stack were selected as 433 K, 443 K, 453 K, and 463 K, and the current densities of the stack were 0.15 A/cm^2, 0.2 A/cm^2, 0.25 A/cm^2, 0.3 A/cm^2, 0.35 A/cm^2 and 0.4 A/cm^2.

As shown in Figure 8a, the mass flow rate of the cooling pump that satisfies the temperature difference between the inlet and outlet of the stack as 5 K is proportional to the current density. As the thermal power of the stack increased as shown in Figure 9d, the required convective heat transfer coefficient also increased, resulting in an increase in the mass flow rate that satisfied the operating conditions. As the inlet temperature of the stack increased, the physical properties of TEG changed, which influenced the formation of the mass flow rate of the cooling pump. The mass flow rate at the evaporator heat source (TEG) side of the ORC subsystem increased as the current density increased, but it decreased as the inlet temperature of the stack increased. The results of the mass flow rate of the cooling pump and the mass flow rate of the evaporator's heat source (TEG) according to the operating conditions, as well as the pressure drop of the stack and the pressure drop of the evaporator are shown in Figure 8c,d, respectively.

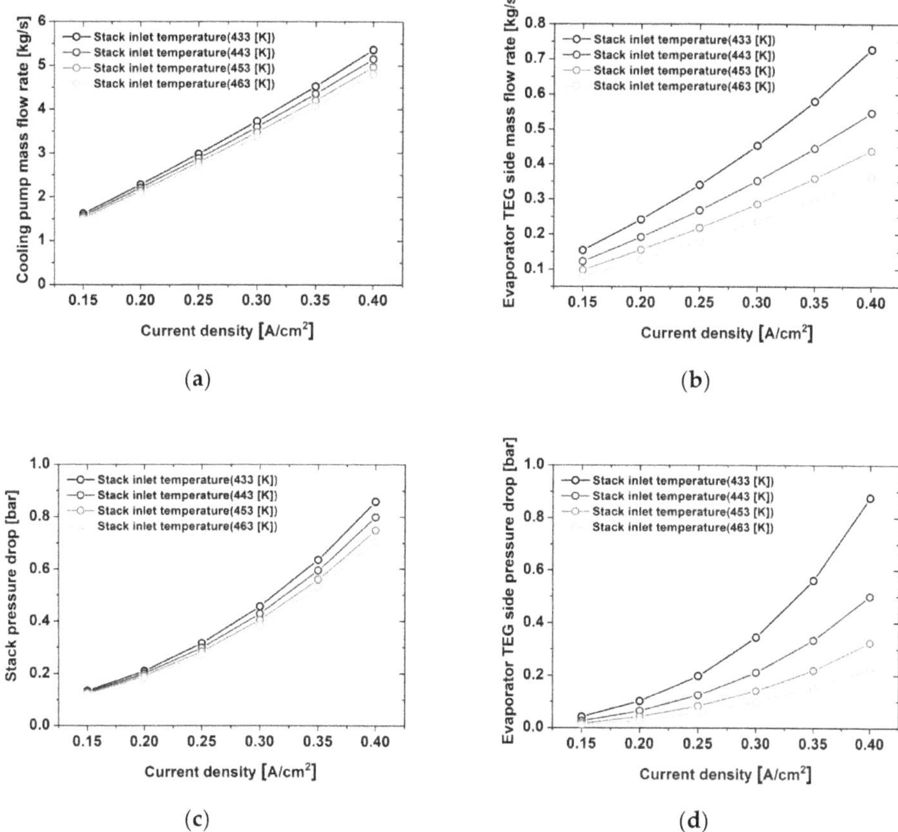

Figure 8. Validation trends of system loss and mass flow rate with stack inlet temperature. (**a**) Cooling pump mass flow rate; (**b**) Evaporator TEG side mass flow rate; (**c**) Stack pressure drop; (**d**) Evaporator TEG side pressure drop.

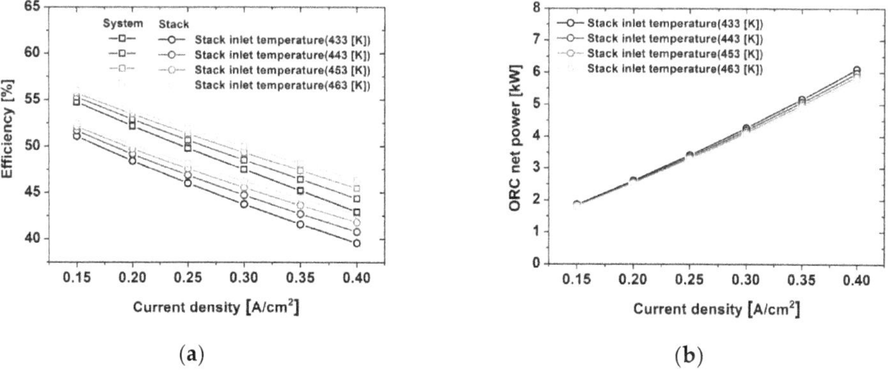

Figure 9. *Cont.*

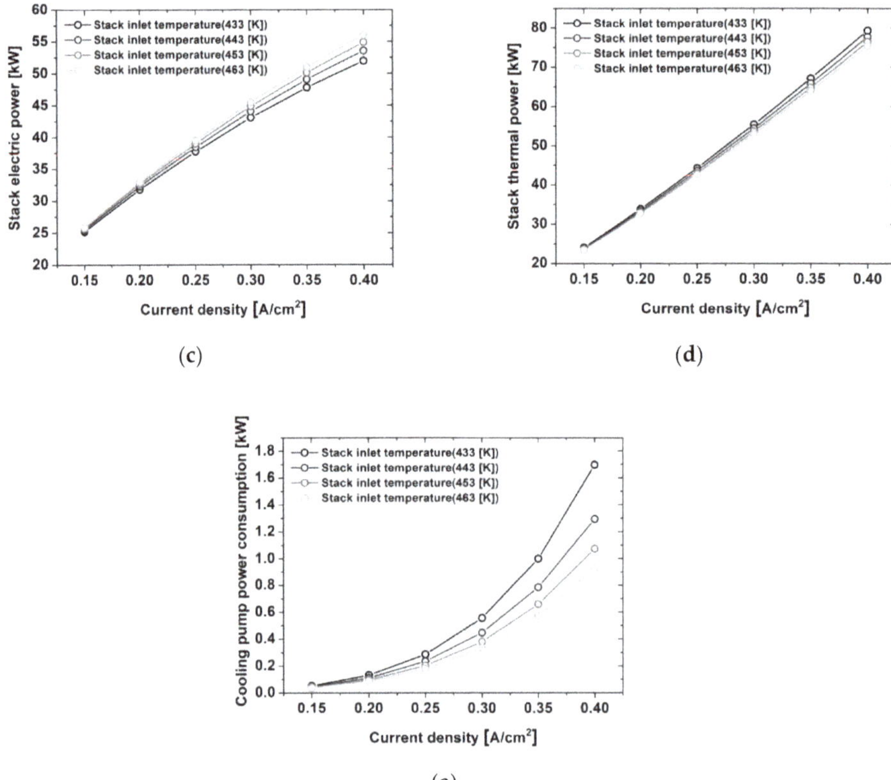

Figure 9. Validation trends of system performance with stack inlet temperature. (**a**) Combined system and stack power efficiency; (**b**) ORC net power; (**c**) Stack electric power; (**d**) Stack thermal power; (**e**) Cooling pump power consumption.

The efficiency of the combined system and the stack power efficiency are shown in Figure 9a, and the highest efficiency was shown as 56.03% and 52.45% at a current density of 0.15 A/cm^2 and a stack inlet temperature of 463 K. Additionally, the percentage increase in the combined system power efficiency compared to the stack power efficiency increased by up to 3.81% at a current density of 0.25 A/cm^2 and a stack inlet temperature of 433 K. In the case of the ORC net power, as the current density increased, the thermal power of the stack and the heat exchange amount of the evaporator increased, resulting in an increase in power generation. However, when the inlet temperature of the stack increased, the power generation decreased, and up to 0.3 kW decreased at a current density of 0.4 A/cm^2. In terms of the stack electric power, it reached a maximum of 55.96 kW at a current density of 0.4 A/cm^2 and 79.36 kW and a stack inlet temperature of 463 K as shown in Figure 9c, whereas in terms of the stack thermal power, it reached a maximum of 79.36 kW at a current density of 0.4 A/cm^2 and a stack inlet temperature of 433 K as shown in Figure 9d. As the current density increased, the power consumption of the cooling pump increased the required mass flow rate on the stack and the TEG side of the evaporator as shown in Figure 8, resulting in the increase in the corresponding pressure drop. As shown in Figure 9e, the power consumption of the cooling pump required a maximum of 1.69 kW at a current density of 0.4 A/cm^2 and a stack inlet temperature of 433 K.

Figure 10a shows the rate of change in the stack power, waste heat generation, and ORC power generation performance according to the difference of the stack coolant inlet temperature for each stack current density. As the current density is relatively higher, the rate of change in the stack power and waste heat generation amount according to difference of the stack inlet temperature clearly increases. Additionally, the rate of change in power generation of stack considering power consumption of cooling pump is increased by up to 20% at current density of 0.4 A·cm^{-2} compared to rate of change in power considering only the stack model. On the other hand, the rate of change in the stack power, heat generation, and ORC power generation performance according to the current density for each inlet temperature showed a relatively low difference as shown in Figure 10b. Based on these system analysis results, the stack inlet temperature of the HT-PEMFC power generation system is judged as an important operating condition that affects the power generation performance change characteristics. In addition, while the effect of the difference of the current density for each the stack inlet temperature is relatively constant, the effect of the difference of the stack inlet temperature is expected to increase as the current density increases. Additionally, when cooling actuators such as a water pump and loss factors are added, the effect of stack operating temperature is expected to increase. In the case of the ORC system, the rate of change in power generation performance according to the temperature and current density was relatively low in the operating temperature range of this stack model.

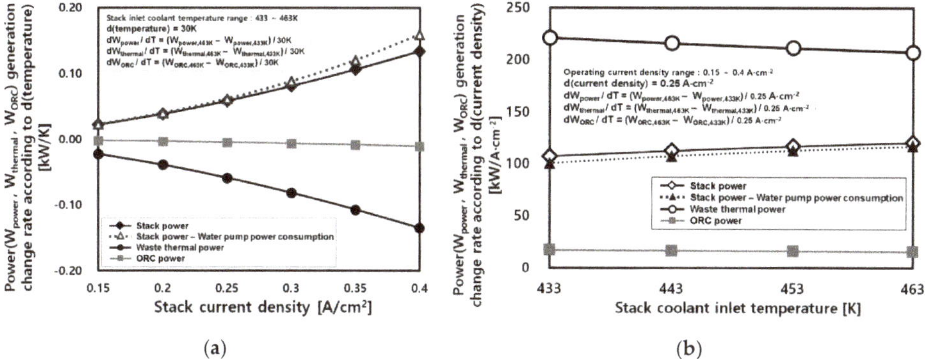

Figure 10. (**a**) The rate of change of stack power, waste heat generation, and ORC power generation according to operating temperature for each stack current density; (**b**) The rate of change in stack power, heat generation, and ORC power generation performance according to the current density for each stack coolant inlet temperature.

As shown in Figure 11, the system efficiency was compared excluding the stack power efficiency as a result of the coolant inlet temperature of the stack and condensing temperature of the working fluid formed at the condenser outlet for the ORC system. This is shown by excluding only the stack power generation from the overall efficiency of the combined power generation system. Through this, the efficiency changes of the ORC power generation system by pumps and heat exchangers excluding the stack were compared, and within the current density range, overall system efficiency except the stack tended to increase as the working fluid condensing temperature decreased. When the working fluid condensing temperature was 20 °C, the maximum efficiency was about 4.75%, which was a 25% increase compared to the case where the working fluid condensing temperature was 35 °C. Additionally, as the operating temperature of the stack increased, the deviation of the system efficiency except the stack tended to decrease relatively according to the change in current density. When the current density was 0.4 A/cm^2, the change in the system efficiency except the stack appeared to be the biggest according to the change of the stack operation temperature. This is believed to indicate that the influence of the operating temperature gradually increases under the power generation condition with the stack high load.

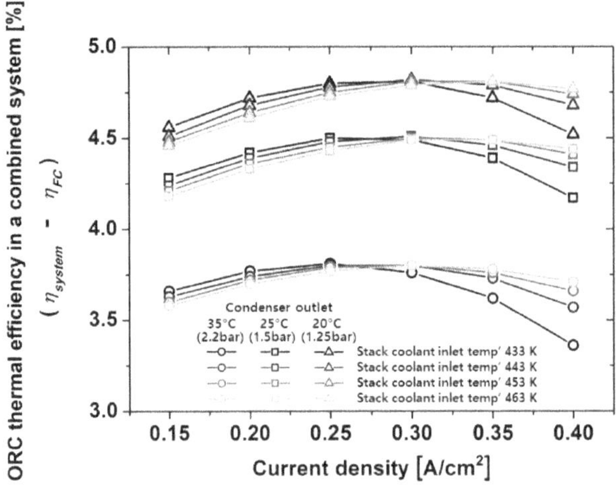

Figure 11. Difference between system and stack power efficiency with stack inlet temperature and condensing temperature.

5. Conclusions

In the case of HT-PEMFC, thermal management is formed as important as a relatively high operating temperature, and in the case of a heat exchange system that utilizes waste heat, since the operating range and strategies taking account of the thermal management of the stack must be selected, it is necessary to predict the power generation performance and efficiency according to the operating conditions. In this study, a model for a combined power generation system composed of a HT-PEMFC stack and an ORC power generation system was established, and the power generation performance and system efficiency were analytically compared according to the stack and ORC operating conditions. Each system was made of a model using the existing research contents and in the case of the plate heat exchanger for the Evaporator of the ORC system, which is the most important element for stack thermal management and waste heat recovery, reliability was secured by comparing the experimental results and the analysis results of the model. Through the analysis using the final combined power generation system model, the system power generation performance and efficiency were compared and predicted according to the operating temperature of the stack, the power generation load, and the ORC system working fluid condensing temperature, and the results are summarized as follows.

(1) For the analytical comparative study, modeling of each of the HT-PEMFC stack and ORC combined power generation system was conducted. In particular, in order to secure the reliability of the plate heat exchanger for the ORC power generation system, previous experimental results under the same operating conditions and the model-based analysis results established in this study were compared. The reliability of the combined power generation system model was secured through this process.

(2) Using the established combined power generation system model, the power generation performance and system efficiency of each stack and ORC system according to the power generation load and operating temperature of the HT-PEMFC stack were compared analytically. It is believed that the model has a higher degree utilization in a HT-PEMFC stack in which the higher the operating temperature within the allowable range, the higher the power generation and efficiency is. It also has higher degree utilization at a stack operating temperature where the waste heat that is proportional to the power generation load is relatively high. Furthermore, it is determined that the stack capacity and rated power generation section (current density range during power generation) need to be selected considering the target subjects as the amount

of waste heat becomes greater than the amount of power generated at points above a certain current density.

(3) And as the current density is relatively higher, the rate of change in the stack power and waste heat generation amount according to the stack operating temperature clearly increases. Additionally, the rate of change in power generation by operating temperature of stack with cooling pump is increased by up to 20% at the current density of 0.4 A·cm^{-2} compared to rate of change in power considering only stack model. Therefore, the operating temperature of the HT-PEMFC stack generation system is able to be considered as an important operating condition that affects the power generation performance change characteristics.

(4) In the model of the HT-PEMFC stack and ORC combined power generation system, comparative analysis was performed according to the operating temperature, power generation load (current density), and working fluid condensing temperature of the ORC system in order to compare the system efficiency excluding the stack, that is, the thermal efficiency of the ORC and subsystem that includes the stack cooling pump and heat exchanger, which change according to the operating conditions. As the operating temperature of the stack increased, the efficiency deviation of ORC and subsystem excluding the stack by the change in current density tended to decrease. Considering the energy load consumed by the thermal management part, it was shown that, under a certain current density, the lower the stack operation temperature was, and the more the efficiency of the ORC and subsystem except the stack improved. Moreover, as the working fluid condensing temperature decreased, the efficiency of the combined power generation system except for the stack tended to increase as well.

The HT-PEMFC stack and ORC combined power generation system require an appropriate operation strategy according to the target subjects and operating environment. To this end, this study constructed a combined power generation system model that considered the thermal management of the stack and the heat exchange process of waste heat and verified the operation range according to operating. It is believed that the results of this study will contribute to the selection of stacks and establishment of the strategies according to the target subjects and operation environments of the HT-PEMFC stack and ORC combined power generation system. In the future, an improvement on the model will be made regarding the target subjects of specific combined power generation, and analytical and experimental comparative studies will be conducted.

Author Contributions: H.S.K. designed the research, H.S.K., M.-H.K. and Y.H.S. discussed the results and contributed to writing the paper. All authors have read and agreed to the published version of the manuscript.

Funding: This research was supported by the Technology Development Program to Solve Climate Changes of the National Research Foundation (NRF) funded by the Ministry of Science, ICT & Future Planning (NRF-2016M1A2A2937158).

Conflicts of Interest: The authors declare no conflict of interest.

References

1. Arpino, F.; Massarotti, N.; Mauro, A.; Vanoli, L. Metrological analysis of the measurement system for a micro-cogenerative SOFC module. *Int. J. Hydrogen Energy* **2011**, *36*, 10228–10234. [CrossRef]
2. Arpino, F.; Dell'Isola, M.; Maugeri, D.; Massarotti, N.; Mauro, A. A new model for the analysis of operating conditions of micro-cogenerative SOFC unit. *Int. J. Hydrogen Energy* **2013**, *38*, 336–344. [CrossRef]
3. Duhn, J.D.; Jensen, A.D.; Wedel, S.; Wix, C. Optimization of a new flow design for solid oxide cells using computational fluid dynamics modelling. *J. Power Sources* **2016**, *336*, 261–271. [CrossRef]
4. Supra, J.; Janßen, H.; Lehnert, W.; Stolten, D. Temperature distribution in a liquid-cooled HT-PEFC stack. *Int. J. Hydrogen Energy* **2013**, *38*, 1943–1951. [CrossRef]
5. Lüke, L.; Janßen, H.; Kvesić, M.; Lehnert, W.; Stolten, D. Performance analysis of HT-PEFC stacks. *Int. J. Hydrogen Energy* **2012**, *37*, 9171–9181. [CrossRef]
6. Kandidayeni, M.; Macias, A.; Boulon, L.; Trovão, J.P.F. Online Modeling of a Fuel Cell System for an Energy Management Strategy Design. *Energies* **2020**, *13*, 3713. [CrossRef]

7. Dudek, M.; Raźniak, A.; Rosół, M.; Siwek, T.; Dudek, P. Design, Development, and Performance of a 10 kW Polymer Exchange Membrane Fuel Cell Stack as Part of a Hybrid Power Source Designed to Supply a Motor Glider. *Energies* **2020**, *13*, 4393. [CrossRef]
8. Jannelli, E.; Minutillo, M.; Perna, A. Analyzing microcogeneration systems based on LT-PEMFC and HT-PEMFC by energy balances. *Appl. Energy* **2013**, *108*, 82–91. [CrossRef]
9. Chen, X.; Li, W.; Gong, G.; Wan, Z.; Tu, Z. Parametric analysis and optimization of PEMFC system for maximum power and efficiency using MOEA/D. *Appl. Therm. Eng.* **2017**, *121*, 400–409. [CrossRef]
10. Bargal, M.H.; Abdelkareem, M.A.; Tao, Q.; Li, J.; Shi, J.; Wang, Y. Liquid cooling techniques in proton exchange membrane fuel cell stacks: A detailed survey. *Alex. Eng. J.* **2020**, *59*, 635–655. [CrossRef]
11. Najafi, B.; Mamaghani, A.H.; Rinaldi, F.; Casalegno, A. Long-term performance analysis of an HT-PEM fuel cell based micro-CHP system: Operational strategies. *Appl. Energy* **2015**, *147*, 582–592. [CrossRef]
12. Cardona, E.; Piacentino, A. A methodology for sizing a trigeneration plant in Mediterranean areas. *Appl. Therm. Eng.* **2003**, *23*, 1665–1680. [CrossRef]
13. Ziher, D.; Poredos, A. Economics of a trigeneration system in a hospital. *Appl. Therm. Eng.* **2006**, *26*, 680–687. [CrossRef]
14. He, T.; Shi, R.; Peng, J.; Zhuge, W.; Zhang, Y. Waste heat recovery of a PEMFC system by using organic rankine cycle. *Energies* **2016**, *9*, 267. [CrossRef]
15. Dickes, R.; Dumont, O.; Lemort, V. Experimental assessment of the fluid charge distribution in an organic Rankine cycle (ORC) power system. *Appl. Therm. Eng.* **2020**, *179*, 115689. [CrossRef]
16. Jang, Y.; Lee, J. Comprehensive assessment of the impact of operating parameters on sub 1-kW compact ORC performance. *Energy Convers. Manag.* **2019**, *182*, 369–382. [CrossRef]
17. Jeong, H.; Oh, J.; Lee, H. Experimental investigation of performance of plate heat exchanger as organic Rankine cycle evaporator. *Int. J. Heat Mass Transf.* **2020**, *159*, 120158. [CrossRef]
18. Yang, S.C.; Hung, T.C.; Feng, Y.Q.; Wu, C.J.; Wong, K.W.; Huang, K.C. Experimental investigation on a 3 kW organic Rankine cycle for low-grade waste heat under different operation parameters. *Appl. Therm. Eng.* **2017**, *113*, 756–764. [CrossRef]
19. Rosli, R.E.; Sulong, A.B.; Daud, W.R.W.; Zulkifley, M.A.; Husaini, T.; Rosli, M.I.; Majlan, E.H.; Haque, M.A. A review of high-temperature proton exchange membrane fuel cell (HT-PEMFC) system. *Int. J. Hydrogen Energy* **2017**, *42*, 9293–9314. [CrossRef]
20. Korsgaard, A.R.; Refshauge, R.; Nielsen, M.P.; Bang, M.; Kær, S.K. Experimental characterization and modeling of commercial polybenzimidazole-based MEA performance. *J. Power Sources* **2006**, *162*, 239–245. [CrossRef]
21. Korsgaard, A.R.; Nielsen, M.P.; Kær, S.K. Part one: A novel model of HTPEM-based micro-combined heat and power fuel cell system. *Int. J. Hydrogen Energy* **2008**, *33*, 1909–1920. [CrossRef]
22. Kang, H.S.; Shin, Y.H. Analytical Study of Tri-Generation System Integrated with Thermal Management Using HT-PEMFC Stack. *Energies* **2019**, *12*, 3145. [CrossRef]
23. Pukrushpan, J.T. Modeling and control of fuel cell systems and fuel processors. Ph.D. Thesis, University of Michigan, Ann Arbor, MI, USA, 2003.
24. Chang, H.; Wan, Z.; Zheng, Y.; Chen, X.; Shu, S.; Tu, Z.; Chan, S.H.; Chen, R.; Wang, X. Energy- and exergy-based working fluid selection and performance analysis of a high-temperature PEMFC-based micro combined cooling heating and power system. *Appl. Energy* **2017**, *204*, 446–458. [CrossRef]
25. Lee, H.S.; Lee, M.Y.; Cho, C.W. Analytic study on thermal management operating conditions of balance of 100 kW fuel cell power plant for a fuel cell electric vehicle. *J. Korea Acad. Ind. Coop. Soc.* **2019**, *16*, 1–6.
26. Muley, A.; Manglik, R.M. Experimental study of turbulent flow heat transfer and pressure drop in a plate heat exchanger with chevron plates. *J. Heat Transf.* **1999**, *121*, 110–117. [CrossRef]
27. Yan, Y.Y.; Lin, T.F. Evaporation Heat Transfer and Pressure Drop of Refrigerant R-134a in a Plate Heat Exchanger. *J. Heat Transf.* **1999**, *121*, 118–127. [CrossRef]

Review

Smart Asset Management for District Heating Systems in the Baltic Sea Region

Anna Grzegórska [1], Piotr Rybarczyk [1,*], Valdas Lukoševičius [2], Joanna Sobczak [3] and Andrzej Rogala [1]

1 Department of Process Engineering and Chemical Technology, Faculty of Chemistry, Gdansk University of Technology, Narutowicza 11/12 Street, 80-233 Gdansk, Poland; anna.grzegorska@pg.edu.pl (A.G.); andrzej.rogala@pg.edu.pl (A.R.)

2 Department of Energy, Faculty of Mechanical Engineering and Design, Kaunas University of Technology, 51424 Kaunas, Lithuania; valdas.lukosevicius@ktu.edu

3 Research and Development Joanna Sobczak, Różnowo 8, 14-240 Susz, Poland; joasobczak@outlook.com

* Correspondence: piotr.rybarczyk@pg.edu.pl

Received: 17 December 2020; Accepted: 6 January 2021; Published: 8 January 2021

Abstract: The purpose of this review is to provide insight and a comparison of the current status of district heating (DH) systems for selected Baltic Sea countries (Denmark, Germany, Finland, Latvia, Lithuania, Poland, and Sweden), especially from viewpoints of application and solutions of novel smart asset management (SAM) approaches. Furthermore, this paper considers European projects ongoing from 2016, involving participants from the Baltic Sea Region, concerning various aspects of DH systems. The review presents the energy sources with particular attention to renewable energy sources (RES), district heating generations, and the exploitation problems of DH systems. The essential point is a comparison of traditional maintenance systems versus SAM solutions for optimal design, operating conditions, and controlling of the DH networks. The main conclusions regarding DH systems in Baltic Sea countries are commitment towards a transition to 4th generation DH, raising the quality and efficiency of heat supply systems, and simultaneously minimizing the costs. The overall trends show that applied technologies aim to increase the share of renewable energy sources and reduce greenhouse gas emissions. Furthermore, examples presented in this review underline the importance of the implementation of a smart asset management concept to modern DH systems.

Keywords: Baltic Sea Region; district heating; DH network; smart asset management; smart grid

---

1. Introduction

The district heating sector constitutes an important segment of energy generation in all developed countries. The history of district heating (DH) systems starts in the 19th century when so-called 1st generation DH systems were introduced, based on the combustion of coal for steam generation. Together with industrial devolvement, district heating systems have been undergoing a transition from using fossil fuels as heat sources towards a continuously rising share of renewable, solar, and geothermal energy sources. Parallel to the development of heat generation systems, heat distribution systems, as well as their maintenance techniques, have been modernized and improved. One of the most important milestones in the management of DH systems was the introduction of pre-insulated pipelines coupled with electric wires connected to alarm systems, which enabled the monitoring of damage in the DH system. In fact, the damage to heat distribution systems, related mainly to the aging of the infrastructure, is posing the biggest social and economic threat related to the exploitation of DH systems. What is more, old infrastructure, e.g., channel pipes with damaged insulation, leads to heat losses, affecting the economy and efficiency of heat distribution. Thus, further development of modern and reliable management systems, allowing for the prediction of possible failures as well as the application of novel durable materials and solutions during the installation of new DH networks, are crucial factors determining their safe, reliable, and long-term exploitation.

Assuring the safety of heat distribution is a legal requirement resulting from energetic policies in all modern countries. Safe and reliable usage of such networks, from the perspective of the consumer, means the lowest possible number of incidents leading to interruption in the heat supply. From the perspective of DH systems' owners, reliable, safe, and efficient networks are those without heat loss and unexpected damage, as well as those that ensure possible high incomes. In order to ensure the implementation of such an approach, so-called smart asset management (SAM) methods need to be considered. By definition, smart asset management is an approach to maintenance via control, prediction, optimization, and selective refurbishment of assets with the aid of novel hardware and software solutions. The SAM approach is universal, e.g., it may be applied in the transportation sector [1]; however, it is well suited to the management of district heating systems. Nowadays, the SAM concept in the DH sector is in the early stage of research and development. However, in the next few years, DH systems are likely to become smart grids. According to Gao et al. [2], smart DH has more prominent benefits than the traditional DH system in numerous aspects, i.e., energy savings or troubleshooting, therefore it has high expansion potential and broad market perspectives in the future. Smart thermal grids, corresponding to smart electricity grids, exploit the potential of renewable energy sources (RES), thermal energy storage (TES), and prosumers and bi-directional networks, as well as the integration of intelligent management [3–5]. Smart heating and cooling use algorithm and model-based control methods. Currently, the SAM methods in DH mainly concern predictive control of the network, including forecasting of heat consumption and thermal comfort optimization [6,7]. Furthermore, smart optimization of pipe dimensions, insulation, and pipeline layout have been applied. Previous studies have emphasized the potential usage of nanofluids instead of water as a solution to improve heat-transfer efficiency for smart thermal grids [8–10]. Moreover, the implementation of intelligent metering and information and communications technology (ICT) may allow for counteracting and quick elimination of failures in DH networks [11]. However, this approach remains only briefly addressed in the literature. This review paper intends to present how the ideas of smart asset management may successfully be applied to district heating systems with a focus on selected Baltic Sea countries. Due to moderate and harsh winter conditions in several Baltic Sea countries, the selection of the proposed location as a field for in-depth analysis of heat distribution systems seem to serve as an excellent reference example for other locations in the world that have similar climate conditions.

In this paper, the search for reference information and data in further sections was done using the following databases: ScienceDirect, Web of Science, and Google Scholar. The following keywords were used during the search: District Heating System, District Heating Network, District Heating Generation, Smart Grid, Smart Asset Management, Smart Maintenance, Intelligent System, etc. To the best knowledge of the authors, this is the first review paper on the application of the smart asset management concept in district heating systems.

2. District Heating Systems

2.1. Energy Sources

The European Union's goal to address climate change and the global demand for energy is the provision of climate-neutral energy systems by 2050, which requires shifting from the utilization of fossil fuels towards the application of renewable energy sources [12]. District heating systems are classified as one of the most important sectors to be optimized with respect to profitability and environmental impact, meaning implementation of sustainable and decarbonized energy [13]. Even though heat generation exceeds 40% of the total energy use in Europe, only a small percentage, approximately 11–12%, of heat demand was delivered by district heating systems (DHS) in 2017 [14]. Nowadays, district heating systems are mainly supported by combined heat and power plants (CHP), commonly based on the combustion of fossil fuels, for instance, coal, natural gas, or petroleum [15]. Nevertheless, DHS presents flexibility in terms of feedstock; thus, incineration of municipal solid waste (MSW) or integration of waste heat from industrial sites is also possible [16,17]. A new direction is the implementation and extension of using renewable energy sources, including solar thermal energy and geothermal energy, as well as biomass fuels [18–20].

One of the ideas is "waste to energy" technology, which implies the use of waste as a feedstock to produce energy in the form of power and heat. This is achieved by means of pyrolysis, gasification, and incineration [21]. Thermal treatment of municipal solid waste is not only a way to minimize the number and area of landfills but also to develop technology to recover energy accumulated therein into valuable heat for DH applications. Another advantage of the "waste to energy" approach is the reduction of both fossil fuel consumption and greenhouse gas (GHG) emissions [22]. Tsai [23] estimated that MSW containing combustible material such as paper, textiles, woods, food waste, and plastic with 50% moisture possess a higher heating value (HHV) equal to about 2500 kcal/kg. A higher value was obtained by Rudra et al. [24], and the MSW energy content was about 22 MJ/kg, depending on the moisture content and amount of incombustible materials. Thus, a proper sorting process is needed to provide sufficiently high efficiency of the process. Sun et al. [17] evaluated the efficiency of solid waste energy recovery and CO_2 emission reduction. Researchers stated that the efficiency of steam production from waste incineration is better than power generation, with fossil fuel reduction being two-times greater for steam than for power generation. They predicted that energy recovery efficiency might be about 66% with a corresponding reduction of CO_2 emissions of 6.58×10 tons. However, they pointed out that the key aspect is the localization of incinerators in the vicinity of DH grids.

Developing biomass district heating (BioDH) systems is a potential method to take advantage of agriculture and forestry residues or energy crops as an alternative to fossil fuels [25]. Due to the fact that biomass has lower energy density compared to fossil fuels, it is not economically feasible to transport the biomass over 80 km [26]. Therefore, this technology is especially encouraging as a local source for small towns. Lower mass density requires higher storage volumes [27]. Additionally, seasonal availability and uncertain quantity are a barrier to expanding its usage. However, the obvious benefit is the reduction of carbon emissions. The biomass implemented in DH systems consists mainly of wood chips, sawdust, wood pellets, briquettes, and bio-oil [28,29]. Biomass may be converted into energy via thermochemical or biochemical processes [30]. Thermochemical processes generate heat, electricity, or biofuels, while biochemical operations provide liquid and gaseous fuels [31–33]. An important parameter determining the possible application is the biomass quality, i.e., moisture content, heating value, and bulk density, which all affect the energy generation efficiency. Generally, as in the case of MSW, biomass with lower moisture content is desirable. According to Quirion-Blais et al. [28], the average heating value of wood residues received by the plant is 19.3 GJ/kg, which is significantly higher than for municipal solid waste. Shahbaz et al. [34] concluded that synthetic natural gas (SNG) from 100 MW biomass gasification incorporated into a CHP plant provides about 27.35 MW for DH and 3.4 MW for power production.

The integration of industrial waste heat in DHS is a different approach to decrease the consumption of fossil fuels and GHG emissions. Because a significant amount of low-grade waste heat from manufacturers (mainly in the temperature range between 30 and 120 °C [35]) is dissipated into the atmosphere during the production processes, huge quantities of potential energy are lost [22,36,37]. Thus, it is deeply advisable to recover, collect, and reuse such industrial waste heat in DH systems. In 28 countries of the European Union, the entire waste heat potential in industry is predicted to be about 300 TWh/year, while one-third of the waste heat has a temperature level below 200 °C [36]. Waste heat at a high-temperature level is reused directly for production processes. Meanwhile, heat at an intermediate temperature level may be recovered by heat exchangers for district heating [35]. However, using this source of energy is principally limited to small districts and the energy is regularly reused by the factory itself or in buildings in the vicinity, without long-distance transfer. The size of the heating area is often insufficient and therefore the heat demand does not rise enough to match the waste heat available from the factory, resulting in an insufficient recovery ratio. Moreover, the waste heat recovery process is straightforward only if a single waste heat source is applied [38]. Fitó et al. [39] evaluated valorization of waste heat from the French National Laboratory of Intense Magnetic Fields. The researchers suggested that the demand-oriented design led to recovering waste heat at 35 °C with a heat pump and storage of 40-MWh, increasing the coverage of residential needs by up to 49%. Meanwhile, the source-oriented design with recovering waste heat at 85 °C without a heat pump and with storage of 40-MWh reached the highest recovery of waste heat of approximately 55%. Nowadays, utilizing geothermal energy (GE) as a renewable source for DH systems is being extensively investigated. GE is the cleanest form of energy obtained from the radioactive decay of minerals such as U, Th, and K, and has vast worldwide resources [40]. Geothermal resources provide base-load energy owing to their non-intermittent character, which is not affected by weather conditions [41]. In Europe, GE is directly utilized in 32 countries, accounting for more than 40% of the world's direct utilization [42]. Geothermal energy sources already produce more than the equivalent of four million tons of oil annually for DHS in the EU, equivalent to more than 15 GWth installed capacity, with geothermal heat pump systems contributing the biggest share [43]. Geothermal sources may be classified into three groups, according to the available temperature. These include shallow GE with low temperature, hydro-geothermal energy with low-medium temperature, and hot dry rock with higher temperature and greater energy density [41]. CHP systems might directly apply low-medium and high-temperature sources, while using low-temperature GE accumulated in-ground soil requires geothermal heat pump systems.

Solar thermal energy might be a notable source of renewable energy for DH systems. Solar district heating (SDH) refers to large-scale district heating systems equipped with solar collectors as part of the integrated heat supply. According to the analysis by Weiss et al. [44], up to 2016 the World installed area of solar collectors included in the DH system was about 1.65 million m^2, which represents only about 0.3% of all global solar collectors. Moreover, annual solar thermal energy yield in 2017 amounted to 388 TWh, which led to saving about 41.7 million tons of oil and 134.7 million tons of CO_2 [45]. The main advantages of SDH include the reduction of operating costs and greenhouse gas emissions, and is a potentially pollution-free method with 100% renewable source. Huang et al. [46] proposed several rules that guarantee the effective performance of SDH. The most significant included installation of SDH plants in a smaller town with the distance from the collector field to the DH network being within 200 m. Furthermore, the lower temperature of about 70–80 °C for DH supply and 35–40 °C for return are important parameters that are needed in order to reach a high efficiency of solar collector integration with the DH network. The authors proposed the connection of SDH with seasonal energy storage as an optimal solution.

2.2. District Heating Generations

District heating systems constantly evolves, and the driving force strives for lower heat and maintenance costs and higher efficiency with the sustainable development of non-fossil and renewable-based energy systems, simultaneously reducing greenhouse gas emissions. Figure 1 shows the timeline of development of DH generations with characteristic parameters.

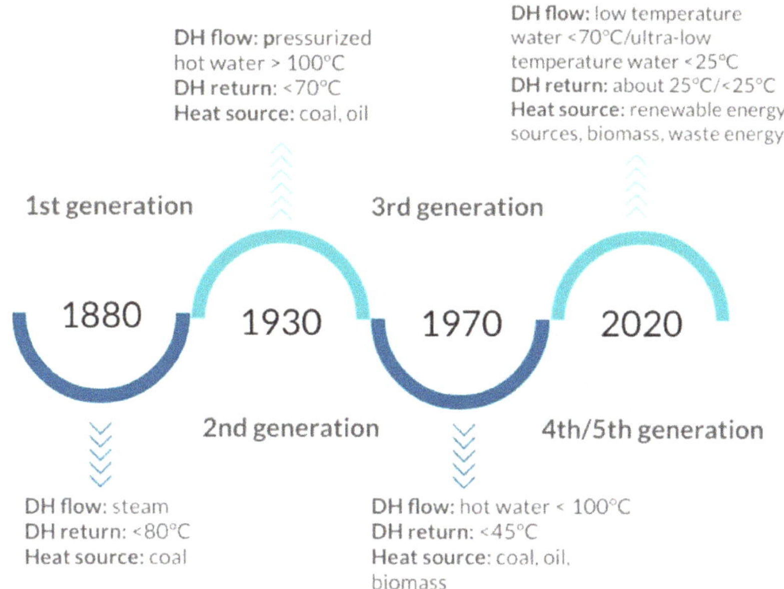

Figure 1. Timeline of district heating (DH) generations.

First generation DH systems were developed in the USA in the 1880s and were exploited until the 1930s. These DHS involved the combustion of coal for steam generation, which was the first heat transfer medium carried by the network to the consumer. Steam-based systems required a complex condensate structure; thus, servicing and maintenance costs were high. Moreover, high operating temperatures caused significant heat losses, especially with weak-insulated pipe systems, and therefore long-distance transport was not possible [47,48].

Second generation DH were introduced in the 1930s and was developed until the 1970s. These systems applied pressurized hot water with a temperature above 100 °C, instead of steam as a heat carrier. With this, heat transfer became safer and, together with better-insulated pipelines, bigger DH networks could develop. This system relied on a two-pipe closed-loop system, where the returning condensate could be reutilized or applied for lower grade heat demand. The significant aspect of the 2nd DH generation is the development of the combined heat and power plant (CHP), both for heat and electricity production, which guarantees substantial savings in energy loss and exploitation costs [47,48].

Developing the third generation DH systems started in the 1970s and nowadays is the most commonly used technology. This mode uses high-pressure water but with lower temperature than in the 2nd generation DH, i.e., below 100 °C. A standard 3rd generation DH network consists of underground pre-fabricated pipes, thinner than in the 2nd generation, with thermal insulation. Moreover, shell and tube heat exchangers were replaced by plate heat exchangers. Additionally, renewable energy sources started to be exploited for heat generation, including biomass [47,48]. These notable improvements led to boosting efficiency and lowering the costs of heat.

According to literature data, the fourth generation DH (4GDH) is being introduced in 2020 and it is predicted to dominate up to the year 2050. Lund et al. [47] formulated five major requirements for 4GDH systems. These are the ability to provide heat with low grid losses and the potential to recycle heat from low-temperature sources and incorporate renewable energy sources. Moreover, other important aspects are the integration of district heating technology to smart energy systems and the provision of suitable planning, cost, and management functions associated with its transformation into a future sustainable energy system. The 4th generation DH concept provides the development of CHP systems with the utilization of renewable energy sources such as biomass, solar thermal energy, geothermal sources, and more [49,50]. Fourth-generation district heating is able to provide low-temperature systems with supply temperatures below 50–60 °C along with significantly smaller pipe dimensions and improved insulation. Thus, the heat losses of the network may be reduced and the heat transmission capacity of the network may be increased [51]. Lowering the supply temperature is possible due to improved heat transfer in heat exchangers. Additionally, thermal energy storage is said to play an important role in 4GDH; therefore, the shift in time between thermal energy production and its utilization is not problematic [52]. Furthermore, the crucial innovation deals with intelligent control and metering of DH network performance as well as heat demand forecasting [53].

The idea of evolving the fifth generation district heating and cooling (5GDHC) networks is relatively new and has not been widely described in the literature. The proposed innovative method assumes a bidirectional low-temperature network supplying water with a temperature lower than 25 °C, close to ground temperature, in a so-called "ambient loop", in order to reduce heat losses to a minimum [54]. In this system, thermal networks include warm and cold pipes, whereby the temperature of water in a warm pipe is about 5–10 °C higher with respect to a cold pipe. Decentralized heat pumps using water from a warm pipe as a heat source enable sufficient water temperature for space heating purposes or domestic hot water. Then, cooled water from the evaporator is evacuated to a cold pipe. Conversely, chillers utilize water from the cold pipe and remove heated water to a warm pipe [55]. Furthermore, the system is facilitated with seasonal heat storage. Buffa et al. [14] suggested that negligible thermal losses, recovering low-temperature excess heat, and usage of renewable energy sources are one of the most essential characteristics of the 5GDHC system. Moreover, these 5th generation systems involve integrated smart network control using artificial intelligence [54]. This might be helpful to promote optimal control and maximize the benefits of exchanging heat or cold between connected buildings instead of direct transport of heat or cold from the supplier. Each consumer may transfer the excess of heat or cold to the network, and hence act as a producer [56]. In this way, the heat losses are limited, resulting in economic and ecological profits. Artificial intelligence solutions may be needed for the optimization of pipelines network configuration, connection of new customers, prediction of heat demand, or matching of the heat sources. Fifth generation DH has already been introduced in Plymount (UK) and Heerlen (Netherlands). Additionally, pilot projects are being tested in Germany. The Mijnwater DHC system in Heerlen (Netherlands) is one of the most advanced 5th generation DHC systems from a technological point of view. This system includes an urban smart DHC grid, incorporating several decentralized heat sources, a data processing center, and residual heat from supermarket refrigerators and from small scale industrial processes, along with the warm return flow from space cooling in the connected buildings. Nowadays, Mijnwater supports over 200,000 m^2 of building floor area [57].

3. Exploitation Issues of DH Networks

3.1. Low Efficiency of Heat Supply Systems

The principal exploitation issue associated with district heating is the low efficiency of the system. One of the circumstances determining this low-efficiency is very often the relatively high difference between the heat demand and the heat load of buildings connected to the DH network, which may arise from variable external temperatures during the heating season or differences in heat demand and accumulation between the buildings [2,58–60]. In this case, the main solution is the forecasting of the heat demand based on weather conditions [61] or application of an individual metering system. A different obstacle with DH efficiency refers to the low number of thermal storage systems [60]. The increase of the heat storage system enables storage of the extra energy produced by the DH plant during the low heat requirement period. Another limiting factor for the efficiency of DH systems is heat loss. Reduction of the temperature supply, application of pre-insulated pipes, the optimal length and diameter of the pipes (preventing over-dimensioning of pipes), and short heat transmission lines from heat plants to consumers have been proven to limit the level of heat losses form the DH network [62–64].

3.2. Heat Supply Problems

The heat supply problems are related to an excessively high flow of heat transfer media in the DH network relative to the pipe diameter. This results in problems with maintaining the sufficient differential pressure in the system [65,66]. Generally, this phenomenon occurs with too small diameter pipelines when more consumers are connected to the DH network. Elimination of this problem may result in lowering the supply temperature or reduction of the pumping power. Earlier, the before-mentioned difficulties were solved by an increase in temperature supply or an increase in the pipe diameter. However, nowadays, different solutions are taken into accounts, including the application of pumps, a local heat supply, or demand-side management.

3.3. Failures in DH Networks and Customers Installations

Another problem regarding the DH system are failures in the DH networks or customer installations, usually arising from outdated DH networks or lack of DH network inspections, as well as poor system monitoring [59]. Leakages and blockages are significant problems related to the DH pipeline system [51,67]. The main reasons for leakage are corrosion, equipment aging, the impact of ground loads, or mechanical factors [68]. Leakages cause loss of medium or energy, which affects the efficiency of heat transfer and related economic aspects. Furthermore, in situations using steam or hot water as a heat transfer medium, leakages are posing a risk to consumer safety and have a negative environmental impact. Blockage of flow is caused, for example, by local chemical or physical deposition or water freezing in cold climates. The blocking of pipelines causes energy losses and flow separation, which decreases the efficiency of the heat supply [69]. Because it is not possible to eliminate them completely, systems for early detection are needed to provide data about the location and size of the blockage.

Meanwhile, customer installation faults are described as the fault of the heat exchangers, control systems, actuators, control valves, or internal heating system of the customer. First of all, it is the fouling of the heat exchanger—deposits on heat transferring elements, that increase the resistance to heat transfer and flow in the heat exchanger. Moreover, the failures may be related to incorrect installation of heat exchangers, where the water flow pattern is co-current instead of counter-current. A common problem is related to leakages from heat exchangers. Defects may also concern temperature sensors. This may result in an excessive return temperature level or losses of heat transfer medium. Furthermore, problems with valves appear when these are oversized, thus affecting the flow when the valve position changes, making it difficult to set small flows through the valve [59].

3.4. Environmental Issues

Carbon dioxide released from the combustion of fossil fuels is a dominant reason for global warming. Combustion of coal emits more than 350 kg CO_2/MWh energy while natural gas emits more than 200 kg CO_2/MWh. According to data from 2018, in the EU, about 75% of heating and cooling is still produced from fossil fuels [70]. Therefore, following the European Union long-term objective of reducing greenhouse gas emissions by 80–95% by 2050 and the target of a 40% reduction by 2030 compared with the 1990 level [71], it is essential to implement renewable energy sources such as solar, wind, and geothermal energy or utilize waste heat with zero CO_2 emissions. Moreover, increasing the efficiency of DH by thermal storage may help to limit CO_2 emissions.

3.5. Occurrence of Legionella in LTDH Systems

Low-temperature water supply, i.e., below 55 °C, constitutes a favorable environment for the growth of *Legionella* bacteria, which cause Legionnaire's disease [72]. The simplest solution for preventing Legionella growth is a high-temperature regime and hot water circulation in households. However, in low-temperature DH generation systems, this solution is not satisfactory. Hence, Yang et al. [73] proposed an alternative design using decentralized substation and micro heat pumps or other sterilization techniques such as ionization, UV light, chlorine addition, or the application of filters.

4. Traditional Maintenance Approach Versus Smart Asset Management for District Heating Networks—If "Prevention is Better than Cure"?

Significant heat losses due to higher heat load than heat demand, leakages in the case of system damage, and difficulties with pipeline inspections leading to problems with finding the location of faults—this is how old, traditional DH Systems without monitoring, controlling, or alarm equipment might be described. In this system, maintenance and repair are based on the reactive (corrective) approach, which means that actions are taken when failure occurs. The new approach is called a smart district heat network—i.e., systems with a proactive and condition-based strategy. Such systems are characterized by optimization of heat load and intelligent control of the network, including information and communication technologies; effective automatic control in real-time in the areas of heat generation, distribution, and consumption; interactions with customers; early detection of errors; and a series of measurements and modeling operations. These solutions provide safe, efficient, and cost-effective DH systems. The most commonly proposed ideas and innovations for Smart DHS are presented below.

4.1. Smart District Heating (sDH) Bi-Directional System with Prosumers Idea and Thermal Energy Storage

In smart DH, different parts of the system, i.e., thermal source, pipelines, substations, and consumers, are connected and integrated into a long-distance management and intelligent system. According to Gao et al. [2], sDH consists of a control center, a communication network, a geographic information system (GIS), supervisory control, and a data acquisition system.

The sDH development strategy involves design optimization of the distribution system, i.e., network layout, pipe size and insulation, or underground depth [74]. All the above-mentioned features directly affect the construction cost, performance, heat loss, and a resulting differential pressure of the pipeline. In a sustainable system, consumers should be close to the site of heat generation to reduce the length of the distribution line. The depth of the DH pipe should be about 0.6–1.2 m underground, where the soil temperature is relatively stable. Furthermore, the application of the so-called hybrid insulation of pipelines with polyurethane and vacuum insulation panels are said to decrease the heat loss by 15–20% in comparison to pipe insulated with pure polyurethane. A reliable design mode applies optimization algorithms to determine the most suitable pipe diameters, insulation layer, or layout [74].

Novel bi-directional substations using a single circuit for both district heating and cooling proposed by 5th generation DH ideas are also part of the sDH strategy [4,75]. In the case of higher demand in the network for cooling than heating, the system circulates from a central plant in one direction. When more heat is required, the system circulates in the opposite direction.

Smart DH includes the approach of so-called prosumers—households that can be both consumers and producers of heat [5,76]. Prosumers possess heat sources (HS) and energy storage devices that allow them, based on the balance between their needs and capabilities, to regulate the amount of heat, depending on the energy received from the system [77,78]. These units, which collect heat or cold, are used to level off the constraints of short-term variation and to provide continuous energy supply. Thermal energy storage provides peak shaving and time-varying management, relieving renewable energy intermittence and lowering operating costs.

4.2. Demand Site Management with Heat Demand/Load Forecasting

Demand site management (DSM), so-called demand response, is one of the strategies used for the optimization of DH system operation. This technology is based on the forecast of the thermal request profile at the building level and the management of optimal schedules for heating systems or control strategies, thus reducing consumption and peak demand. Thermal request profiles illustrate how the heat is consumed by households, and they may be compared for a different location or time. DSM measures should be beneficial both for DH companies (heat consumption variations are balanced) and for consumers (lower energy costs) [79,80].

Guelpa et al. [81] proposed a multi-level thermal request prediction for buildings in the DH network for the case study in Turin. The method consists of the following two steps. The first is a black-box model—a smart prediction of the thermal demand evolution of the buildings based on data available in the buildings connected to the DHS, i.e., inlet and outlet temperatures and mass flow rate on the primary side. The second step is a network physical model, combining the demand of the buildings with fluid flow and heat transfer in the DH network. The results showed that the applied prediction model for buildings and the DH network was able to determine the request profiles with an error below 10%. Furthermore, the authors stated that the proposed model is appropriate for large DH networks due to its compactness—the low number of input data and simplicity of application and implementation together with low computational costs. Meanwhile, Idowu et al. [82] proposed machine learning for the prediction of heat demand in buildings connected to the DHS located in Skellefteå, Sweden. The heat load forecast models were generated using supervised machine learning techniques, in comparison to other methods, including support vector machine, regression tree, feed-forward neural network, and multiple linear regression. The proposed model takes into account parameters such as outdoor temperature, historical values of heat load, time factor variables, and physical parameters of DHS as its input. The obtained results revealed that the support vector machine was the most suitable machine learning method and has the least normalized root mean square error of 0.07 for the forecast horizon of 24 h.

From the theoretical viewpoint, application of DSM for thermal peak shaving according to Guelpa et al. [83] might have many advantages. These including the possibility of exploiting DH system capabilities, increasing potential for the connection of other buildings to DH networks without modifying the pipeline by making the request as low as possible, or a more effective combination of thermal and electrical energy production. Kontu et al. [84] presented an example of a case study that compared households with implemented smart control actions and customers who did not apply any DSM solutions, in two Finnish cities—Espoo and Vantaa. The benefits of DSM were evaluated from the perspectives of a point customer and a DH company. The results confirmed that customers with implemented DSM solutions decreased heat consumption and thus lowered their energy costs. However, the benefits were not observed by the DH companies, and higher short-term variations in the heat load were observed.

4.3. Multi-Heat Source (MHS) for DH with Integration of RES

Traditional DH systems are usually based on one heat source that provides heat to a single district. Other principals are implemented for Smart DH with the employment of multi-heat a source strategy. The use of intelligent control systems led to the selection and matching of the most profitable heat sources. These MHS systems can efficiently promote the utilization of local renewable energy sources and waste heat. Generally, MHS consist of a CHP plant as a major heat source with maximum heat capacity and additional, supporting heat sources such as solar thermal energy, wind energy, or waste heat, which are regulated to meet the different heat demands of consumers. In such a way, the effectiveness of DH systems may be improved. This complex system is a flexible and reliable solution, with a limited number of possible failures. If any problem occurs with one heat source, the system should still be able to operate [2,85–87].

4.4. DH Network and Household Integrated into Intelligent System

Smart DH systems are based on sensors and devices such as IT hardware and software, smart grid integrated communications, and metering hardware and software [88]. Various devices measure diagnostic signals that carry information about the actual state of the equipment. The signal goes through the primary selection of diagnostic information and then is fully processed and analyzed by the control center. Intelligent control systems might supervise all elements integrated into heat production, distribution, and consumption. DH plants and households are connected to one smart network [2]. This smart supervision manages to diagnose the technical condition of plants, pipelines, and equipment, identify all possible defects, faults, or dangerous situations in a continuous manner with a rapid response of alarm signals. Moreover, the prediction and planning of heat loads with respect to weather conditions and historical data are also included [89–91].

Zhong et al. [92] described the smart saliency analysis method for DHS pipeline leakage detection, applying remotely multi-source data from sensed infrared imagery, visible imagery, and geographic information system (GIS). Pipeline leakages are salient in the infrared imagery as the leaked material creates a local high-temperature area. The false alarm rate might be limited by combining multisource data analysis. GIS data from pipelines led to the removal of the potential of false leakage locating outside the pipeline area.

Ahn and Cho [93] proposed an intelligent controller to improve thermal comfort and reduce the peak demand in a DH system. The model, based on artificial intelligence with temperature and thermal comfort detectors, was used to maintain the desired room temperature. Application of the model resulted in a decrease of peak demands in order to optimize the DH distribution capacity. Comparative analysis proved that the model maintained thermal comfort level and decreased peak demands by 30%, compared to a conventional on/off controller.

Wang et al. [94] proposed a control model for the detailed estimation of heat loss for buried pipes and thermal deterioration of pipe insulations. The temperature, pressure, and flow rate meters are included to provide real-time data at the inlets and outlets of each heat source and substation. The complex heat loss profile along pipes might be defined by the presented model, and thus the accuracy of the location of damaged insulation might be enhanced. Moreover, the thermal deterioration of buried pipes may be evaluated conveniently; thus, aging or corrosion are more easily detected.

4.5. Selected Examples of Smart Solutions for DH Systems

Smart active box (SAB), proposed by the Arne Jensen AB company (Sweden), is an example of a smart solution for DH systems. It is a controller that enables leak detection by Delta-t® measurement based on a unique acoustic module. The SAB unit is installed in a DH chamber. Measurement data are collected from the sensors according to the desired frequency, ranging from every six minutes to once a day. The sound and signal analysis provide information about temperature and humidity in the chamber, contact temperature on the supply and return line, the flood alarms, and recordable audio files [95]. Such an approach allows for monitoring of the DH network between the adjacent chambers.

STORM controller (Belgium) is a smart district heating and cooling network management system based on self-learning algorithms. This controller includes three main modules, i.e., an energy forecaster, the operational optimization planner, and the demand-side management tracker, and uses three different control strategies: peak shaving, market interactions, and cell balancing [96].

FLEXYNETS (Germany) is a single system control with a centralized intelligent control platform. This platform includes a high-level controller that interacts with a strategic control platform, supporting the control decisions and strategies by weather forecast data in combination with simulation models of the entire system. Weather forecast data and simulation tools help to optimize the control strategies a day in advance [97].

Danfoss Link™ app (Denmark) enables the control of room temperature via an application on a smartphone connected to the central controller, room, and floor thermostats. The application enables intelligent adjustment of room temperatures to the schedule of the day with a lower temperature at night, including a weekly plan, holiday, or absent mode [98].

Vexve intelligent valve solution (Finland) operates underground without an external power supply and transmits the information to a cloud service. Through the application of an intelligent valve, the temperature, humidity, and network pressure variations are monitored continuously. Thus, leakage points linked to leak detection cables can be immediately localized [99].

4.6. Estimated Revenues from Using SAM Methods in District Heating Networks

It is generally assumed that there are three areas in which application of smart asset management may be beneficial to both district heating companies and consumers. These are social, technical, and economical areas and include increased durability and usage safety of the networks, reduction of number and frequency of failures, and economical benefits. The installation of SAM solutions in DH networks is aimed at prolonging the exploitational time of the network and can be used to plan retrofitting and maintenance operations. This, in turn, results in a much lower number of failures and shortens the time when consumers are deprived of a heat supply. The following analysis on the possible revenues may be proposed. According to operational data from the Lithuanian District Heating Association, the annual maintenance cost per kilometer of network is about 950–1150 EUR. Without proper preventive monitoring methods, failures in the network are hardly foreseen. Assuming the failure of a typical 100 mm tube, its repair may cost up to several thousand EUR. This is because the failure must be initially localized, then the pipe must be excavated, drained, repaired, and buried, and the original state of the landscape must be restored. It must be noted that this analysis excludes the costs of depreciation as well as the potential costs related to disappointed heat consumers.

The average maintenance costs related to the pipe replacement are in the range of 250–2350 EUR per 1 km, for a diameter of roughly 70–600 mm, respectively. In Lithuania, for example, the normative depreciation period of a DH section is 35 years. The savings from applying SAM solutions may be calculated on the basis of prolongation of the operational time. Thus, if the operational time is prolonged by 1 year, the resulting savings are about 1/35 of the initial cost. Swedish district heating company, Örresundskraft, reports that installing the moisture bands on the pipes in the Helsingborg region results in about 250,000 EUR saving per year [100]. However, the precise evaluation of the possible prolongation period of network exploitation and savings is very hard and requires new experimental data, coming from, e.g., already installed SAM solutions. Databases on the development of damages in DH networks (regarding both different types of failures as well as damage to pipe materials) as a function of their lifetime are missing. Currently, the German Energy Efficiency Association for heating, cooling, and CHP (AGFW) is building up such a database for Germany, Austria, and Switzerland (the so called DACH region) [101].

5. Overview of District Heating Systems in Selected Baltic Sea Countries

This section provides an overall overview of the current status of DH systems and strategies for the future development in seven Baltic Sea Countries, i.e., Poland, Sweden, Finland, Latvia, Lithuania, Germany, and Denmark. Table 1 presents a brief summary of the current status of DH systems in the above listed countries, while Table 2 presents selected projects in the field of district heating systems involving partners from the Baltic Sea Region, both completed and still on-going since 2016.

Table 1. Current status of district heating systems in selected Baltic Sea countries—a summary.

Country	Share of District Heating in Residential Sector, %	Main Heat Source/Heat Production	Share of RES, %	Current Status	Typical Supply/Return Temepratures, °C	Goals
Poland	42	Hard coal and coal product cogeneration plants	7.4	Transformation into 3rd generation system	135/70	By 2030: reduction of CO_2 emissions by 42%, increase of share of RES up to 50%; by 2050: 100% reduction of CO_2 emissions (with respect to reference year 1990)
Denmark	65	Biomass and solar	58.9	Transformation from 3rd to 4th generation system	80/40	By 2050: zero CO_2 emissions and fossil independence
Sweden	50.4	Biomass and waste heat	>65	3rd generation system	86/47	By 2045: CO_2—neutrality
Finland	38	Coal and wood-derived fuels	55	2nd and 3rd generation systems	65–115/40–60	Development of nuclear power and solar heating
Lithuania	>65	Biomass and municipal waste	>65	3rd and 4th generation systems	~85/45	By 2050: zero CO_2 emissions, 80% of RES share
Germany	13.8	Combined heat and power plants	12	3rd and 4th generation systems	~85/45	By 2050: 100% of RES share and 80% reduction of CO_2 emissions (with respect to reference year 1990)
Latvia	30	Central heat pumps, natural gas and biomass	47	2nd and 3rd generation sysmtems	~85–130/45–70	By 2050: CO_2 reduction by 80–95% (with respect to reference year 1990), fossil independence

RES—renewable energy sources.

Table 2. Summary of district heating (DH) projects ongoing from 2016 with participating countries from the Baltic Sea Region.

Project	Baltic Sea Region Project Partners	Years of the Project	Main Goals	Reference
LowTEMP (Low Temperature District Heating for the Baltic Sea Region)	Denmark, Finland, Germany, Latvia, Lithuania, Poland, Sweden	2017–2021	(1) Combining low-temperature DH solutions with energy supply systems (2) Promoting energy efficiency and the application of renewable energy sources (RES) with the reduction of CO_2 emission	[163]
BSAM (Baltic Smart Asset Management)	Lithuania, Poland, Sweden	2019–2022	(1) Distinguishing limitations and profits for the implementation of smart asset management (SAM) and digitalization of DH distribution networks (2) Developing nationally adapted methods for condition monitoring of district heating systems (DHS).	[164]

Table 2. *Cont.*

Project	Baltic Sea Region Project Partners	Years of the Project	Main Goals	Reference
HRE4 (Heat Roadmap Europe 4)	Finland, Germany, Poland, Sweden	2016–2019	(1)Improving at least 15 new policies at the local, national, or EU level for decarbonization of district heating and cooling (DHC) systems (2) The setting of how to save approximately 3,000,000 GWh/year of fossil fuels in Europe.	[165]
InDeal (Innovative Technology for District Heating and Cooling)	Finland, Poland	2017–2019	(1) Offering a novel platform that will manage a fair distribution of heating and cooling in the DHC network by real-time energy demand data collected using artificial intelligent meters and classifying the network's building demand. (2) Turning the current DHC systems into innovative high-level computerized DHC systems.	[166]
RELaTED (REnewable Low TEmperature District)	Denmark, Poland, Sweden	2017–2021	(1) Providing a novel idea of decentralized ultra low-temperature district heating (LTDH), which enables the adding of low-grade heat reservoirs (2) Following the approach of the smart grids, where heat production is decentralized, and consumers become prosumers.	[167]
THERMOS (Thermal Energy Resource Modelling and Optimisation System)	Denmark, Germany, Latvia, Poland	2016–2020	(1) Developing the methods, data, and tools to facilitate public authorities and other stakeholders to undertake a more sophisticated thermal energy system planning to incorporate real-world cost, benefit, and performance data	[168]
TEMPO (TEMPerature Optimisation for Low Temperature District Heating across Europe)	Germany, Sweden	2017–2021	(1) Creating low temperature (LT) networks for raised network efficiency with the integration of renewable and residual heat sources. (2) Developing innovative business models to promote LTDH network competitiveness.	[169]
W.E. District (Smart and local reneWable Energy DISTRICT heating and cooling solutions for sustainable living)	Denmark, Germany, Poland, Sweden	2019–2023	(1) Demonstrating DHC as a joined solution that employs the combination of RES, thermal storage, and waste heat recovery to provide 100% of the heating and cooling requirement. (2) Incorporating integrated control technologies and decision making to DHC systems.	[170]
ReuseHeat (Recovery of Urban Excess Heat)	Denmark, Germany, Sweden	2017–2021	(1) Demonstrating high-level systems allowing the recovery and reuse of excess heat available at the municipal level. (2) Development, monitoring, and evaluation of four large scale cases showing the technical utility and financial profitability of waste heat recovery.	[171]
Cool DH (Cool ways of using low grade Heat Sources from Cooling and Surplus Heat for heating of Energy Efficient Buildings with new Low Temperature District Heating Solutions)	Denmark, Sweden	2017–2021	(1) Supporting municipalities in planning and expanding new, effective district DHC systems, extend and renovate existing units to higher standards. (2) Designing LTDH systems with non-conventional pipe materials and controls inside buildings that link LTDH with the integration of locally generated renewable energy for low DH temperatures (40–65 °C).	[172]
Keep Warm (Improving the performance of district heating systems in Central and East Europe)	Germany, Latvia	2018–2020	(1) The growing energy effectiveness of DH systems by an acceleration of cost-effective investments in the modernization of DHS. (2) Reducing greenhouse gas (GHG) emissions by promoting a switch from fossil fuel to renewable sources.	[173]

5.1. Poland

Poland is the second, after Germany, greatest district heat producer in the European Union [102]. In 2017, the DH system satisfied 42% of the heat demand in the residential sector [103]. Polish DH networks provide heat for space heating for at least six months per year, depending on weather conditions. Nowadays, the highest share in DH is covered by hard coal and coal product cogeneration plants [103]. In 2015, the share of RES in district heating systems reached only about 7.4% [104]. Many Polish citizens still use individual coal-fired boilers and stoves. However, The Clean Air Program, implemented by the Polish Government, assumed thermal modernization of Polish households with the simultaneous replacement of heat sources during the period 2018–2029 to enhance energy efficiency and diminish the emission of dust and other pollutants related to the problem of smog [105].

Nowadays, the major challenge for Polish DH is a step by step transformation into modern 3rd generation systems, with temperature supply below 100 °C. The typical DH network supply temperature is 135 °C, while the return is 70 °C for a large district heating system [106]. Cenian et al. [107] described the case study in a Polish town, Łomża, where DH system thermal modernization was introduced. The first stage of modernization led to decreasing the DH supply temperature from 121 °C to 109.8 °C, with a reduction of about 14% of heat losses compared to the previous heating system and improvement in the hydraulic stability of the DH system. The oversized radiators, heat exchangers, and grids after thermal modernization of buildings are advantageous. An important element in Polish DH systems is thermal energy storage (TES) tanks that support ten Polish CHP plants, which provide heat source load-leveling for variable heat demand periods [108].

One of the directions of DH development in Poland is the use of geothermal energy as a heat source. Poland is one of the richest countries in Europe in terms of low-temperature geothermal resources. In Poland, four geothermal areas are distinguished, i.e., Polish Lowlands, Carpathian Province, Carpathian Foredeep, and Sudety Region [109,110]. The most valuable, from a heating point of view, are the areas of Carpathian Province as well as the Lower Cretaceous and Lower Jurassic reservoirs in the Polish Lowlands. The total installed capacity from six geothermal DH plants is 76.2 MWth [110]. The outflow water temperatures vary from 20 to 97 °C, while water reserves amount from several L/s to 150 L/s, and heat flux values range from 20 to 90 mW/m^2 [109]. The implementation of geothermal heat sources has played an important role in the improvement of air quality in Poland. Moreover, geothermal DH plants could operate at lower temperatures, offering a higher efficiency of heat transfer.

The major Polish DH goals assumed two time limits, the first by the year 2030, with a reduction of CO_2 emissions by 42% (compared to levels from 2016) and a share of 40% RES. The second limit is set at the year 2050, with the reduction of CO_2 emissions by 100% (compared to levels from 2016) and utilization of 100% renewable energy [111].

5.2. Denmark

Denmark is one of the most advanced Baltic Sea Region countries in terms of DH systems. In 2017, about 65% of its citizens' heat demand was satisfied by DH systems. Currently, various Danish DH networks are going through the transformation from 3rd generation (80/40 °C) to 4th generation (50–55 °C/25 °C) networks [112].

The share of renewable energy sources for DH in 2017 was equal to 58.9%, principally in the form of biomass [113]. However, Denmark is also a leader in both total installed capacity and number of large-scale solar DH plants. Furthermore, it is the first and only country with commercial market-driven solar DH plants. Until 2017, 110 solar heating plants, with more than 1.3 million m^2 of collector area, had been operating in Denmark, which represents more than 70% of the total worldwide large scale solar DH plants. Solar DH production is expected to achieve 6000 TJ in 2025 [114]. Moreover, an important part of the Danish solar DH systems is long term large heat storage, consisting of water pit storage, constructed in Marstal, Dronninglund, Vojens, and Gram, and boreholes in Brædstrup. The share of solar heating in a district heating system without heat storage is about 5–8% of the annual heat demand, while an application of diurnal heat storage reaches 20–25%. Application of solar energy is also an economically efficient solution, because the cost of solar heat is in the range of 20–40 EU/MWh, while the heating price of natural gas boilers is higher than 60EU/MWh [114].

According to Buhler et al. [115], excess heat from industry might be applied as a resource for the Danish DH systems. The results for the case study for Denmark revealed that about 1.36 TWh of DH demand could be covered annually with industrial excess heat from thermal processes, which is equal to about 5.1% of the current usage. Additionally, more than half of the heat might be used directly, without the application of heat pumps.

Another option used in Denmark is the application of large scale heat pumps in DH systems. This solution may provide stable and efficient heat supply, especially when wind or solar electricity production is high, and might be converted into heat, and thus replace the heat produced by fuel boilers. According to Lund et al. [116], Denmark is well able to introduce heat pumps, with potential heat sources located close to all DH areas and, moreover, seawater will play a significant role as a heat source for Denmark in the future.

A smart approach in DH management in Denmark is presented by Foteinaki et al. [117], who described the cooperation between three DH companies in the Copenhagen metropolitan area, coordinated by Varmelast Company. The main task of Varmelast Company is related to the preparation of the day-before heating plan, based on the DH forecast disclosed by companies, which considers fuel prices, operating and maintenance costs, hydraulic bottlenecks in the network, and other parameters. Moreover, the combined optimization of heat and power generation is aimed at providing maximum economic efficiency for the whole system. Each day, adaptation of the heating system is done, based on the consumption forecast, capacities, and power costs.

The main goals for DH systems approved by the Danish Government concern net-zero emissions of carbon dioxide by 2050 and fossil-independent DH systems before the year 2050 [118].

5.3. Sweden

DH systems were introduced in Sweden in 1948 [119]. District heating is a large industry in Sweden, satisfying about 50.4% of heat demand in the residential sector, according to data from 2017 [120]. The major competition for DH is using individual heat pumps, which, in 2014, constituted about 25% of the market share. Swedish DH systems may be classified as a 3rd generation DH system, regarding their temperature levels. Recently, average distribution temperatures have been 86 °C for supply pipes and 47 °C for return pipes [119]. Nowadays, about two-thirds of the heat supply to DH is based on biomass and waste heat, which is exceptional among high-income countries. Sweden was probably the first country that utilized industrial waste heat, and nowadays it is a country with the highest percentage share of industrial waste heat recovery. In 2015, waste heat represented about 8% of the total energy supplied to Swedish DH [121]. The idea presented in a case study performed by Brange et al. [5] in the Hyllie area in Malmö, Sweden, concerned DH prosumers—customers who both use and supply energy. The excess heat might be directly used in the DH network, or the temperature may be raised with a heat pump. According to obtained data for the Hyllie region, about 50% of the annual heat demand could be covered by prosumer heat. However, the majority of the excess heat is generated during the summer season; thus, thermal storage systems are required. Sweden supports the construction of "near-zero energy buildings". The case study described by Joly et al. [19] presents the approach, combining a near-zero energy house with a solar-assisted 100% renewable heating solution in the residential area of "Vallda Heberg" in Kungsbacka, Sweden. Extruded polyethylene insulated pipes are used as a more economic and technically preferable solution, in comparison to traditional steel pipes. The solar installation is able to produce 37% of the thermal energy requirement.

The Swedish parliament has set, as a goal for the country, the attainment of a carbon dioxide-neutral state by the year 2045. Additionally, total energy use per heated area in Swedish buildings should be 50% lower by 2050 (compared to levels from 1995) [122,123].

5.4. Finland

In 2017, about 38% of heat demand was supplied by district heating systems in Finland [124]. The temperature of district heating water in the supply pipe changes according to the weather conditions and varies from 65 to 115 °C for the supply and from 40 to 60 °C for the return; thus, the DHS might be classified as a 2nd or 3rd generation system [125]. In 2016, the main DH resources in 2016 were coal and wood-derived fuels, and now about 80% of DH is produced by CHP [126]. However, coal-fired heat generation will be prohibited from May 2029 [127]. In 2016, the utilization of renewable energy sources amounted to 32%, while, in 2018, this amount increased to 55%. Nowadays, Finland has the third highest share, after Sweden and Latvia, of renewable energy in the DHC sector in the European Union [128]. Recently, heat pumps and utilization of industrial waste heat on a large scale became a promising source for Finnish DH [129]. The growing interest in large HPs in Finnish DH systems has been considered according to data provided by Kontu et al. [130]. The heat pumps in Finland have a potential of 10–25% of DH sales (while currently it is about 3%). The analysis revealed that HP has the highest potential in small DH systems for replacing fossil fuel-fired boilers. Wahlroos et al. [131] investigated the possibility of using waste heat for Finnish DH systems in Espoo. The utilization of the waste heat led to savings in the total operational costs of the DH systems of between 0.6% and 7.3%. Therefore, from an economic, environmental, and technical point of view, it is a sustainable solution. In Finland, one important future solution for the replacement of fossil fuel is the use of nuclear power as a heat source, resulting in a major reduction of pollutant emissions. The potential of small modular nuclear reactors (SMR) for CO_2-free DH production has been evaluated. The case study by Varri and Syri [132] includes application of NuScale SMR reactors with a total heat capacity of 300 MWdh in the Helsinki region. Teräsvirta et al. [133] investigated replacing biomass cogeneration plants using heat-only small modular reactors of 24–200 MW for a mid-sized city in eastern Finland. The most promising solution assumed the use of five 24 MW heat-only SMR units combined with a 100 MWt cogeneration plant. Both studies confirmed that the application of SMR might be profitable. However, the hesitance concerns the investment costs for rising SMR technologies.

A new interesting trend in Finland is solar heating. In Finland, most of the community lives in regions that reach above 5.3 GJ/m^2 total solar radiation per year [134]. The seasonal storage of solar heat is crucial in solar district heating plants as a solution to the mismatch between energy supply and demand. According to Rehman et al. [134], Finland is located in an area suitable for borehol TES (BTES). One of the examples of an early large-scale, high-temperature solar-heating system with seasonal storage of energy was constructed in Finland is the Kerava solar village [135]. The total heat demand of the community, including 44 apartments with flat-plate solar panels, was 495 MWh annually. Considering heat loss, the solar heating system was planned to produce 550 MWh per year. The borehole TES is a combined system using 54 tilted boreholes filled with water. The holes provide layered thermal storage, with a water temperature of 55–65 °C at the top and 8 °C at the bottom during the winter season. The top water may be applied for short-term storage or for household hot water generation. Moreover, aquifer thermal energy storage (ATES), combined with ground-source heat pumps (GSHP), is considered to be an attractive technology for Finnish DHS. The case study was performed by Todorov et al. [128] for the integration of ATES–GSHP into the existing DH system in the Finnish Urban District. The use of a pre-cooling exchanger led to an increase in the heating and cooling demand covered by ATES of 13% and 15%, respectively. Furthermore, a decrease in the heat generation cost by 5.2% has been reported.

5.5. Lithuania

Lithuania has a well-developed DH system. The heating season in Lithuania starts when the outdoor temperatures, after three consecutive days, are below +10 °C. Nowadays, the heating season lasts from October until April [136]. Meanwhile, heat demand for domestic heating is almost 1.8 times higher in Lithuania in comparison to other EU countries with comparable weather conditions [137]. According to data in 2017, DH satisfied about 56% of heat demand in the residential sector [138]. Other Lithuanian households use individual heating systems, fed with firewood, natural gas, coal, or oil products as fuel [139].

According to the literature, Lithuania's potential for solar energy production is approximately 1000 kWh/m^2 per year. The daily potential is variable depending on the season, from about 0.55 kWh in January to 5.8 kWh in June [140]. Nowadays, the potential of solar heat energy generation in Lithuania reaches 1.5 TWh/year [140] and, up until 2016, about 14800 m^2 of solar collectors were installed in the country, while the application refers mainly to a single-family houses [141].

Šliaupa et al. [142] noted that Lithuania contains large hydro-geothermal resources, mainly in the largest Cambrian, Lower Devonian, and Upper-Middle Devonian reservoirs. Despite very good geothermal conditions, there is only the Klaipėda geothermal district heating plant in Lithuania, which was opened in 2001. However, in 2017, due to decreasing injectivity, the operation of the plant was ceased.

Lithuania's national goal is to produce heat with zero CO_2 emissions by the year 2050. The National Energy Independence Strategy of the Republic of Lithuania, accepted in 2018, assumed a goal to increase the RES share in final energy consumption by up to 30% by 2020, 45% by 2030, and 80% by 2050 [143]. Furthermore, Lithuania's energy strategy is to promote the development of DH systems and cogeneration plants and to start district cooling of buildings. According to Kveselis et al. [144], Lithuania shows a positive trend towards the fulfillment of these energy goals. The most significant factors stimulating the progress of DH in Lithuania include international responsibilities to enhance the participation of CHP plants using RES in energy generation and the development of small CHP plants, as well as the establishment of an energy strategy that supports the use of renewable energy sources in DH systems. Based on the research performed by Šiupšinskas and Rogoža [145], in Lithuania, DH systems have been combined with renewable energy sources; biomass and solar energy are environmentally favorable solutions for heat production. The utilization of these sources may help the country to reach nearly zero CO_2 emissions.

5.6. Germany

In 2017, DH satisfied about 14–24% of heat demand in the residential sector (for existing buildings, while it is about 44% for newly built houses) [146,147]. Currently, more than 80% of district heating in Germany is generated by CHP generation [147]. However, in 2017, the share of renewable energy in district heating consumption in Germany was only 12%, including energy from biomass and geothermal sources [147]. Additionally, a significant source is MSW energy, which, in 2017, reached 11.6%, while waste industrial energy constituted about 2% of the share [147]. An essential aspect of German DH is also geothermal energy. Thirty geothermal plants in Germany generate about 155 GWh of electricity and 1.3 TWh of heat, annually [148].

Petersen [149] described the idea of self-supplying communities based on local and renewable energy sources. A case study using oat peel biomass CHP in the community of Krückau-Vormstegen in Elmshorn, Germany, was developed. For the investigated community, the local heating energy potential based on renewable energies is up to 44.64 GWh/year. This surpasses the local heating demand, ranging between 4.22 and 8.42 GWh/year. However, the implementation process of this approach might be time-consuming and will require initial investment and operational costs.

Pelda et al. [150] analyzed the potential for Germany to incorporate waste industrial heat and solar thermal energy into DH systems. The obtained results showed that solar thermal energy and waste industrial heat potential have been used only to a very limited extent. The calculated theoretical potential of solar thermal power is about 377 $\times 10^3$ TWh/year, and the theoretical potential of waste heat from the industry sector is between 43 TWh/year and 193 TWh/year. Thus, implementation of both strategies into DH systems might significantly reduce CO_2 emissions. However, according to the predictions of the German industry association for district heating and cooling, solar thermal systems with 800,000 m^2 of collector field area in the district heating systems exist in 2020 in Germany [151].

Germany has several seasonal thermal energy storage systems (STES), especially tank thermal energy storage (TTES) and water gravel thermal energy storage (WGTES). Additionally, four aquifer thermal energy storages (ATES) exist in Germany. The investigation performed by Schüppler et al. [152] evaluated the techno-economic and environmental impact of the developed ATES in Karlsruhe. The examined ATES has a cooling capacity of 3.0 MW and a heating capacity of 1.8 MW. The most productive supply preference is that the direct cooling by the ATES results in a decrease in electricity cost of 80%. Moreover, concerning the reference system, the ATES achieves CO_2 savings of around 600 tons annually.

The scenario is to achieve a 100% renewable German energy system by 2050. The energy transition concept of Germany specifies the reduction of CO_2 emissions by 40% by 2020, 55% by 2030, and at least 80% by the year 2050, compared with the year 1990. To achieve these goals, according to Hansen et al. [20], DH systems in Germany should comply with the following requirements: increase the expansion and share of DH for heat demand, heat savings, and increase the number of individual heat pumps.

5.7. Latvia

In Latvia, the duration of the heating season exceeds half a year; thus, high-quality heat supply is needed. In 2015, the share of DH in Latvia satisfied about 30% of the residential sector [153]. Most of the heat energy is produced by boiler houses and CHP plants (the share of CHP in DH generation according to data from 2019 was 73%) [154]. DH is mostly charged by natural gas and biomass (renewables constitute about 47–50% of energy source [155]. However, the goal is to increase the share of renewable energy sources in DH to 60% by 2023 [156].

In 2019, the first large–scale solar district heating plant was opened in Salaspis, Latvia [157]. The plant includes 1720 collectors, can generate about 12,000 kWh of power per year, and is the biggest solar DH installation in Europe. Moreover, the plant is equipped with a 28 m high storage tank and a 3MW wood chip boiler.

Povilanskas et al. [158] analyzed Latvia's potential for using geothermal energy for DH systems. The case study involved Nīca, a small municipality in southwest Latvia. The investigation revealed that the temperature of the Lower Cambrian sandy siltstone aquifer was confirmed to be not less than 43 °C, and it was possible to obtain about 2 l/s of the hot brine to provide sufficient heat to the DHS of Nīca throughout the year. The authors stated that three main requirements had to be met to reach high efficiency. These were the satisfactory extractable heat resources of the aquifer, immediate availability of a nearby deep well, and a plan for the environmentally acceptable transfer of the cooled geothermal water to the surface waters.

The majority of the DH grids work in 2nd and 3rd generation temperature regimes [159]. However, according to the analysis by Ziemele et al. [160], the transition to 4th generation DH is possible in Latvia and could provide a 68% share of RES, and the heat tariff could be reduced by 48% until 2030. Moreover, CO_2 emissions could be reduced by 58% by 2030. The pilot case study performed by Feofilovs et al. [161] confirmed that transition to low-temperature district heating (LTDH) with Solar PV will improve the environmental performance of DH; however, significant modernization or reconstruction will be required, including changes of valves, pipelines, insulations, and boiler houses to improve the heat production effectiveness. However, Cirule et al. [162] suggested that legislation in Latvia does not provide a proper basis to develop or safeguard 4th generation DH. The authors pointed out that it is necessary to raise the competitiveness of the DH system in comparison to other heating solutions, thereby making DH systems more interesting for consumers.

Independence from fossil fuel and the reduction of carbon dioxide emissions by 80–95% by 2050 compared to 1990 levels are the main targets for the Latvian community [156].

6. Conclusions and Recommendations for Future Research

Nowadays, according to European Union regulations, the DH Sector in the Baltic Sea Region countries is constantly undergoing a transition related to increasing the share of renewable energy sources and the reduction of greenhouse gases emissions. From an economic point of view, DH companies strive for an increase in heat supply efficiency, becoming unrivaled, and extending the number of customers. Thus, the intention to establish 4th generation DH systems is fully justified. Lowering the temperature of the water supply, integration of RES such as solar, thermal, wind, or geothermal energy, and application of better insulated and optimized pipeline layouts are significant steps towards improving the effectiveness of current DH systems. Moreover, the introduction of smart management tools into DH systems may overcome both the challenges in existing DH networks and be an essential aspect to ensure profitability for both heat producers and consumers. Intelligent networks with demand site management are an inherent part of DHS development and could have advantages in providing safe and efficient DH systems. Nowadays, implementation of the SAM concept in the DH sector mostly involves the forecasting of heat supply and demand-side management, which may significantly reduce the amount of energy used for heating. However, in the current state-of-the-art, there is a key research gap in the application of smart predictive and preventive methods for DH network performance control and maintenance. An interesting approach of future research may be related to the development of novel smart methods to predict and detect possible failures and damages, minimize the risk of failures, and introduce a system of preventive response. Future studies may examine and model the techno-economic aspects of SAM. A clear indication of the SAM solution's competitiveness in the DH sector is also needed. Moreover, the design and development of innovative multi-source DH systems instead of them being charged by a single fuel type should be proposed. This solution should lead to the heat demand being covered and the ability to select a preferable heat source at any given time, as well as the capability to maintain a continuous heat supply when one of the sources fails. Furthermore, an important aspect is assuring the sustainable development of DH systems, especially related to thermal energy storage. Research is proposed in the area of more dynamic prosumer involvement and the development of bi-directional DHC networks. Implementation of these solutions may help to reduce CO_2 emissions as well as energy costs in the near future.

Author Contributions: Conceptualization, P.R., A.G., and A.R.; writing—original draft preparation, A.G., P.R., and J.S.; formal analysis, A.R. and V.L., technical consultation, V.L.; writing—review and editing, P.R. and A.G.; funding acquisition, A.R. All authors have read and agreed to the published version of the manuscript.

Funding: This research received no external funding.

Institutional Review Board Statement: Not applicable.

Informed Consent Statement: Not applicable.

Data Availability Statement: Data available in a publicly accessible repository as well as in a publicly accessible repository that does not issue DOIs. All data referred to in this paper are accessible according to the list of references.

Acknowledgments: This manuscript was prepared within a scope of a Baltic smart asset management (BSAM) project, co-financed by the Interreg South Baltic Program 2014–2020.

Conflicts of Interest: The authors declare no conflict of interest.

References

1. Van Dongen, L.A.M.; Frunt, L.; Martinetti, A. Smart Asset Management or Smart Operation Management? The Netherlands Railways Case. In *Transportation Systems*; Singh, S., Martinetti, A., Majumdar, A., van Dongen, L.A.M., Eds.; Springer: Singapore, 2019; pp. 113–132. [CrossRef]
2. Gao, L.; Cui, X.; Ni, J.; Lei, W.; Huang, T.; Bai, C.; Yang, J. Technologies in Smart District Heating System. *Energy Procedia* **2017**, *142*, 1829–1834. [CrossRef]
3. Dahash, A.; Ochs, F.; Tosatto, A.; Streicher, W. Toward efficient numerical modeling and analysis of large-scale thermal energy storage for renewable district heating. *Appl. Energy* **2020**, *279*, 115840. [CrossRef]
4. Bünning, F.; Wetter, M.; Fuchs, M.; Müller, D. Bidirectional low temperature district energy systems with agent-based control: Performance comparison and operation optimization. *Appl. Energy* **2018**, 502–515. [CrossRef]
5. Brange, L.; Englund, J.; Lauenburg, P. Prosumers in district heating networks—A Swedish case study. *Appl. Energy* **2016**, *164*, 492–500. [CrossRef]
6. Guelpa, E.; Verda, V. Demand Response and other Demand Side Management techniques for District Heating: A review. *Energy* **2020**, 119440. [CrossRef]
7. Kristensen, M.; Hedegaard, R.; Petersen, S. Long-term forecasting of hourly district heating loads in urban areas using hierarchical archetype modeling. *Energy* **2020**, *201*, 117687. [CrossRef]
8. Sarafraz, M.M.; Tlili, I.; Tian, Z.; Bakouri, M.; Safaei, M.R. Smart optimization of a thermosyphon heat pipe for an evacuated tube solar collector using response surface methodology (RSM). *Physica A* **2019**, *534*, 122146. [CrossRef]
9. Sarafraz, M.M.; Tlili, I.; Tian, Z.; Bakouri, M.; Safaei, M.R.; Goodarzi, M. Thermal Evaluation of Graphene Nanoplatelets Nanofluid in a Fast-Responding HP with the Potential Use in Solar Systems in Smart Cities. *Appl. Sci.* **2019**, *9*, 2101. [CrossRef]
10. Sarafraz, M.M.; Tlili, I.; Baseer, M.A.; Safaei, M.R. Potential of Solar Collectors for Clean Thermal Energy Production in Smart Cities using Nanofluids: Experimental Assessment and Efficiency Improvement. *Appl. Sci.* **2019**, *8*, 1877. [CrossRef]
11. Hong, M.S.; Young-Shin, J.; Woo-Cheol, K.; Joon, J.C.; Jung-Gu, K. Optimization of Cathodic Protection System for River-Crossing District Heating Pipeline using Computational Analysis: Part I. The Basic Model. *Int. J. Electrochem. Sci.* **2020**, *15*, 7013–7026. [CrossRef]
12. Mandley, S.J.; Daioglou, V.; Junginger, H.M.; van Vuuren, D.P.; Wicke, B. EU bioenergy development to 2050. *Renew. Sustain. Energy Rev.* **2020**, *127*, 109858. [CrossRef]
13. Hast, A.; Syri, S.; Lekavičius, V.; Galinis, A. District heating in cities as a part of low-carbon energy system. *Energy* **2018**, *152*, 627–639. [CrossRef]
14. Buffa, S.; Cozzini, M.; D'Antoni, M.; Baratieri, M.; Fedrizzi, R. 5th generation district heating and cooling systems: A review of existing cases in Europe. *Renew. Sustain. Energy Rev.* **2019**, *104*, 504–522. [CrossRef]
15. Sayegh, M.A.; Danielewicz, J.; Nannou, T.; Miniewicz, M.; Jadwiszczak, P.; Piekarska, K.; Jouhara, H. Trends of European research and development in district heating technologies. *Renew. Sustain. Energy Rev.* **2017**, *68*, 1183–1192. [CrossRef]
16. Köfinger, M.; Schmidt, R.R.; Basciotti, D.; Terreros, O.; Baldvinsson, I.; Mayrhofer, J.; Moser, S.; Tichler, R.; Pauli, H. Simulation based evaluation of large scale waste heat utilization in urban district heating networks: Optimized integration and operation of a seasonal storage. *Energy* **2018**, *159*, 1161–1174. [CrossRef]
17. Sun, L.; Fujii, M.; Tasaki, T.; Dong, H.; Ohnishi, S. Improving waste to energy rate by promoting an integrated municipal solid-waste management system. *Resour. Conserv. Recycl.* **2018**, *136*, 289–296. [CrossRef]
18. Francisco Pinto, J.; Carrilho da Graça, G. Comparison between geothermal district heating and deep energy refurbishment of residential building districts. *Sustain. Cities Soc.* **2018**, *38*, 309–324. [CrossRef]

19. Joly, M.; Ruiz, G.; Mauthner, F.; Bourdoukan, P.; Emery, M.; Anderson, M. A methodology to integrate solar thermal energy in district heating networks confronted with a Swedish real case study. *Energy Procedia* **2017**, *122*, 865–870. [CrossRef]
20. Hansen, K.; Mathiesen, B.V.; Skov, I.R. Full energy system transition towards 100% renewable energy in Germany in 2050. *Renew. Sustain. Energy Rev.* **2019**, *102*, 1–13. [CrossRef]
21. Dong, J.; Tang, Y.; Nzihou, A.; Chi, Y. Key factors influencing the environmental performance of pyrolysis, gasification and incineration Waste-to-Energy technologies. *Energy Convers. Manag.* **2019**, *196*, 497–512. [CrossRef]
22. Simeoni, P.; Ciotti, G.; Cottes, M.; Meneghetti, A. Integrating industrial waste heat recovery into sustainable smart energy systems. *Energy* **2019**, *175*, 941–951. [CrossRef]
23. Tsai, W.T. Analysis of municipal solid waste incineration plants for promoting power generation efficiency in Taiwan. *J. Mater. Cycles Waste Manag.* **2016**, *18*, 393–398. [CrossRef]
24. Rudra, S.; Tesfagaber, Y.K. Future district heating plant integrated with municipal solid waste (MSW) gasification for hydrogen production. *Energy* **2019**, *180*, 881–892. [CrossRef]
25. Jonynas, R.; Puida, E.; Poškas, R.; Paukštaitis, L.; Jouhara, H.; Gudzinskas, J.; Miliauskas, G.; Lukoševičius, V. Renewables for district heating: The case of Lithuania. *Energy* **2020**, *211*, 119064. [CrossRef]
26. Hendricks, A.M.; Wagner, J.E.; Volk, T.A.; Newman, D.H.; Brown, T.R. A cost-effective evaluation of biomass district heating in rural communities. *Appl. Energy* **2016**, *162*, 561–569. [CrossRef]
27. Sartor, K.; Dewallef, P. Integration of heat storage system into district heating networks fed by a biomass CHP plant. *J. Energy Storage* **2018**, *15*, 350–358. [CrossRef]
28. Quirion-Blais, O.; Malladi, K.T.; Sowlati, T.; Gao, E.; Mui, C. Analysis of feedstock requirement for the expansion of a biomass-fed district heating system considering daily variations in heat demand and biomass quality. *Energy Convers. Manag.* **2019**, *187*, 554–564. [CrossRef]
29. Ericsson, K.; Werner, S. The introduction and expansion of biomass use in Swedish district heating systems. *Biomass Bioenergy* **2016**, *94*, 57–65. [CrossRef]
30. Kucharska, K.; Rybarczyk, P.; Hołowacz, I.; Konopacka-Łyskawa, D.; Słupek, E.; Makoś, P.; Cieslński, H.; Kamiński, M. Influence of alkaline and oxidative pre-treatment of waste corn cobs on biohydrogen generation efficiency via dark fermentation. *Biomass Bioenergy* **2020**, *141*, 105691. [CrossRef]
31. Karimi Alavijeh, M.; Yaghmaei, S. Biochemical production of bioenergy from agricultural crops and residue in Iran. *Waste Manag.* **2016**, *52*, 375–394. [CrossRef]
32. García-Velásquez, C.A.; Cardona, C.A. Comparison of the biochemical and thermochemical routes for bioenergy production: A techno-economic (TEA), energetic and environmental assessment. *Energy* **2019**, *172*, 232–242. [CrossRef]
33. Clausen, L.R. Energy efficient thermochemical conversion of very wet biomass to biofuels by integration of steam drying, steam electrolysis and gasification. *Energy* **2017**, *125*, 327–336. [CrossRef]
34. Shahbaz, M.; Yusup, S.; Inayat, A.; Patrick, D.O.; Partama, A. System analysis of poly-generation of SNG, power and district heating from biomass gasification system. *Chem. Eng. Trans.* **2016**, *52*, 781–786. [CrossRef]
35. Wang, J.; Wang, Z.; Zhou, D.; Sun, K. Key issues and novel optimization approaches of industrial waste heat recovery in district heating systems. *Energy* **2019**, *188*, 116005. [CrossRef]
36. Xu, Z.Y.; Wang, R.Z.; Yang, C. Perspectives for low-temperature waste heat recovery. *Energy* **2019**, *176*, 1037–1043. [CrossRef]
37. Dagilis, V.; Vaitkus, L.; Balcius, A.; Gudzinskas, J.; Lukoševičius, V. Grade heat recovery system for woodfuel cogeneration plant using water vapour regeneration. *Therm. Sci.* **2018**, *22*, 2667–2677. [CrossRef]
38. Fang, H.; Xia, J.; Jiang, Y. Key issues and solutions in a district heating system using low-grade industrial waste heat. *Energy* **2015**, *86*, 589–602. [CrossRef]
39. Fitó, J.; Hodencq, S.; Ramousse, J.; Wurtz, F.; Stutz, B.; Debray, F.; Vincent, B. Energy- and exergy-based optimal designs of a low-temperature industrial waste heat recovery system in district heating. *Energy Convers. Manag.* **2020**, *211*, 112753. [CrossRef]

40. Lund, J.W. Development and utilization of geothermal resources. In Proceedings of the ISES World Congress 2007, Beijing, China, 18–21 September 2007; Volume 1, pp. 87–95. [CrossRef]
41. Chen, Y.; Wang, J.; Lund, P.D. Sustainability evaluation and sensitivity analysis of district heating systems coupled to geothermal and solar resources. *Energy Convers. Manag.* **2020**, *220*, 113084. [CrossRef]
42. Geothermal Energy. Available online: http://www.geothermaleranet.is/about-geothermal-era-net/geothermal-energy-/ (accessed on 10 September 2020).
43. Dumas, P. A European perspective of the development of deep geothermal in urban areas: Smart thermal grids, geothermal integration into smart cities. *Geomech. Tunnelbau* **2016**, *9*, 447–450. [CrossRef]
44. Weiss, W.; Spörk-Dür, M. Solar Heat Worldwide 2018. Global Market Development and Trends in 2017. Detailed Market Figures 2016. *IEA Sol. Heat. Cool. Program.* **2018**, 94. [CrossRef]
45. Chung, K.M.; Chen, C.C.; Chang, K.C. Effect of diffuse solar radiation on the thermal performance of solar collectors. *Case Stud. Therm. Eng.* **2018**, *12*, 759–764. [CrossRef]
46. Huang, J.; Fan, J.; Furbo, S. Feasibility study on solar district heating in China. *Renew. Sustain. Energy Rev.* **2019**, *108*, 53–64. [CrossRef]
47. Lund, H.; Werner, S.; Wiltshire, R.; Svendsen, S.; Thorsen, J.E.; Hvelplund, F.; Mathiesen, B.V. 4th Generation District Heating (4GDH). Integrating smart thermal grids into future sustainable energy systems. *Energy* **2014**, *68*, 1–11. [CrossRef]
48. Millar, M.A.; Burnside, N.M.; Yu, Z. District heating challenges for the UK. *Energies* **2019**, *12*, 310. [CrossRef]
49. Averfalk, H.; Werner, S. Economic benefits of fourth generation district heating. *Energy* **2020**, *193*, 116727. [CrossRef]
50. Askeland, K.; Rygg, B.J.; Sperling, K. The role of 4th generation district heating (4GDH) in a highly electrified hydropower dominated energy system—The case of Norway. *Int. J. Sustain. Energy Plan. Manag.* **2020**, *27*, 17–34. [CrossRef]
51. Li, H.; Nord, N. Transition to the 4th generation district heating—Possibilities, bottlenecks, and challenges. *Energy Procedia* **2018**, *149*, 483–498. [CrossRef]
52. Ziemele, J.; Cilinskis, E.; Blumberga, D. Pathway and restriction in district heating systems development towards 4th generation district heating. *Energy* **2018**, *152*, 108–118. [CrossRef]
53. Schweiger, G.; Kuttin, F.; Posch, A. District heating systems: An analysis of strengths, weaknesses, opportunities, and threats of the 4GDH. *Energies* **2019**, *12*, 4748. [CrossRef]
54. Revesz, A.; Jones, P.; Dunham, C.; Davies, G.; Marques, C.; Matabuena, R.; Scott, J.; Maidment, G. Developing novel 5th generation district energy networks. *Energy* **2020**, *201*, 117389. [CrossRef]
55. Wirtz, M.; Kivilip, L.; Remmen, P.; Müller, D. 5th Generation District Heating: A novel design approach based on mathematical optimization. *Appl. Energy* **2020**, *260*, 114158. [CrossRef]
56. von Rhein, J.; Henze, G.P.; Long, N.; Fu, Y. Development of a topology analysis tool for fifth-generation district heating and cooling networks. *Energy Convers. Manag.* **2019**, *196*, 705–716. [CrossRef]
57. Boesten, S.; Ivens, W.; Dekker, S.C.; Eijdems, H. 5th Generation District Heating and Cooling Systems as a Solution for Renewable Urban Thermal Energy Supply. *Adv. Geosci.* **2019**, *49*, 129–136. [CrossRef]
58. Li, H.; Wang, S.J. Challenges in smart low-T emperature district heating development. *Energy Procedia* **2014**, *61*, 1472–1475. [CrossRef]
59. Månsson, S.; Johansson Kallioniemi, P.O.; Thern, M.; Van Oevelen, T.; Sernhed, K. Faults in district heating customer installations and ways to approach them: Experiences from Swedish utilities. *Energy* **2019**, *180*, 163–174. [CrossRef]
60. Gustafsson, J.; Sandin, F. *District Heating Monitoring and Control Systems*; Elsevier Ltd.: Amsterdam, The Netherlands, 2015; ISBN 9781782423959.
61. Turski, M.; Nogaj, K.; Sekret, R. The use of a PCM heat accumulator to improve the efficiency of the district heating substation. *Energy* **2019**, *187*. [CrossRef]
62. Masatin, V.; Latõšev, E.; Volkova, A. Evaluation Factor for District Heating Network Heat Loss with Respect to Network Geometry. *Energy Procedia* **2016**, *95*, 279–285. [CrossRef]

63. Yarahmadi, N.; Sällström, J.H. Improved maintenance strategies for district heating pipe-lines. In Proceedings of the 14th International Symposium on District Heating and Cooling, Stockholm, Sweden, 6–10 September 2014.
64. Franco, A.; Bellina, F. Methods for optimized design and management of CHP systems for district heating networks (DHN). *Energy Convers. Manag.* **2018**, *172*, 21–31. [CrossRef]
65. Brange, L.; Englund, J.; Sernhed, K.; Thern, M.; Lauenburg, P. Bottlenecks in district heating systems and how to address them. *Energy Procedia* **2017**, *116*, 249–259. [CrossRef]
66. Brange, L.; Lauenburg, P.; Sernhed, K.; Thern, M. Bottlenecks in district heating networks and how to eliminate them—A simulation and cost study. *Energy* **2017**, *137*, 607–616. [CrossRef]
67. Sattar, A.M.; Chaudhry, M.H.; Kassem, A.A. Partial blockage detection in pipelines by frequency response method. *J. Hydraul. Eng.* **2008**, *134*, 76–89. [CrossRef]
68. Zhou, S.; O'Neill, Z.; O'Neill, C. A review of leakage detection methods for district heating networks. *Appl. Therm. Eng.* **2018**, *137*, 567–574. [CrossRef]
69. Mohapatra, P.K.; Chaudhry, M.H.; Kassem, A.A.; Moloo, J. Detection of partial blockage in single pipelines. *J. Hydraul. Eng.* **2006**, *132*, 200–206. [CrossRef]
70. Heating and Cooling. Available online: https://ec.europa.eu/energy/topics/energy-efficiency/heating-and-cooling_en (accessed on 10 September 2020).
71. Greenhous Gas Emissions. Available online: https://www.eea.europa.eu/airs/2018/resource-efficiency-and-low-carbon-economy/greenhouse-gas-emission (accessed on 10 September 2020).
72. Yang, X.; Li, H.; Svendsen, S. Decentralized substations for low-temperature district heating with no Legionella risk, and low return temperatures. *Energy* **2016**, *110*, 65–74. [CrossRef]
73. Yang, X.; Li, H.; Svendsen, S. Alternative solutions for inhibiting Legionella in domestic hot water systems based on low-temperature district heating. *Build. Serv. Eng. Res. Technol.* **2016**, *37*, 468–478. [CrossRef]
74. Li, Y.; Rezgui, Y.; Zhu, H. District heating and cooling optimization and enhancement—Towards integration of renewables, storage and smart grid. *Renew. Sustain. Energy Rev.* **2017**, *72*, 281–294. [CrossRef]
75. Zarin Pass, R.; Wetter, M.; Piette, M.A. A thermodynamic analysis of a novel bidirectional district heating and cooling network. *Energy* **2018**, *144*, 20–30. [CrossRef]
76. Postnikov, I.; Stennikov, V.; Penkovskii, A. Prosumer in the district heating systems: Operating and reliability modeling. *Energy Procedia* **2019**, *158*, 2530–2535. [CrossRef]
77. van der Heijde, B.; Vandermeulen, A.; Salenbien, R.; Helsen, L. Integrated optimal design and control of fourth generation district heating networks with thermal energy storage. *Energies* **2019**, *12*, 2766. [CrossRef]
78. Tang, R.; Wang, S. Model predictive control for thermal energy storage and thermal comfort optimization of building demand response in smart grids. *Appl. Energy* **2019**, *242*, 873–882. [CrossRef]
79. Guelpa, E.; Marincioni, L. Demand side management in district heating systems by innovative control. *Energy* **2019**, *188*, 116037. [CrossRef]
80. Wang, Z.; Crawley, J.; Li, F.G.N.; Lowe, R. Sizing of district heating systems based on smart meter data: Quantifying the aggregated domestic energy demand and demand diversity in the UK. *Energy* **2020**, *193*, 116780. [CrossRef]
81. Guelpa, E.; Marincioni, L.; Deputato, S.; Capone, M.; Amelio, S.; Pochettino, E.; Verda, V. Demand side management in district heating networks: A real application. *Energy* **2019**, *182*, 433–442. [CrossRef]
82. Idowu, S.; Saguna, S.; Åhlund, C.; Schelén, O. Applied machine learning: Forecasting heat load in district heating system. *Energy Build.* **2016**, *133*, 478–488. [CrossRef]
83. Guelpa, E.; Marincioni, L.; Capone, M.; Deputato, S.; Verda, V. Thermal load prediction in district heating systems. *Energy* **2019**, *176*, 693–703. [CrossRef]
84. Kontu, K.; Vimpari, J.; Penttinen, P.; Junnila, S. City scale demand side management in three different-sized district heating systems. *Energies* **2018**, *11*, 3370. [CrossRef]
85. Wang, H.; Wang, H.; Haijian, Z.; Zhu, T. Optimization modeling for smart operation of multi-source district heating with distributed variable-speed pumps. *Energy* **2017**, *138*, 1247–1262. [CrossRef]
86. Wang, H.; Wang, H.; Zhou, H.; Zhu, T. Modeling and optimization for hydraulic performance design in multi-source district heating with fluctuating renewables. *Energy Convers. Manag.* **2018**, *156*, 113–129. [CrossRef]

87. Lu, S.; Gu, W.; Zhou, J.; Zhang, X.; Wu, C. Coordinated dispatch of multi-energy system with district heating network: Modeling and solution strategy. *Energy* **2018**, *152*, 358–370. [CrossRef]
88. Zaporozhets, A.; Eremenko, V.; Serhiienko, R.; Ivanov, S. *Methods and Hardware for Diagnosing Thermal Power Equipment Based on Smart Grid Technology*; Springer International Publishing: Berlin/Heidelberg, Germany, 2019; Volume 871, ISBN 9783030010683.
89. De Lorenzi, A.; Gambarotta, A.; Morini, M.; Rossi, M.; Saletti, C. Setup and testing of smart controllers for small-scale district heating networks: An integrated framework. *Energy* **2020**, *205*, 118054. [CrossRef]
90. Novitsky, N.N.; Shalaginova, Z.I.; Alekseev, A.A.; Tokarev, V.V.; Grebneva, O.A.; Lutsenko, A.V.; Vanteeva, O.V.; Mikhailovsky, E.A.; Pop, R.; Vorobev, P.; et al. Smarter Smart District Heating. *Proc. IEEE* **2020**, *108*, 1596–1611. [CrossRef]
91. Vivian, J.; Jobard, X.; Hassine, I.B.; Hurink, J. Smart Control of a District Heating Network with High Share of Low Temperature Waste Heat Direct Use of Low Temperature Heat in District Heating Networks with Booster Heat Pumps View project i_city: Intelligent city View project. In Proceedings of the 12th Conference on Sustainable Development of Energy, Water and Environmental Systems-SDEWES, Dubrovnik, Croatia, 4–8 October 2017.
92. Zhong, Y.; Xu, Y.; Wang, X.; Jia, T.; Xia, G.; Ma, A.; Zhang, L. Pipeline leakage detection for district heating systems using multisource data in mid- and high-latitude regions. *ISPRS J. Photogramm. Remote Sens.* **2019**, *151*, 207–222. [CrossRef]
93. Ahn, J.; Cho, S. Development of an intelligent building controller to mitigate indoor thermal dissatisfaction and peak energy demands in a district heating system. *Build. Environ.* **2017**, *124*, 57–68. [CrossRef]
94. Wang, H.; Meng, H.; Zhu, T. New model for onsite heat loss state estimation of general district heating network with hourly measurements. *Energy Convers. Manag.* **2018**, *157*, 71–85. [CrossRef]
95. Smart Active Box. Available online: https://www.smartactivebox.com (accessed on 10 October 2020).
96. The Storm Controller. Available online: https://storm-dhc.eu/en/storm-controller (accessed on 10 October 2020).
97. Fifth Generation, Low Temperature, High EXergY District Heating and Cooling NETworkS. Available online: https://cordis.europa.eu/project/id/649820 (accessed on 10 October 2020).
98. Smart Heating—Danfoss Link. Available online: https://www.smartheating.danfoss.com/pl (accessed on 20 September 2020).
99. Vexve Has Developed the First Smart Valve Solution for Underground District Energy Networks. Available online: https://www.vexve.com/ (accessed on 20 September 2020).
100. Ohlsson, M. Take Care of Your Network (Presentation). Available online: https://lsta.lt/wp-content/uploads/2020/09/3_Magnus-Ohlsson.pdf (accessed on 20 November 2020).
101. Bestandsdaten und Schadensstatistik. Available online: www.agfw-bestand.online (accessed on 3 January 2021).
102. Wojdyga, K.; Chorzelski, M. Chances for Polish district heating systems. *Energy Procedia* **2017**, *116*, 106–118. [CrossRef]
103. District Energy in Poland. Available online: https://www.euroheat.org/knowledge-hub/district-energy-poland/ (accessed on 12 September 2020).
104. Polish District Heating Sector—From Coal to Solar? Available online: https://www.euroheat.org/news/polish-district-heating-sector-coal-solar/ (accessed on 12 September 2020).
105. Burchard-Dziubińska, M. Air pollution and health in Poland: Anti-smog movement in the most polluted Polish cities. *Ekon. Sr.* **2019**, *2*, 76–90. [CrossRef]
106. Wojdyga, K. An influence of weather conditions on heat demand in district heating systems. *Energy Build.* **2008**, *40*, 2009–2014. [CrossRef]
107. Cenian, A.; Dzierzgowski, M.; Pietrzykowski, B. On the road to low temperature district heating. *J. Phys. Conf. Ser.* **2019**, *1398*. [CrossRef]
108. Zwierzchowski, R. Characteristics of large thermal energy storage systems in Poland. *E3S Web Conf.* **2017**, *22*. [CrossRef]
109. Kępińska, B. Geothermal Energy Use, Country Update for Poland, 2013–2015. In Proceedings of the EGC Proceedings, Strasbourg, France, 19–24 September 2016; pp. 1–10.

110. Kaczmarczyk, M. Potential of existing and newly designed geothermal heating plants in limiting of low emissions in Poland. *E3S Web Conf.* **2018**, *44*. [CrossRef]
111. Renewables in Heating. Available online: https://www.forum-energii.eu/en/analizy/oze-w-cieplownictwie (accessed on 1 September 2020).
112. Yang, X.; Li, H.; Svendsen, S. Energy, economy and exergy evaluations of the solutions for supplying domestic hot water from low-temperature district heating in Denmark. *Energy Convers. Manag.* **2016**, *122*, 142–152. [CrossRef]
113. District Energy in Denmark. Available online: https://www.euroheat.org/knowledge-hub/district-energy-denmark/ (accessed on 1 September 2020).
114. Tian, Z.; Zhang, S.; Deng, J.; Fan, J.; Huang, J.; Kong, W.; Perers, B.; Furbo, S. Large-scale solar district heating plants in Danish smart thermal grid: Developments and recent trends. *Energy Convers. Manag.* **2019**, *189*, 67–80. [CrossRef]
115. Bühler, F.; Petrović, S.; Karlsson, K.; Elmegaard, B. Industrial excess heat for district heating in Denmark. *Appl. Energy* **2017**, *205*, 991–1001. [CrossRef]
116. Lund, R.; Persson, U. Mapping of potential heat sources for heat pumps for district heating in Denmark. *Energy* **2016**, *110*, 129–138. [CrossRef]
117. Foteinaki, K.; Li, R.; Péan, T.; Rode, C.; Salom, J. Evaluation of energy flexibility of low-energy residential buildings connected to district heating. *Energy Build.* **2020**, *213*, 109804. [CrossRef]
118. Denmark's Integrated National Energy and Climate Plan. Available online: https://ec.europa.eu/energy/sites/ener/files/documents/dk_final_necp_main_en.pdf (accessed on 7 September 2020).
119. Werner, S. District heating and cooling in Sweden. *Energy* **2017**, *126*, 419–429. [CrossRef]
120. District Energy in Sweden. Available online: https://www.euroheat.org/knowledge-hub/district-energy-sweden/ (accessed on 1 September 2020).
121. Söderholm, K. Pioneering industry/municipal district heating collaboration in Sweden in the 1970s. *Energy Policy* **2018**, *112*, 328–333. [CrossRef]
122. Börjesson, P.; Hansson, J.; Berndes, G. Future demand for forest-based biomass for energy purposes in Sweden. *For. Ecol. Manag.* **2017**, *383*, 17–26. [CrossRef]
123. Savvidou, G.; Nykvist, B. Heat demand in the Swedish residential building stock—Pathways on demand reduction potential based on socio-technical analysis. *Energy Policy* **2020**, *144*. [CrossRef]
124. District Energy in Finland. Available online: https://www.euroheat.org/knowledge-hub/district-energy-finland/ (accessed on 1 September 2020).
125. Energy Sector in Finland. Available online: https://energia.fi/en/energy_sector_in_finland/energy_networks/district_heating_networks (accessed on 1 September 2020).
126. District Heating. Available online: http://gebwell.fi/en/district-heating/ (accessed on 2 September 2020).
127. The Act Banning the Use of Coal for Energy Generation in 2029 to Enter into Force in Early April. Available online: https://valtioneuvosto.fi/en/-/1410877/kivihiilen-energiakayton-vuonna-2029-kieltava-laki-voimaan-huhtikuun-alussa?_101_INSTANCE_YZfcyWxQB2Me_languageId=en_US (accessed on 2 September 2020).
128. Todorov, O.; Alanne, K.; Virtanen, M.; Kosonen, R. Aquifer thermal energy storage (ATES) for district heating and cooling: A novel modeling approach applied in a case study of a Finnish urban district. *Energies* **2020**, *13*, 2478. [CrossRef]
129. Abdurafikov, R.; Grahn, E.; Kannari, L.; Ypyä, J.; Kaukonen, S.; Heimonen, I.; Paiho, S. An analysis of heating energy scenarios of a Finnish case district. *Sustain. Cities Soc.* **2017**, *32*, 56–66. [CrossRef]
130. Kontu, K.; Rinne, S.; Junnila, S. Introducing modern heat pumps to existing district heating systems—Global lessons from viable decarbonizing of district heating in Finland. *Energy* **2019**, *166*, 862–870. [CrossRef]
131. Wahlroos, M.; Pärssinen, M.; Manner, J.; Syri, S. Utilizing data center waste heat in district heating—Impacts on energy efficiency and prospects for low-temperature district heating networks. *Energy* **2017**, *140*, 1228–1238. [CrossRef]
132. Värri, K.; Syri, S. The possible role of modular nuclear reactors in district heating: Case Helsinki region. *Energies* **2019**, *12*, 2195. [CrossRef]
133. Teräsvirta, A.; Syri, S.; Hiltunen, P. Small Nuclear Reactor—Nordic District Heating Case Study. *Energies* **2020**, *13*, 3782. [CrossRef]

134. Ur Rehman, H.; Hirvonen, J.; Sirén, K. A long-term performance analysis of three different configurations for community-sized solar heating systems in high latitudes. *Renew. Energy* **2017**, *113*, 479–493. [CrossRef]
135. Gehlin, S. Borehole Thermal Energy Storage. In *Advances in Ground-Source Heat Pump System*; Ress, S.J., Ed.; Woodhead Publishing (Elsevier): Cambridge, UK, 2016; pp. 295–327. [CrossRef]
136. Varnagirytė-Kabašinskienė, I.; Lukminė, D.; Mizaras, S.; Beniušienė, L.; Armolaitis, K. Lithuanian forest biomass resources: Legal, economic and ecological aspects of their use and potential. *Energy. Sustain. Soc.* **2019**, *9*, 1–19. [CrossRef]
137. Valancius, R.; Singh, R.M.; Jurelionis, A.; Vaiciunas, J. A review of heat pump systems and applications in cold climates: Evidence from Lithuania. *Energies* **2019**, *12*, 4331. [CrossRef]
138. District Energy in Lithuania. Available online: https://www.euroheat.org/knowledge-hub/district-energy-lithuania/ (accessed on 1 September 2020).
139. Jonkutė, G.; Norvaišienė, R.; Banionis, K.; Monstvilas, E.; Bliūdžius, R. Analysis of carbon dioxide emissions in residential buildings through energy performance certification in Lithuania. *Energy Sources Part B Econ. Plan. Policy* **2020**, 1–18. [CrossRef]
140. Valancius, R.; Jurelionis, A.; Vaiciunas, J. Review of Solar Thermal Systems and Their Potential in Lithuania. In Proceedings of the International Conference on Solar Energy and Buildings, Palma de Mallorca, Spain, 11–14 October 2016; pp. 1–7. [CrossRef]
141. Valancius, R.; Cerneckiene, J.; Singh, R.M. Review of Combined Solar Thermal and Heat Pump Systems Installations in Lithuanian Hospitals. In Proceedings of the EuroSun 2018 Conference of ISES Europe, Rapperswil, Switzerland, 10–13 September 2018. [CrossRef]
142. Geothermal Energy Use, Country Update for Sweden. Available online: http://europeangeothermalcongress.eu/wp-content/uploads/2019/07/CUR-28-Sweden.pdf (accessed on 30 October 2020).
143. National Energy Independence Strategy Executive Summary—Energy for Competitive Lithuania. Available online: https://enmin.lrv.lt/uploads/enmin/documents/files/National_energy_independence_strategy_2018.pdf (accessed on 1 September 2020).
144. Kveselis, V.; Dzenajavičienė, E.F.; Masaitis, S. Analysis of energy development sustainability: The example of the lithuanian district heating sector. *Energy Policy* **2017**, *100*, 227–236. [CrossRef]
145. Šiupšinskas, G.; Rogoža, A. Modernisation of the existing energy supply system to CO2 neutral system: Lithuanian town case study. In Proceedings of the International Conference on Environmental Engineering, ICEE 2014, Vilnius, Lithuania, 22–24 May 2014. [CrossRef]
146. Krikser, T.; Profeta, A.; Grimm, S.; Huther, H. Willingness-to-Pay for District Heating from Renewables of Private Households in Germany. *Sustainability* **2020**, *12*, 4129. [CrossRef]
147. District Energy in Germany. Available online: https://www.euroheat.org/knowledge-hub/district-energy-germany/ (accessed on 1 September 2020).
148. Weinand, J.M.; Kleinebrahm, M.; McKenna, R.; Mainzer, K.; Fichtner, W. Developing a combinatorial optimisation approach to design district heating networks based on deep geothermal energy. *Appl. Energy* **2019**, *251*, 113367. [CrossRef]
149. Petersen, J.P. Energy concepts for self-supplying communities based on local and renewable energy sources: A case study from northern Germany. *Sustain. Cities Soc.* **2016**, *26*, 1–8. [CrossRef]
150. Pelda, J.; Stelter, F.; Holler, S. Potential of integrating industrial waste heat and solar thermal energy into district heating networks in Germany. *Energy* **2020**, *203*. [CrossRef]
151. Shrestha, N.L.; Urbaneck, T.; Oppelt, T.; Platzer, B.; Göschel, T.; Uhlig, U.; Frey, H. Implementation of large solar thermal system into district heating network in Chemnitz (Germany). *ISES Sol. World Conf.* **2017**, 322–332. [CrossRef]
152. Schüppler, S.; Fleuchaus, P.; Blum, P. Techno-economic and environmental analysis of an Aquifer Thermal Energy Storage (ATES) in Germany. *Geotherm. Energy* **2019**, *7*. [CrossRef]
153. District Energy in Latvia. Available online: https://www.euroheat.org/knowledge-hub/district-energy-latvia/ (accessed on 5 September 2020).
154. Latvian City Invests in Solar District Heating. Available online: https://www.euroheat.org/knowledge-hub/latvian-city-invests-solar-district-heating/ (accessed on 5 September 2020).

155. Kamenders, A.; Vilcane, L.; Indzere, Z.; Blumberga, D. Heat Demand and Energy Resources Balance Change in Latvia. *Energy Procedia* **2017**, *113*, 411–416. [CrossRef]
156. National Energy and Climate Plan of Latvia 2021–2030. Available online: https://ec.europa.eu/energy/sites/ener/files/documents/ec_courtesy_translation_lv_necp.pdf (accessed on 7 September 2020).
157. Latvian DH Share. Available online: https://www.euroheat.org/knowledge-hub/district-energy-austria/attachment/latvia-dh-share/ (accessed on 7 September 2020).
158. Povilanskas, R.; Satkūnas, J.; Jurkus, E. Conditions for deep geothermal energy utilisation in southwest Latvia: Nīca case study. *Baltica* **2013**, *26*, 193–200. [CrossRef]
159. Ziemele, J.; Gravelsins, A.; Blumberga, A.; Blumberga, D. Combining energy efficiency at source and at consumer to reach 4th generation district heating: Economic and system dynamics analysis. *Energy* **2017**, *137*, 595–606. [CrossRef]
160. Ziemele, J.; Gravelsins, A.; Blumberga, A.; Vigants, G.; Blumberga, D. System dynamics model analysis of pathway to 4th generation district heating in Latvia. *Energy* **2016**, *110*, 85–94. [CrossRef]
161. Feofilovs, M.; Pakere, I.; Romagnoli, F. Life Cycle Assessment of Different Low-Temperature District Heating Development Scenarios: A Case Study of Municipality in Latvia. *Environ. Clim. Technol.* **2019**, *23*, 272–290. [CrossRef]
162. Cirule, D.; Pakere, I.; Blumberga, D. Legislative Framework for Sustainable Development of the 4th Generation District Heating System. *Energy Procedia* **2016**, *95*, 344–350. [CrossRef]
163. Low Temperature District Heating for the Baltic Sea Region. Available online: https://projects.interreg-baltic.eu/projects/lowtemp-112.html (accessed on 12 September 2020).
164. BSAM—Baltic Smart Asset Management. Available online: https://southbaltic.eu/-/bsam (accessed on 12 September 2020).
165. Heat Roadmap Europe. Available online: https://heatroadmap.eu/ (accessed on 12 September 2020).
166. Innovative Technology for District Heating and Cooling. Available online: https://cordis.europa.eu/project/id/696174/pl (accessed on 13 September 2020).
167. REnewable Low TEmperature District. Available online: https://cordis.europa.eu/project/id/768567/pl (accessed on 13 September 2020).
168. THERMOS (Thermal Energy Resource Modelling and Optimisation System). Available online: https://cordis.europa.eu/project/id/723636 (accessed on 13 September 2020).
169. TEMPerature Optimisation for Low Temperature District Heating across Europe. Available online: https://cordis.europa.eu/project/id/768936/pl (accessed on 14 September 2020).
170. Smart and Local Renewable Energy DISTRICT Heating and Cooling Solutions for Sustainable Living. Available online: https://cordis.europa.eu/project/id/857801 (accessed on 14 September 2020).
171. Recovery of Urban Excess Heat. Available online: https://cordis.europa.eu/project/id/767429/pl (accessed on 15 September 2020).
172. Cool Ways of Using Low Grade Heat Sources from Cooling and Surplus Heat for Heating of Energy Efficient Buildings with New Low Temperature District Heating (LTDH) Solutions. Available online: https://cordis.europa.eu/project/id/767799 (accessed on 13 September 2020).
173. Improving the Performance of District Heating Systems in Central and East Europe. Available online: https://cordis.europa.eu/project/id/784966 (accessed on 20 September 2020).

Article

Modeling Artificial Ground Freezing for Construction of Two Tunnels of a Metro Station in Napoli (Italy)

Alessandro Mauro [1], Gennaro Normino [1,*], Filippo Cavuoto [2], Pasquale Marotta [3] and Nicola Massarotti [1]

[1] Dipartimento di Ingegneria, Università degli Studi di Napoli "Parthenope", Centro Direzionale, Isola C4, 80143 Napoli, Italy; alessandro.mauro@uniparthenope.it (A.M.); nicola.massarotti@uniparthenope.it (N.M.)
[2] Studio Cavuoto, Via Benedetto Brin, 63/D2, 80142 Napoli, Italy; dl.cavuoto@mmspa.eu
[3] Consorzio di Ricerca per l'Ambiente i Veicoli l'Energia e i Biocombustibili (CRAVEB), Centro Direzionale, Is. C4–80143 Napoli, Italy; pasquale.marotta@uniparthenope.it
* Correspondence: gennaro.normino@uniparthenope.it; Tel.: +39-081-5476709

Received: 9 January 2020; Accepted: 3 March 2020; Published: 10 March 2020

Abstract: An artificial ground freezing (AGF) technique in the horizontal direction has been employed in Naples (Italy), in order to ensure the stability and waterproofing of soil during the excavation of two tunnels in a real underground station. The artificial freezing technique consists of letting a coolant fluid, with a temperature lower than the surrounding ground, circulate inside probes positioned along the perimeter of the gallery. In this paper, the authors propose an efficient numerical model to analyze heat transfer during the whole excavation process for which this AGF technique was used. The model takes into account the water phase change process, and has been employed to analyze phenomena occurring in three cross sections of the galleries. The aim of the work is to analyze the thermal behavior of the ground during the freezing phases, to optimize the freezing process, and to evaluate the thickness of frozen wall obtained. The steps to realize the entire excavation of the tunnels, and the evolution of the frozen wall during the working phases, have been considered. In particular, the present model has allowed us to calculate the thickness of the frozen wall equal to 2.1 m after fourteen days of nitrogen feeding.

Keywords: numerical modeling; heat transfer; artificial ground freezing; underground station; metro in Napoli; GEO heating

1. Introduction

Artificial Ground Freezing (AGF) is a consolidation technique adopted in geotechnical engineering, when underground excavations must be executed in granular soils, or below the groundwater level [1,2]. The realization of relevant underground structures in urban areas often involves the management of constructive problems related to avoiding the presence of water in the excavation, especially if the soil has poor geo-mechanical proprieties. The artificial ground freezing technique has been extensively used in the last decades as an effective and powerful construction method, which provides ground support, groundwater control, and structural underpinning during construction. However, the use of this technology requires a good knowledge of frozen soil behavior and a robust numerical model able to predict ground movements around the excavation. This is important, especially in densely urbanized areas, where frost action is detrimental for surrounding structures.

The AGF method consists of letting a refrigerant circulate inside probes located along the perimeter of the excavation, at a temperature significantly lower than the surrounding soil. The water in the soil goes from liquid to solid phase, and it forms a block of frozen ground in the area surrounding the probes.

The process is, in general, divided into two different phases: a first "freezing phase", that ends when the soil achieves the design temperature needed to start the excavation, and a "maintenance phase", which is characterized by heat absorption in order to keep the temperature constant during the excavation [3].

The main advantages of this technique, among the available ground consolidation and waterproofing technologies, are: (i) security and compatibility with the environment, since there is no injection, and the dispersion of products in the ground. Water already present in the ground is, in fact, frozen, using refrigerant fluids that are never directly in contact with the ground and groundwater, avoiding contamination phenomena; (ii) applicability to any type of soil, from coarse to fine grain and rocks [3,4].

Depending on the working fluid used, two types of methods can be identified: (i) the direct method, which is based on the use of liquid nitrogen entering the probes at a temperature of −196 °C and released in the atmosphere in gaseous phase at a temperature between −80 °C and −170 °C; (ii) the indirect method, which is based on the use of a mixture of water and calcium chloride (known as brine), whose circulation temperature can vary between −25 °C and −40 °C. A combination of the two previous methods is known as a mixed-method, which uses the direct method for the freezing phase, and the indirect method for the maintenance phase.

Several numerical and experimental works analyzing the AGF technique are available in the literature. Colombo [5] in the first part of his work invokes a well-known approximate approach for the a priori evaluation of the parameters influencing the technique, such as the time required to reach the target temperatures, or the heat flow rate needed by the plant. The results proposed by the author, applied to Neapolitan tuff, were compared with those obtained from a series of numerical analyses conducted by using the finite element method as a discretization technique, and with experimental data measured on site during freezing operations carried out for the realization of the galleries for the stations of Piazza Dante and Piazza Garibaldi of metro Line 1 in Napoli (Italy). Papakonstantinou et al. [6] first analyzed the experimental data of monitored temperatures in the ground during the freezing process, and then performed a numerical analysis through the FREEZE calculation code, a thermohydraulic software developed at the ETH in Zurich. The authors found that the thermal conductivity of the soil is an important parameter to be taken into account, and can be reasonably estimated by a posterior numerical analysis, if it not known a priori. Subsequently, Pimental et al. [7] analyzed the results of three applications of the AGF technique in urban underground construction projects, comparing the experimental data with the thermohydraulic coupled code model FREEZE. The first case study concerned the construction of a tunnel for the underground in Fürth (Germany) in soft ground with significant infiltration flow. The second case study concerned a platform tunnel in a metro station in Naples, and aimed at the determination of relevant thermal parameters through retrospective analysis, and to compare the results obtained by using the forecasting model with on-site measures. In the third case, regarding a tunnel under the river Limmat in Zurich, numerical simulations were used to identify potential problems caused by geometrical irregularities in the well layout, in combination with infiltration flow. Russo et al. [1] analyzed the experimental data collected during the execution of the excavation with the AGF technique, and developed a numerical model to evaluate stress in the ground during the freezing and defrosting of a frozen wall. The focus of the work was the settlement caused by the tunnel excavation, and the use of the AGF technique to allow the safe digging of a service gallery located half in the silty sand layer, and a half in the yellow tuff layer, below groundwater. The phases of the tunnel construction were accompanied by the monitoring of the measurements and control activities of the effects of the gallery excavation. The measurements collected during the construction process, in fact, allowed us to monitor the freezing-thawing process, and the change in volume related to the excavation, providing useful information for the future implementation of similar projects. Finally, the analysis in the test procedure was conducted using a complete three-dimensional (3D) model implemented in the DFM Flac3D package.

Vitel et al. [8] developed a numerical model considering both the freezing tube and the surrounding ground. The model is based upon the following principles: (i) heat conduction around the well is solved by considering vertical heat transfer processes negligible compared to the trans-horizontal heat transfer; (ii) heat transfer in the freezing probe is reduced to a one-dimensional (1D) calculation. In this study, convection in the ground was not taken into account with respect to the heat conduction, and therefore, the effects of groundwater flow were not considered. Vitel et al. [9] developed a numerical thermohydraulic model in order to simulate artificial ground freezing by considering a saturated and nondeformable porous medium under groundwater flow conditions. Marwan et al. [10] presented a thermohydraulic finite element model integrated into an optimization algorithm, using the Ant Colony Optimization (ACO). This technique allowed researchers to optimize the positions of the freezing probes with respect to the groundwater flow. Kang et al. [11] combined a freezing method and a New Tubular Roof (NTR) simulated by thermomechanical coupling analysis. The temperature range obtained in the freezing process indicated that the thickness of the frozen wall grows of about 2.0 m after 50 days of freezing. Moreover, the stability of the surrounding ground and the support structures in the bench cutting phase were also studied. Panteleev et al. [12] focused on the development of a monitoring system for the artificial ground freezing process for a vertical shaft. The temperature in the wells was measured by using the fiber optic system Silixa, based on the Raman effect. Alzoubi et al. [13] have evaluated the development of a frozen wall between two freezing probes, with and without the presence of groundwater infiltration for 2D geometry, by was using ANSYS. Fan et al. [14] show a case study concerning the monitoring of frozen wall formation during soil freezing using brine, then developed a three-dimensional numerical model to analyze the temperature distribution. The numerical simulation was conducted by using ADINA software.

Based on the analysis of the available literature, the interest of the research community on the AGF technique is evident. Both experimental and numerical works can be found, however, more research effort is needed to numerically analyze the evolution of the process for real cases, considering the geometry of civil works and the development of the freezing probes in the ground. For these reasons, the authors have developed an efficient transient numerical model to effectively analyze heat transfer in the soil, and at the same time, save computing resources. The proposed approach is based on the coupling of a heat transfer model between the freezing probes and the surrounding ground with a heat transfer and phase change model of the soil, and for the first time in the literature, all of the phases of AGF process have been reproduced. The model was used as a preliminary predictive analysis for the construction of two tunnels in Napoli. The model has been validated against the data of Colombo [5]. After validation, the numerical model has been employed to analyze a real case study of two tunnels between Line 1 and Line 6 of the metro station in Piazza Municipio, Napoli, southern Italy. The purpose of the work is to study in detail the heat transfer process during ground freezing for the realization of the tunnels. The analysis is carried out by employing a FEM-based model, using Comsol Multiphysics commercial software, to model artificial ground freezing during the whole processe.

The model developed in the present work allows us to simulate, for the first time in the literature, a mixed-method used for the freezing process, from the first phase based on nitrogen feeding, a maintenance phase, and a third phase that involves the use of brine. The maintenance phase is necessary to avoid the freezing of brine in the probes. The novelty introduced by this work relies on the development of a thermal analysis of the entire artificial ground freezing process, considering all the phases and the influence of the process on the thermodynamic behavior of a second nearby tunnel that was also subject to the AGF. Moreover, the excavation phase has been reproduced, by imposing a convective heat transfer condition related to the presence of men and machines, while the second tunnel was subject to AGF with a mixed method.

This paper is structured as follows: in Section 2, the characteristics of the AGF technique are described, while in Section 3 the numerical model developed is presented. Section 4 reports the validation carried out against literature data. The results of the parametric analysis performed after model validation are reported in Section 5, while conclusions are drawn in Section 6 of the paper.

2. Description of AGF Technique and Case Study

Artificial Ground Freezing (AGF) consists of freezing the ground by means of heat transfer, with a refrigerant fluid circulating inside probes located along the perimeter of the excavation to be realized. In this way, the water contained in the soil undergoes a phase transition from liquid to solid, forming a block of frozen ground called a "frozen wall" in the area surrounding the probes.

The mixed-method AGF process used for the Piazza Municipio galleries can be divided into different phases:

Phase 1-Nitrogen: The probes are fed with nitrogen at an inlet temperature of about −196 °C, while the expected outlet temperature is around −110 °C. The duration of Phase 1 is related to the time required for the formation of the minimum thickness of the frozen wall (1.5 m);

Phase 2-Waiting: At the end of nitrogen feeding, in order to obtain a temperature adequate for brine feeding, and to avoid brine freezing inside the probes;

Phase 3-Brine: Maintaining the ice thickness on the tunnel vault by feeding the probes with brine, at a temperature of about −35 °C. Phase 3 is used during the tunnel excavation, in order to maintain the soil temperature below water freezing over time, and the desired thickness of the frozen wall.

For the realization of two tunnels between Line 1 and Line 6 of the underground station in Piazza Municipio in Napoli, the AGF technique with the mixed method has been used. The choice to use this method is due to the possibility of combining the cryogenic power of nitrogen with the flexibility and safety of freezing with brine. It consists essentially of making complementary direct and indirect methods, using the same freezing probes.

An overview of the nitrogen feeding procedure and system used is reported in Figure 1, which reports the truck and tanks at the construction yard, and the loading phase of liquid nitrogen into the tanks.

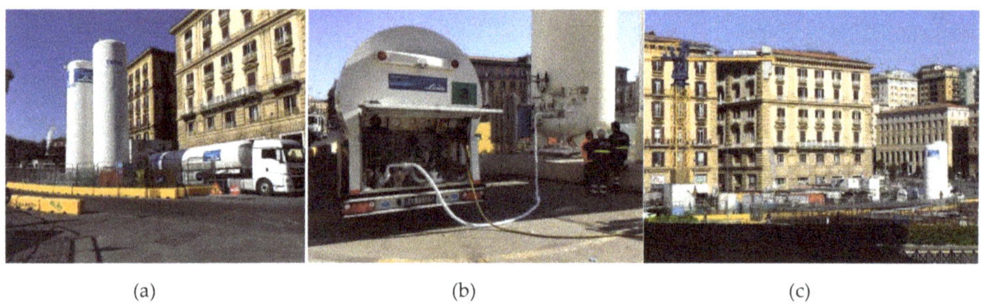

(a) (b) (c)

Figure 1. (**a**) Truck and tanks of liquid nitrogen; (**b**) loading of liquid nitrogen into the tanks; (**c**) view of the construction site.

Figure 2 shows an overview of the construction site, where it is possible to see the nitrogen plant and the two brine refrigeration units. The nitrogen feeding system works by gravity. The brine is refrigerated by one unit, while the second is used as back up in case of the failure of the first one.

The present case study involves the construction of two tunnels for the connection between Line 1 and Line 6 of the Metro station of Piazza Municipio in Napoli, southern Italy. The soil affected by the excavation consists of a layer of pozzolana overlaying a bench of tuff. As shown in Figure 3a, the horizontal distribution of the freezing probes is influenced by the actual development of the two tunnels, that have a slight curvature. Instead, the freezing probes, for technological reasons, have a straight distribution along their axis. Section A-A is located at 5.0 m from the Tunnel Boring Machine (TBM) extraction well, section B-B is in a central position with respect to the tunnels, and section C-C is located at 5.0 m from Line 6 Station well. Figure 3b shows a cross section of the case study with the position of the freezing probes, while Figure 4 shows the axonometry of the two tunnels connecting Line 1 and Line 6.

Figure 2. Overview of the construction site and view of the nitrogen and brine plant.

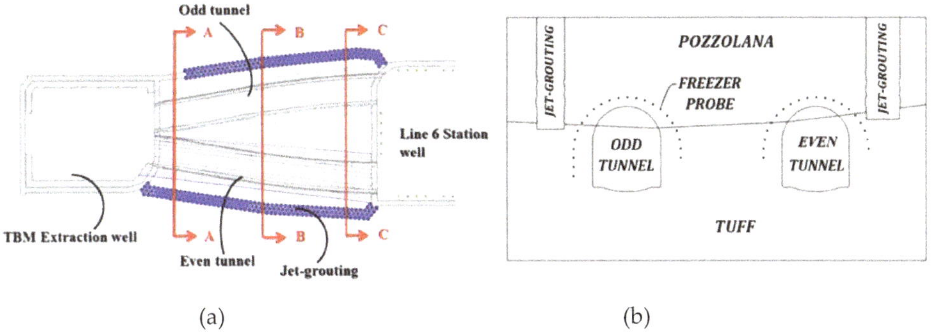

Figure 3. (**a**) Key plan of the tunnels and freezing probes, with indication of the three cross sections considered; (**b**) cross section of the case study with the position of the freezing probes.

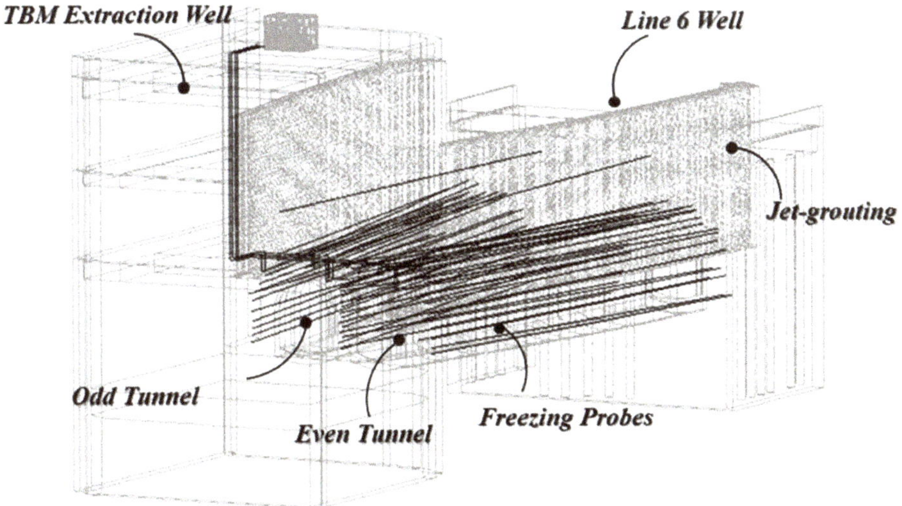

Figure 4. Axonometry of the two tunnels connecting Line 1 and Line 6.

In order to construct the two tunnels connecting Line 1 and Line 6, the excavation of the odd tunnel occurred before the one in the even tunnel. The odd tunnel excavation began after freezing the soil around it, by cutting the diaphragm of the Line 6 station well (see Figure 3), at the opposite point of nitrogen input into the probes, and continuing with the excavation one meter at the time, with the laying of steel ribs and spritz beton until the TBM extraction well was reached. Once the odd tunnel excavation operations were completed, the even tunnel was frozen and constructed.

On the exterior side of the tunnels, two jet grouting walls reaching the depth of the tuff bench had been employed with the purpose of containing fluids motion in the ground. The freezing of the tunnels has been realized using 43 freezing probes (23 for the odd tunnel, and 20 for the even one) with a length of about 40 m, arranged in an arch outside the excavation section with a constant wheelbase equal to 0.75 m. The installed probes are made up of two concentric tubes, as shown in Figure 5. The outer one is made of steel and has a diameter equal to 76 mm, while the inner one is made of copper, with a diameter of 28 mm.

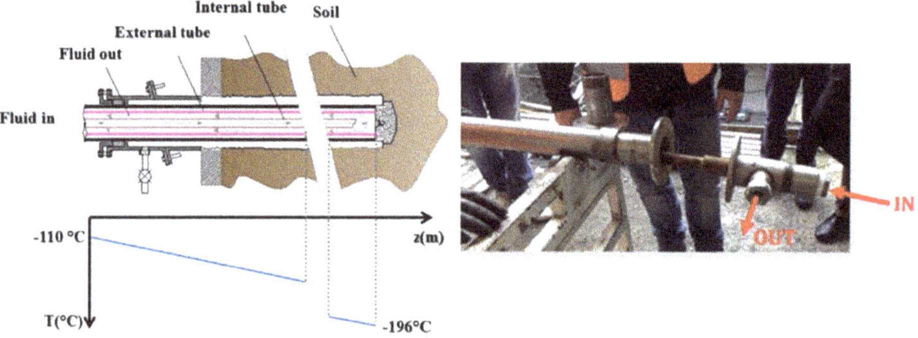

Figure 5. Freezing probe configuration and refrigerant direction: (**left**) scheme of freezing probe and temperature profile along the probes; (**right**) picture of the actual freezing probe.

3. Mathematical Model of the Freezing Process

The model is based on 2D conductive heat transfer in the ground surrounding the probes, and takes into account the phase change phenomenon of water. The mathematical model has been implemented within the commercial software Comsol Multiphysics, based on finite element discretization technique, and has been solved by using the MUltifrontal Massively Parallel sparse direct Solver. The ground subdomain has a depth of 20 m and a length of 35 m, and can be considered sufficiently large to avoid thermal interference with the external environment, and sufficiently deep to assume an undisturbed soil temperature. Figure 6 shows the cross section of the computational domain considered in the present analysis.

The assumptions underlying the present model are the following: (i) homogeneous and isotropic materials in each layer of the computational domain; (ii) thermophysical properties of the soil varying with temperature, between the frozen and unfrozen phases; (iii) for the whole volume of soil, phase transition takes place at a temperature of 0 °C within an interval of 1 °C; (iv) the temperature of the cooling fluid in the probes varies linearly along the axis; (v) heat transfer is purely conductive in the soil, due to the limited convective motion of the water in the ground.

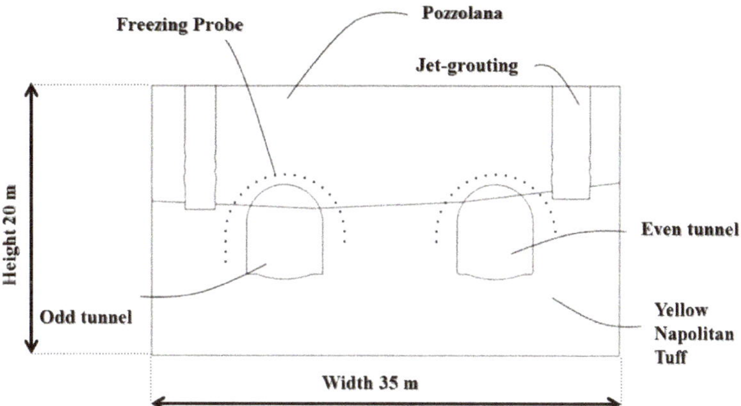

Figure 6. Sketch of the computational domain with the two tunnels considered in the numerical analysis.

3.1. Governing Equations

The problem under investigation has been simulated by means of a dynamic model reproducing the 2D conductive heat transfer in the ground [15,16], taking into account the phase change of water in the soil. The governing equation for heat transfer is reported as follows:

Transient conduction heat transfer

$$\rho_i c_{p_i} \frac{\partial T}{\partial t} = \frac{\partial}{\partial x}\left(k_i \frac{\partial T}{\partial x}\right) + \frac{\partial}{\partial y}\left(k_i \frac{\partial T}{\partial y}\right) + Q \qquad i = 1, \ldots\ldots\ldots, n \tag{1}$$

where ρ_i is the density of the materials constituting the subdomain (kg/m^3), c_{pi} is the specific heat capacity (J/kg·K), k_i is the thermal conductivity (W/m·K), T is the temperature (K), and finally, Q is the heat generated or absorbed per unit of volume (W/m^3). This last term on the right-hand side allows us to model the latent heat of solidification, such as the heat absorbed or released at a constant temperature during the phase change of water in the soil. In fact, this phenomenon is characterized by a significant variation of the thermal diffusion coefficient and the specific heat of the saturated soil, in addition to the absorption of melting latent heat. The time-step employed in the simulations is equal to twelve hours.

3.2. Phase Change in the Soil

The formulation used in the present work provides the latent heat as an additional term in the heat capacity.

Instead of adding the latent heat L in the energy balance equation exactly when the material reaches its phase change temperature, Tpc, it is assumed that the transformation occurs in a temperature interval between Tpc − ΔT/2 and Tpc + ΔT/2. ΔT is the temperature interval which occurs within the phase change of water. In this interval, the material phase is modeled by a smooth function, ϑ, representing the fraction of phase change during transition, which is equal to 1 below Tpc − ΔT/2 and to 0 above Tpc + ΔT/2. The density, ρ, and the specific enthalpy, h, of the ground are then calculated as:

$$\rho = \vartheta \rho_{phase1} + (1 - \vartheta)\rho_{phase2} \tag{2}$$

$$h = \vartheta \rho_{phase1} h_{phase1} + (1 - \vartheta)\rho_{phase2} h_{phase2} \tag{3}$$

where *phase*1 and *phase*2 indicate the characteristics of the material during the different phases of water within the soil. The specific heat at constant pressure can be defined as:

$$\overline{c_p} = \left(\frac{\partial h}{\partial T}\right)_p \tag{4}$$

that becomes, with the product derivatives:

$$\overline{c_p} = \frac{1}{\rho}\left(\vartheta_1 \rho_{phase1} c_{p,phase1} + \vartheta_2 \rho_{phase2} c_{p,phase2}\right) + \left(h_{phase2} - h_{phase1}\right)\frac{\partial \alpha_m}{\partial T} \tag{5}$$

where ϑ_1 and ϑ_2 are, respectively, equal to ϑ and $1 - \vartheta$. The term α_m is defined as:

$$\alpha_m = \frac{1}{2}\frac{\vartheta_2 \rho_{phase2} - \vartheta_1 \rho_{phase1}}{\rho} \tag{6}$$

and it is assumed equal to −1/2, before the phase change process, and 1/2 at the end of the process.

Therefore, the specific heat during the phase change phenomenon is given by the sum of two terms, one proportional to the equivalent thermal capacity C_{eq}:

$$c_{eq} = \frac{1}{\rho}\left(\vartheta_1 \rho_{f1} C_{p,f1} + \vartheta_2 \rho_{f2} c_{p,f2}\right) \tag{7}$$

and the other proportional to the latent heat C_L:

$$c_L(T) = \left(h_{f2} - h_{f1}\right)\frac{d\alpha_m}{dT} = L\frac{d\alpha_m}{dT} \tag{8}$$

so that the total heat per unit of volume released during the phase change process is equal to the latent heat of solidification:

$$Q = \int_{T_{f2+\frac{\Delta T}{2}}}^{T_{f1}+\frac{\Delta T}{2}} C_L(T)dT = L\int_{T_{f2+\frac{\Delta T}{2}}}^{T_{f1}+\frac{\Delta T}{2}} \frac{d\alpha_m}{dT}dT \tag{9}$$

Finally, the apparent thermal capacity C_p used in the heat conservation equation, is given by:

$$c_p = \frac{1}{\rho}\left(\vartheta_1 \rho_{phase1} c_{p,phase1} + \vartheta_2 \rho_{phase2} c_{p,phase2}\right) + c_L \tag{10}$$

The effective thermal conductivity of the portion of soil affected by the phase change is expressed as:

$$k = \vartheta_1 k_{phase1} + \vartheta_2 k_{phase2} \tag{11}$$

while the effective density is calculated as:

$$\rho = \vartheta_1 \rho_{phase1} + \vartheta_2 \rho_{phase2} \tag{12}$$

Finally, continuity of heat flux is assumed on internal interfaces between the materials. To solve the equation of transient heat conduction, appropriate values must be assigned to the coefficients ρ_i, c_{pi}, k_i. The values used in this work have been derived from the literature (Papakonstantinou et al. [6] and Rocca [3]). The thermal characteristics as mineral density, dry density, porosity, wet density of the soil layers and jet-grouting, and the thermal characteristics dependent on the frozen and unfrozen phase, as thermal conductivity and heat capacity, are reported in Table 1.

Table 1. Characteristics of the soil layers and jet-grouting ([3,6]).

Property	Tuff	Pozzolana	Jet-Grouting
Porosity, n	0.5	0.51	
Mineral density, ρ_s (kg/m^3)	2713	2392	3000
Dry density, ρ_d (kg/m^3)	1223	1172	
Wet density, ρ_{wet} (kg/m^3)	1733	1682	
Unfrozen/frozen ground properties (at 16 °C/−50 °C)			
Thermal conductivity, k (W/mK)	1.48/3.14	1.28/2.61	1.40
Heat capacity, c_v (kJ/m^{3}K)	3120/1990	3150/2709	900

3.3. Initial and Boundary Conditions

The initial condition in the whole domain is:

$$T(x, y, 0) = T_0 = 16\ ^\circ\mathrm{C} \qquad \forall (x, y)\ \epsilon\ \Omega \tag{13}$$

where Ω is the computational domain for each of the three sections considered in this work and reported in Figure 3.

The following Dirichlet condition is imposed on the external surface of each probe during Phase 1 and Phase 3 of the AGF process:

$$T(x, y, \vartheta) = T(\vartheta)\ \ \forall (x, y)\ \epsilon\ \Omega_{probes} \tag{14}$$

The boundary conditions employed in the present model refer to the soil and probes domain and are sketched in Figures 7 and 8. In particular, Figure 7 refers to the domain considered before excavation of the first tunnel, while Figure 8 refers to the domain considered after excavation of the first tunnel. The temperature of the top, bottom and lateral surfaces of the soil has been assumed to be constant during the analysis, equal to the average yearly temperature of the site under investigation, $T_0 = 16$ °C.

As previously specified, due to the complexity of nitrogen phase change phenomena occurring inside the freezing probes, a linear temperature profile has been assumed for the refrigerant fluid between the inlet and outlet sections of the probes (refer to Figure 5).

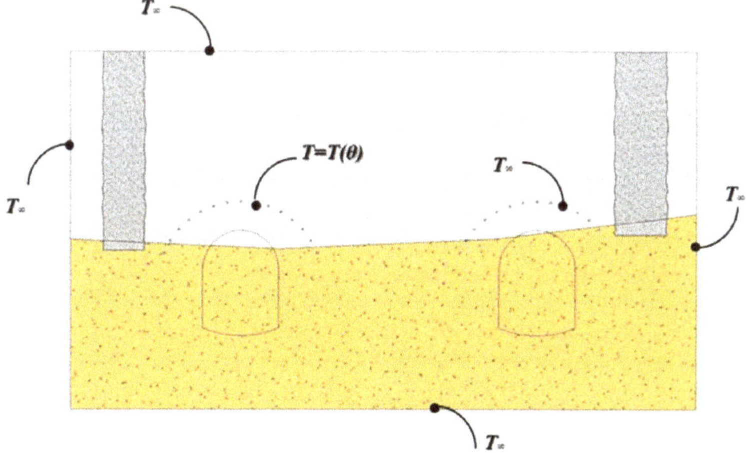

Figure 7. Computational domain and boundary conditions before the excavation.

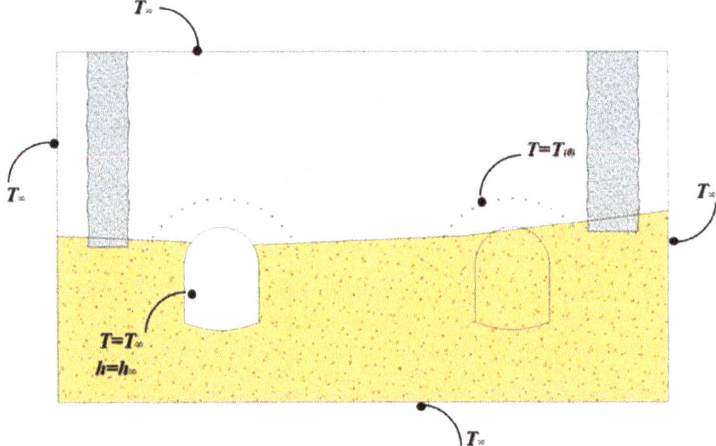

Figure 8. Computational domain and boundary conditions after the excavation of the odd tunnel.

The temperature boundary conditions applied on the external perimeter of each probe depend on the phase of the freezing process. During "Phase 1-Nitrogen", the temperature has a linear variation along the axis, from −196 °C to −110 °C, as shown in Figure 5. During "Phase 2-Waiting", the adiabatic condition, $\overline{\nabla} T{\cdot}n = 0$, has been imposed on the probe boundary.

During "Phase 3-Brine", the temperature of the probe boundary has been imposed equal to the temperature of the brine, −33 °C, in all the sections of the excavation.

In order to simulate the excavation of the even tunnel, the same steps employed for that of the odd tunnel have been considered. During these phases, the excavated odd tunnel (left) has been reproduced by eliminating the corresponding domain of soil and applying a proper boundary condition (refer to Figure 8). This condition takes into account the presence of men, vehicles and air circulation in the excavated tunnel, and is represented by convective heat transfer on the walls of the odd tunnel:

$$-k\overline{\nabla} T{\cdot}n = h(T - T_\infty) \tag{15}$$

$$h(x, y, \vartheta) = h_\infty = 15\ \mathrm{W/m^2K} \tag{16}$$

where T_∞ is equal to 30 °C.

3.4. Mesh Sensitivity Analysis

A mesh sensitivity analysis has been carried out in order to obtain grid-independent numerical results. A domain of 20 × 25 m^2 has been considered. All the computational grids are made by triangular quadratic elements, and are refined near the freezing probes (Figure 9).

Table 2 reports the details of the eight grids considered, together with a summary of the main numerical results. In particular, considering the nitrogen activation (phase 1), the days required for the formation of the frozen wall at 1.5 m have been calculated and reported in Table 2, together with the computing time necessary to reach the convergence. The nitrogen freezing phase is stopped when the desired design value of the frozen wall thickness is reached. The evolution of the frozen wall can be monitored, controlling the temperature of the ground at 0.5 m from the probe axis, which generally must be around −10 °C. Therefore, in the analysis, the temperature at this point has been taken into account.

On the basis of the present sensitivity analysis, the grid employed for the calculations is the one with 103,482 elements (letter f in the table), since the difference between the results obtained by using this grid, and the ones obtained by using the most effective adaptive grid, is around 1%.

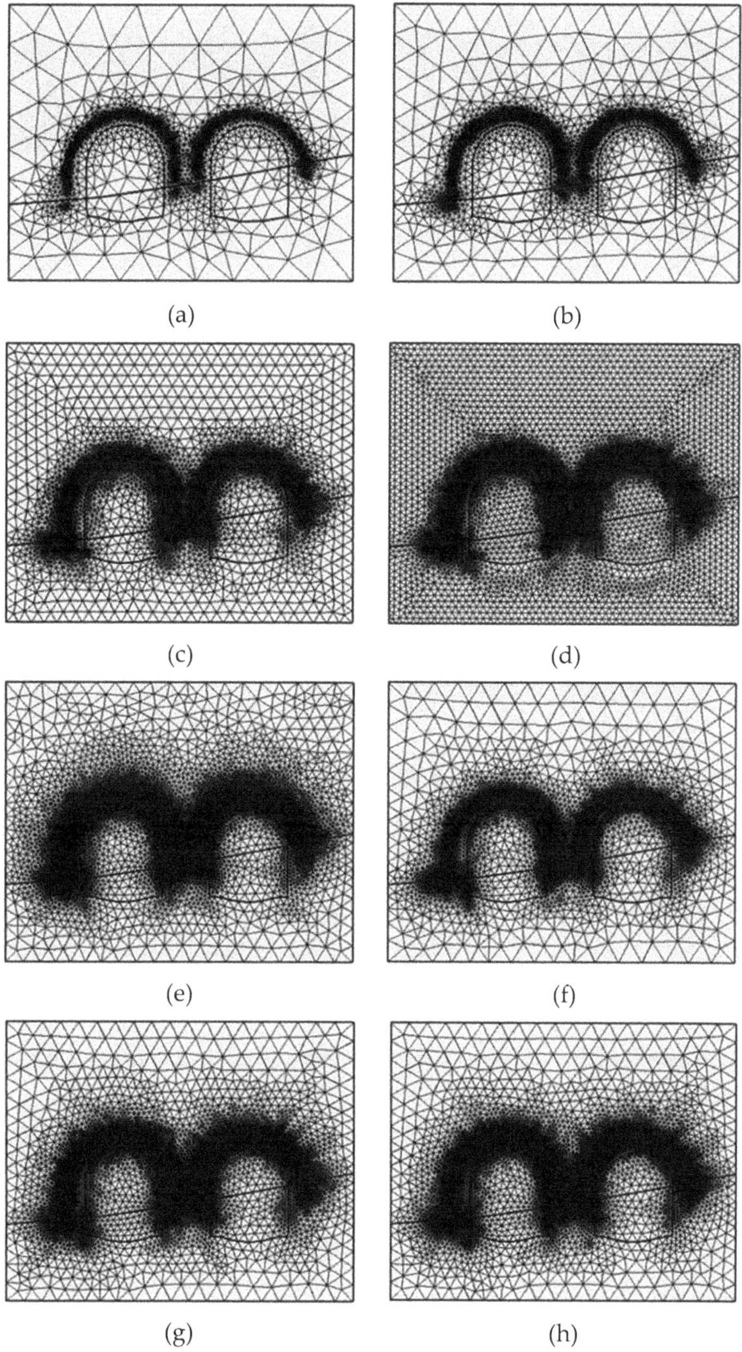

Figure 9. Meshes used for the grid independent study (**a**) 5339 elements; (**b**) 9596 elements; (**c**) 42,671 elements; (**d**) 73,804 elements; (**e**) 82,046 elements; (**f**) 103,482 elements; (**g**) 113,342 elements; (**h**) 177,964 elements.

Table 2. Details of the meshes and obtained results for the grid-independent study.

Mesh Code	Number of Elements (-)	Time for Frozen Wall Formation at 0.6 m (d)	Temperature at 0.5 m (°C)	Computing Time (min)
a	5339	9	−13.1	18
b	9596	9	−14.6	33
c	42,671	8	−12.3	399
d	73,804	8	−12.0	1231
e	82,046	8	−11.3	1820
f	103,482	8	−11.4	4213
g	113,342	8	−11.6	8122
h	177,964	8	−11.6	25,015

4. Model Validation

The present model has been validated against the numerical data reported by Colombo [5], which is validated with on-field data. In that study, the software ABAQUS was used to solve the numerical model, and the computational domain was meshed by using DC2D4 elements. The geometry of the tunnel was symmetric. The present study reproduces the case study of Colombo [5], based on a 2D model, by taking into account a computational domain representative of a portion of land equal to 10×20 m. A mesh consisting of 33,605 triangular elements has been considered, and the thermal characteristics of the materials are those reported in Colombo [5], in particular: volumetric heat capacity of 1910 kJ/m^{3}K and 3100 kJ/m^{3}K for solid and liquid phases, respectively; thermal conductivity of 3.07 W/mK and 1.48 W/mK for the solid and liquid phases, respectively; volumetric latent heat of 179,280 kJ/m^3 and a saturated tuff density of 1550 kg/m^3. The initial temperature has been assumed equal to 18 °C, and the analysis has been carried out, imposing a linear variation of temperature down to −33 °C for the first day, on the nodes representing the perimeter of the freezing probe. In order to compare the results obtained from the present FEM analysis with those reported in Colombo [5], the authors have considered two points located on a line orthogonal to the junction between the probes, at 0.50 m and 0.90 m. Figure 10 shows the conditions that determine the propagation velocity of the freezing front in the tuff, evaluated for two points located at 0.5 m and 0.9 m from the freezing probe. It is evident that the temperature gradually decreases over time, and that a good agreement between the present numerical results and those available in the literature [5] is observed.

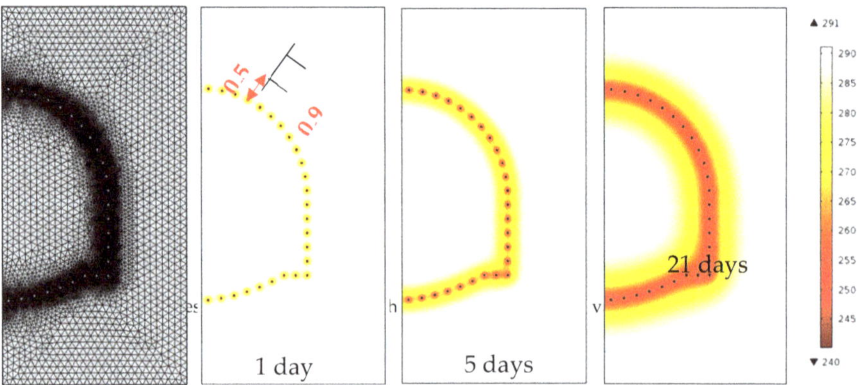

Figure 10. *Cont.*

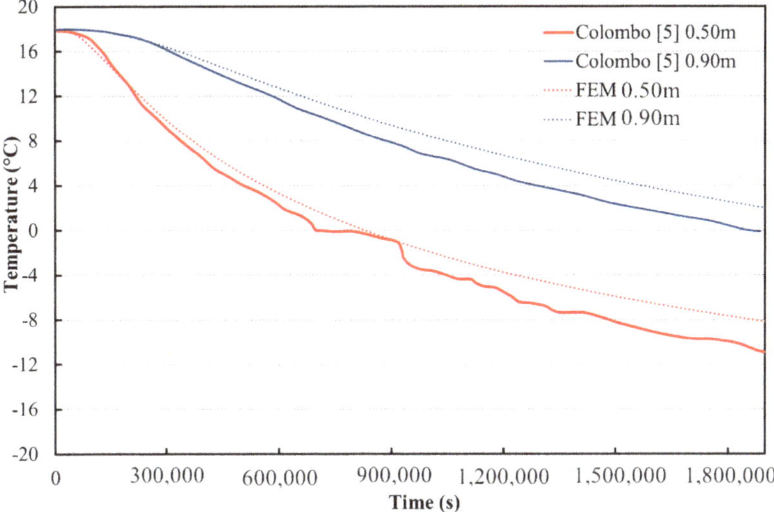

Figure 10. Comparison between field emission microscopy (FEM) analysis and Colombo [5].

5. Results and Discussion

5.1. Odd Tunnel Freezing

The numerical analysis has been developed considering the probes connected in parallel, i.e., all with the same temperature distribution. The boundary conditions on the surface of probes, described in Section 3, are summarized in Table 3, together with the calculated thickness of frozen wall after 14 days.

Table 3. Boundary conditions imposed on the surface of the probes at the considered sections and for the different freezing phases, and frozen wall thickness calculated after 14 days of AGF with liquid nitrogen.

		Phase 1		Phase 2		Phase 3		
Section	z (m)	Duration (days)	Probe Temperature (°C)	Duration (days)	Probe Temperature (°C)	Duration (days)	Probe Temperature (°C)	Thickness of *frozen wall* (m)
A-A	5	14	−180	1	Adiabatic	15	−33	1.50
B-B	20	14	−160	1	Adiabatic	15	−33	1.70
C-C	35	14	−120	1	Adiabatic	15	−33	1.90

The results refer to the freezing process in the odd tunnel (left), for which it has been assumed duration of the direct freezing phase equal to 14 days. The waiting phase, before switching to the indirect method phase with brine, has been imposed equal to 1 day, enough to have suitable temperatures for the brine intake in the pipe system, without freezing it. The third phase of brine retention has been considered to last for 15 days, for a total of 30 days of the whole freezing process in the odd gallery.

Figure 11 reports the temperature field calculated after 14 days of nitrogen feeding in the three sections considered for the odd tunnel (refer to Figure 3), and it is possible to clearly see the formation of the frozen wall, since the figure reports only the temperature values below −2 °C.

In section C-C, the frozen wall reaches the desired thickness of 1.5 m in less time than the other two sections, due to the temperature distribution in the probes. The thickness of 1.5 m ensures the static stability of the ground, according to the design specifications. After 14 days of freezing with nitrogen, the frozen wall reaches the thickness reported in Table 3 for the three sections.

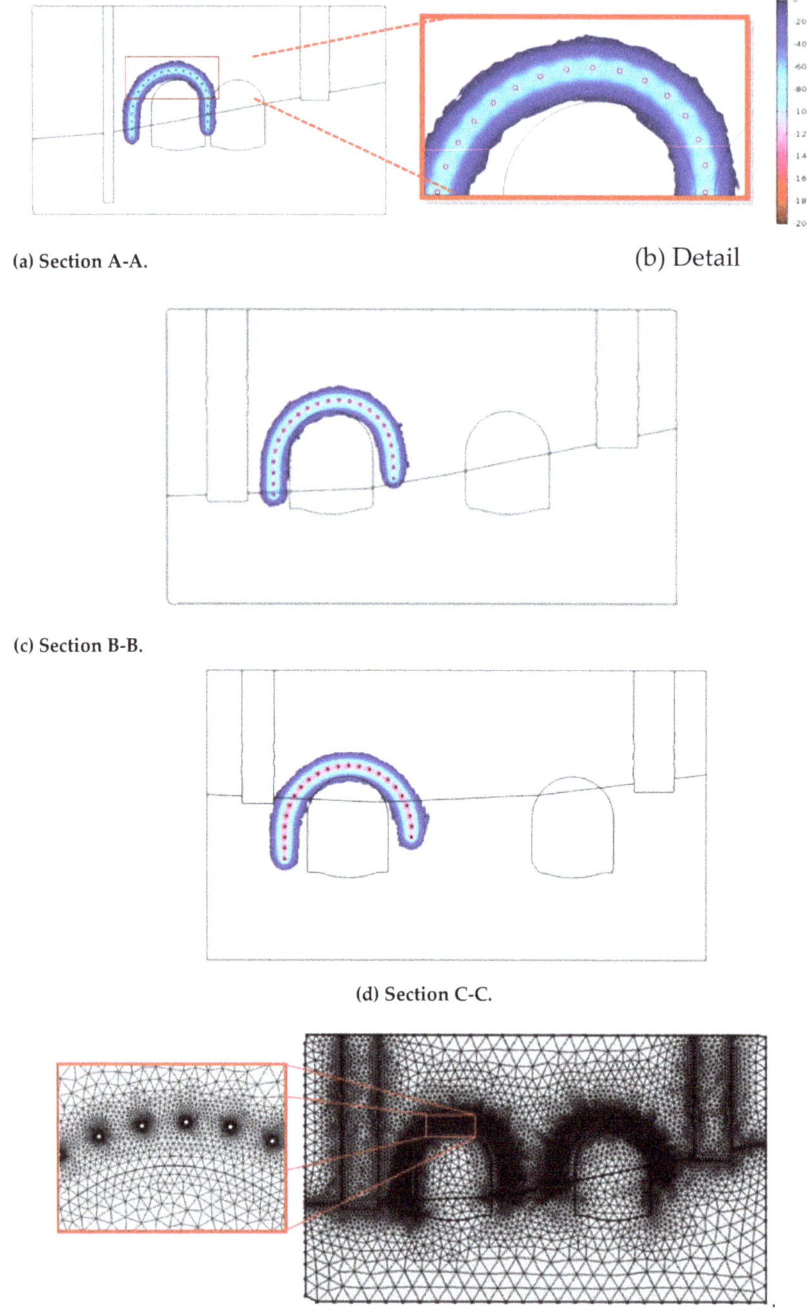

(a) Section A-A.

(b) Detail

(c) Section B-B.

(d) Section C-C.

(e) Detail of the mesh employed in the simulations.

Figure 11. From (**a**–**d**): Temperature field below −2 °C after 14 days of AGF with liquid nitrogen in the three sections of the odd tunnel reported in Figure 3; (**e**) detail of the employed mesh.

Figure 12a shows the frozen wall thickness reached after 14 days of nitrogen freezing phase, by showing the isotherms at −2 °C, and reports a detail of the frozen wall for section A-A. Figures 13b and 12c show the frozen wall thickness reached after 14 days of nitrogen freezing phase, by showing the isotherms at −2 °C, for sections B-B and C-C. Moreover, the figure presents a vertical segment across probe n. 13, that is considered useful for the analysis of the temperature field in the frozen wall.

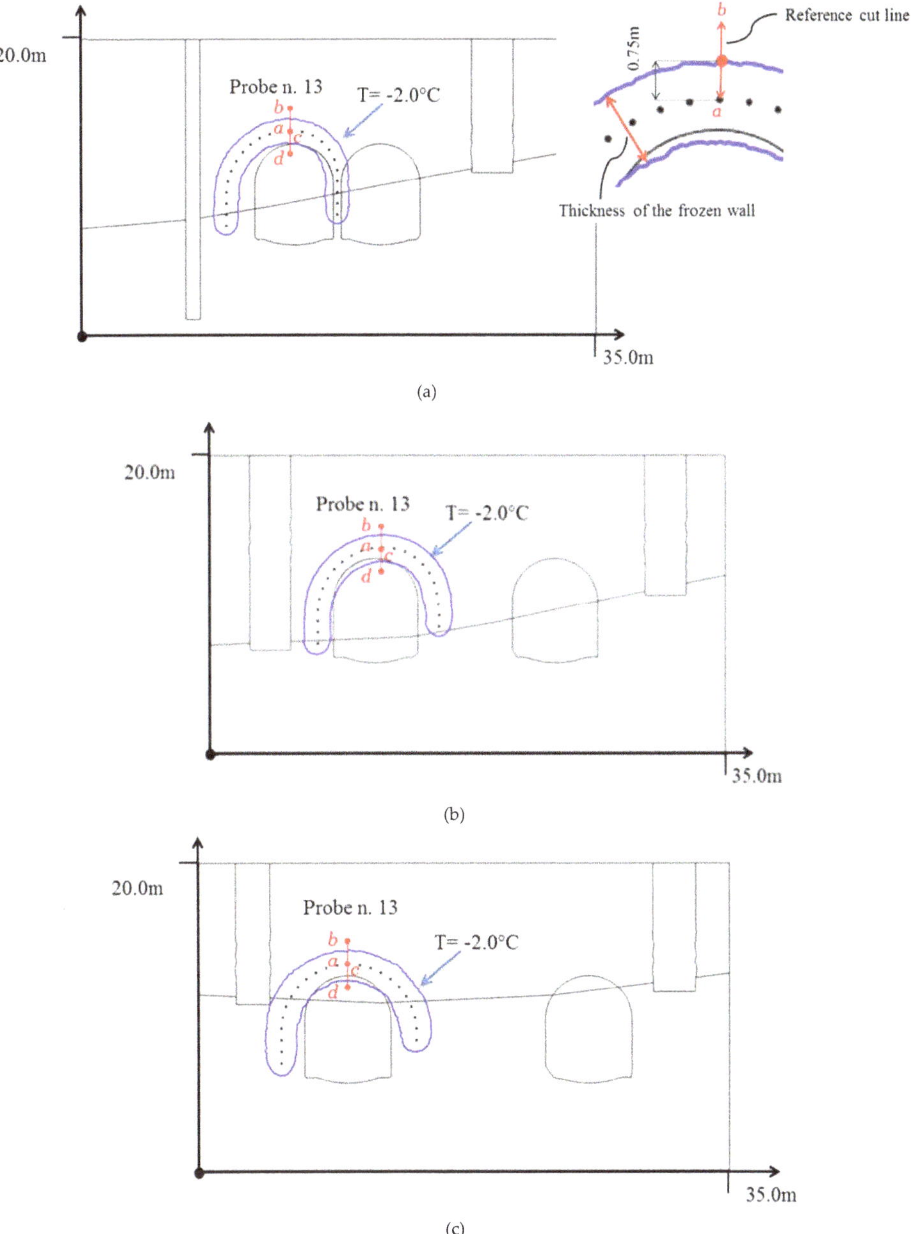

Figure 12. Thickness of the frozen wall represented through isotherm at −2 °C, for section A-A (**a**), section B-B (**b**); section C-C (**c**).

Figure 13 shows the temperature profile calculated on a vertical segment of 1.5 m from probe 13 (point a) to point b for the three sections considered, as shown in the previous figure. From the analysis of the figure, it is evident that at the end of the first phase, the frozen wall has reached the minimum required thickness of 1.5 m.

Figure 14 shows the temperature profile and thickness of the frozen wall with respect to the probe axis, after 14 days of freezing, for sections A-A, B-B and C-C. The graph shows that the frozen wall at section A-A after 14 days of activation with nitrogen has a thickness of 1.5 m for section A-A, 1.7 m for section B-B, and 1.9 m for section C-C.

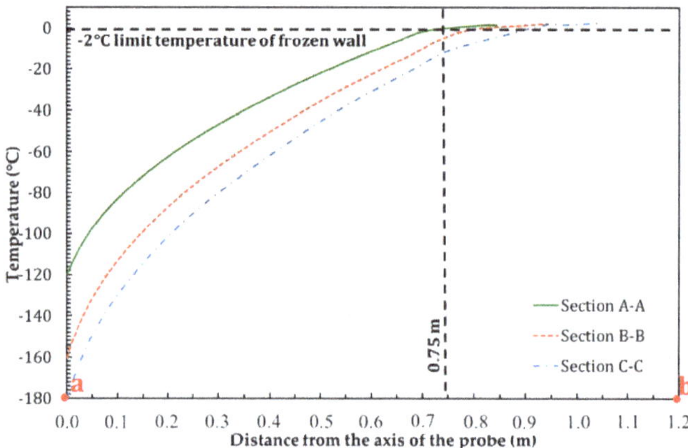

Figure 13. Temperature profile after 14 days of artificial ground freezing (AGF) with liquid nitrogen along the segment a-b.

For the odd tunnel (left), further segments are considered, in order to verify the thickness of the ice wall at the end of Phase 1. Figure 15a shows the cut lines taken into account, while Figure 15b shows the temperature profiles calculated after 14 days for section C-C.

Figure 16a presents the temperature profiles at the sections A-A, B-B, and C-C. From the analysis of this figure, it is possible to notice that the section located 5 m from the TBM well (A-A) is the most disadvantaged for the freezing process. This is due to the nitrogen freezing temperature imposed on the probe perimeter in this section (−120 °C), which is higher than that in the other sections. Consequently, more time is needed for the formation of the frozen wall in this section, defining the total duration of the freezing Phase 1 (nitrogen activation). After almost 5 days of activation with brine, constant temperature values can be reached, but always below the safety value of −5 °C.

Moreover, it is possible to notice that the ground reaches the temperature of 0 °C after about ten days of liquid nitrogen feeding. This finding is in good agreement with the results reported by Colombo [17] and Manassero [18]. Moreover, from the analysis of Figure 16a, it is possible to notice that the defrosting phase for the odd tunnel starts after 60 days and that after 90 days, the temperature rises up to 0 °C. These temperature values are important to evaluate the ground displacement.

Figure 16b shows the temperature profile for the freezing process of the even tunnel. The beginning of the freezing with nitrogen starts at the end of the freezing of the odd tunnel.

Table 4 shows the thickness of the frozen wall after 14 days of artificial ground freezing with nitrogen and is always larger than the design value of 1.5 m for both tunnels.

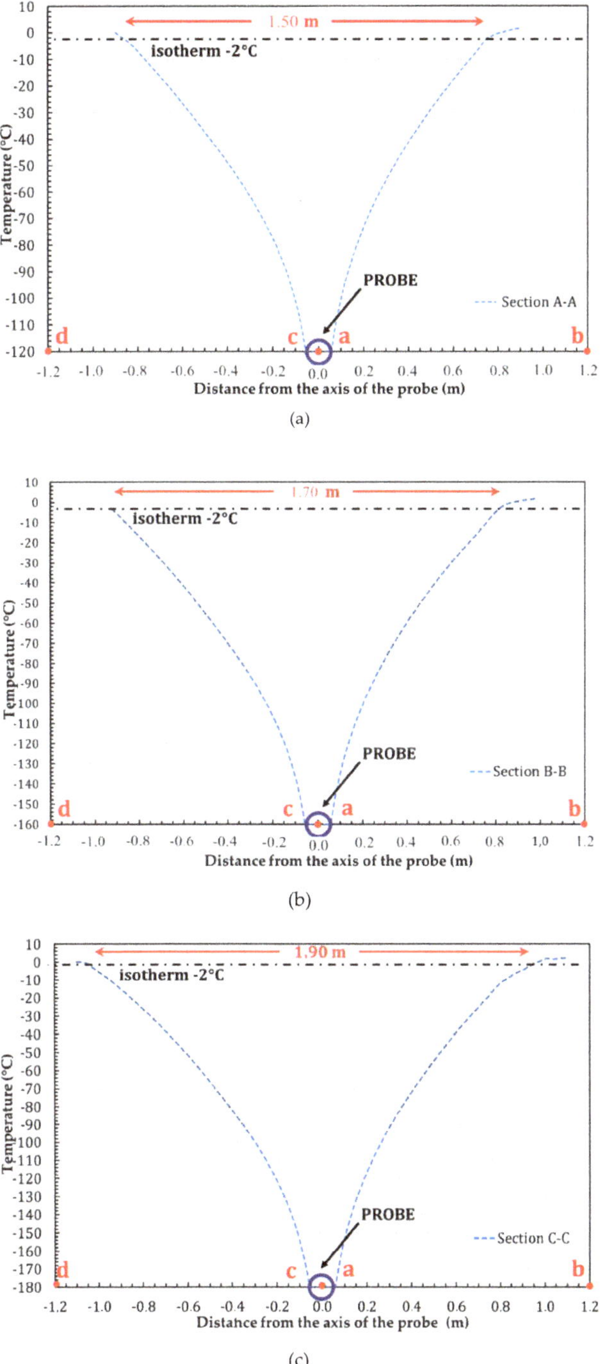

Figure 14. Temperature profile after 14 days of AGF with liquid nitrogen on segment a-b at Section A-A (**a**); section B-B (**b**); section C-C (**c**).

Table 4. Thickness of the frozen wall after 14 days of AGF with nitrogen liquid.

Odd Tunnel Section	z (m)	Thickness *Frozen wall* (m) a-a′	b-b′	c-c′	d-d′	e-e′	f-f′	g-g′
C-C	35	2.0	2.1	2.1	2.1	2.0	2.1	2.0
Even Tunnel Section	**z (m)**	**Thickness *Frozen wall* (m) a-a′**	**b-b′**	**c-c′**				
A-A	5	1.8	1.6	1.6				
B-B	20	1.9	1.9	1.9				
C-C	35	2.1	2.1	2.1				

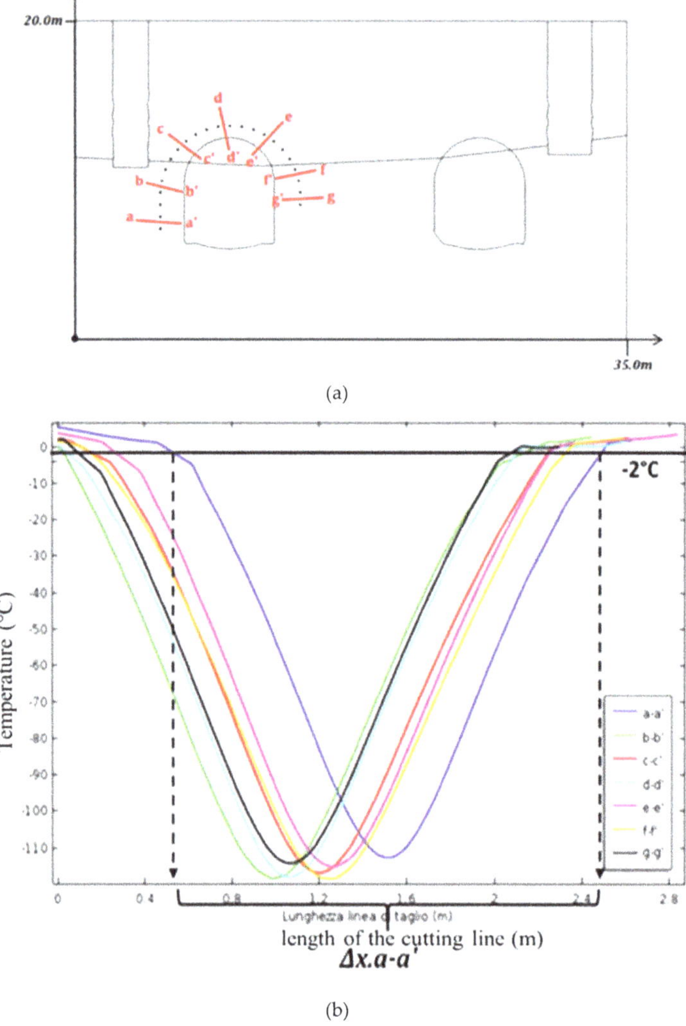

Figure 15. (**a**) Cutline for the section C-C; (**b**) Temperature trend after 14 days for section C-C and frozen wall was formed for all cut lines considered.

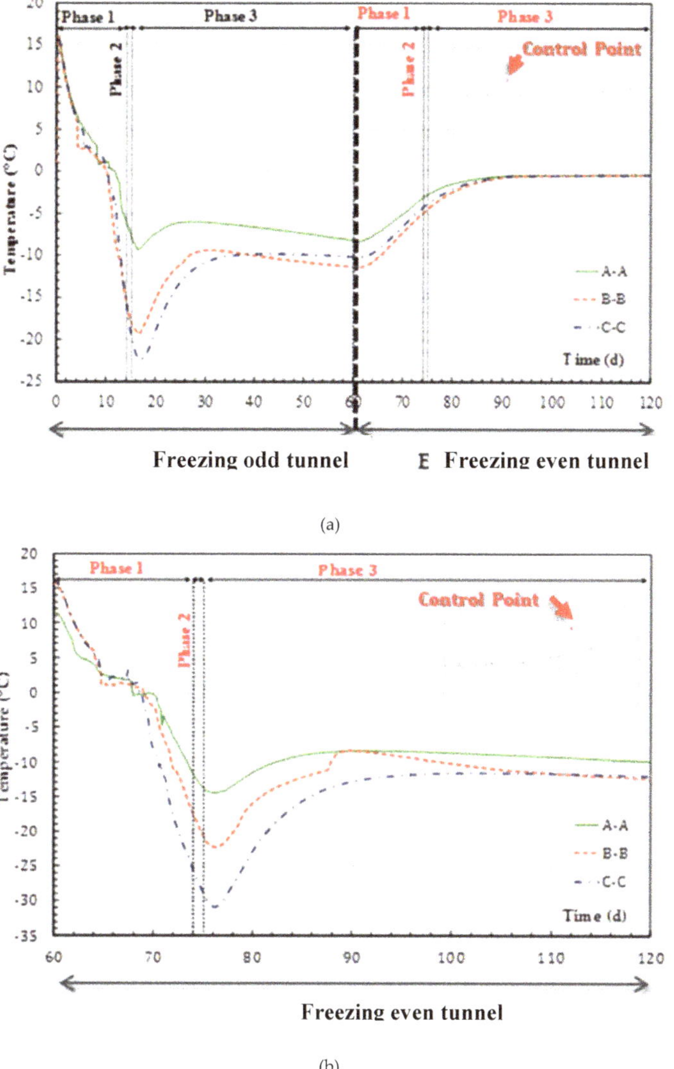

Figure 16. (**a**) Temperature profiles at the control point on the top of the odd tunnel for the three sections; (**b**) Temperature profiles at the control point on the top of the even tunnel for the three sections.

5.2. Even Tunnel Freezing

At the end of the 60 days, after the freezing and excavation activities of the odd tunnel, freezing is repeated with the same conditions for the even tunnel, disabling completely the supply of brine for the maintenance phase of the odd tunnel. The temperature applied on the freezing probes for the three freezing phases is reported in Table 3. The duration of each freezing phase of the even tunnel is equal to fourteen days for Phase 1 with nitrogen, one day for Phase 2, and forty days for Phase 3 (brine maintenance), for a total of sixty days. The results obtained at the end of the first phase with nitrogen for the even tunnel have been reported in Figure 17.

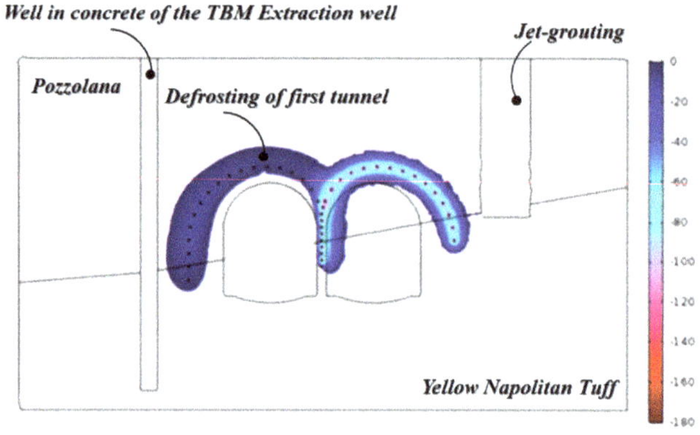

(a) Section A-A

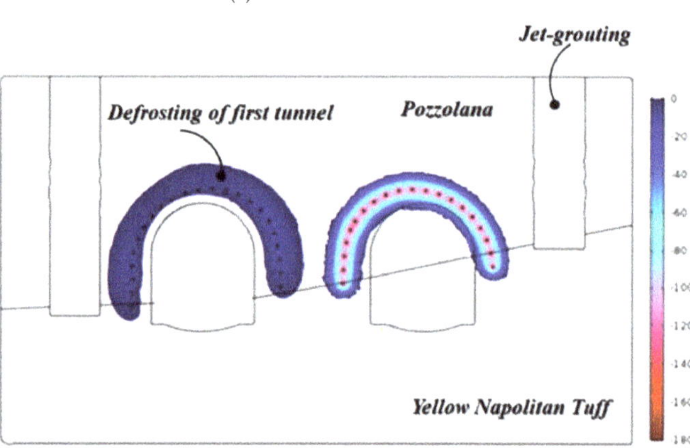

(b) Section B-B

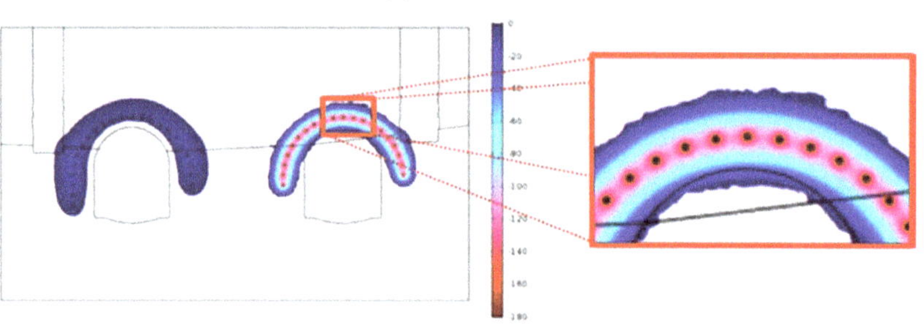

(c) Section C-C

Figure 17. Frozen wall at 0 °C after 14 days.

The results of artificial freezing of soil with the use of liquid nitrogen, after fourteen days from the beginning of the phase, are presented in Figure 18. In particular, the thickness of the frozen wall in the three sections is reported, identified by the isotherm at −2 °C. The minimum thickness of 1.5 m is used as a design target necessary to consider Phase 1 completed.

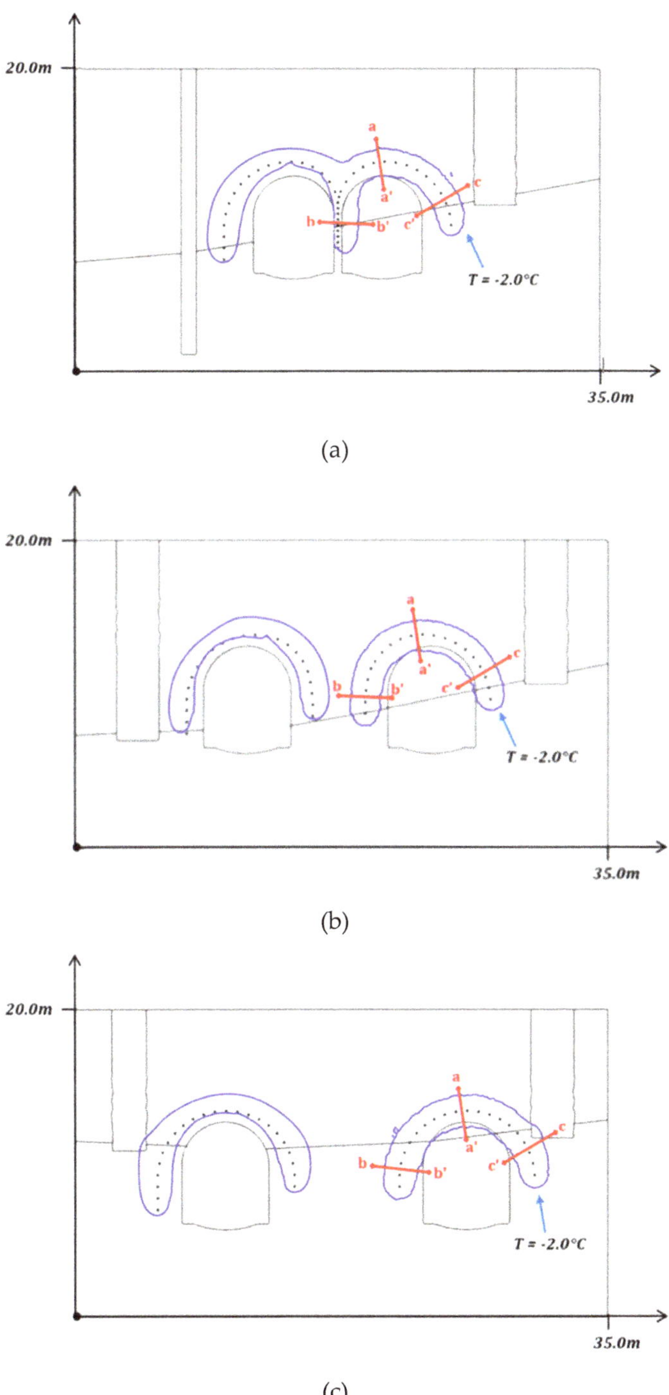

Figure 18. *Cont.*

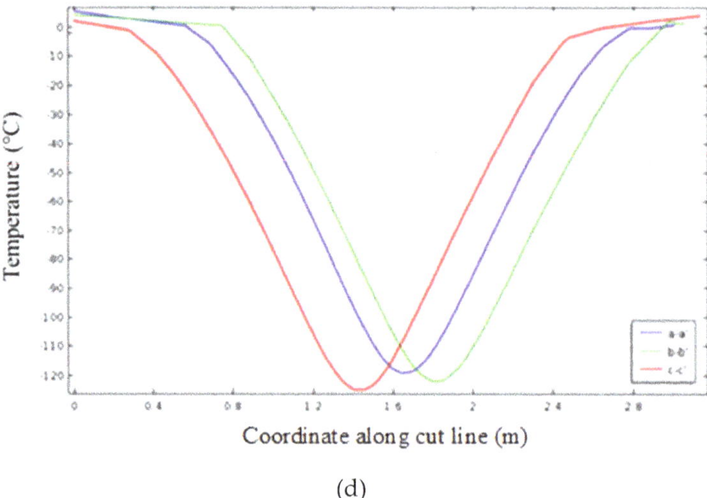

(d)

Figure 18. The frozen wall after 14 days of nitrogen feeding for section A-A (**a**); for section B-B (**b**); for section C-C (**c**); Temperature profile for section C-C (**d**).

Figure 18d shows the temperature profile at three cut lines of section C-C during the freezing process of the even tunnel. From the analysis of this figure, it is possible to observe that the temperature values below −2 °C involve a thickness of ground around 2.1 m, in all the three cut lines considered, as reported in Table 4. This is a relevant result, proving that the minimum ice wall limit (1.5 m) is respected in the whole ice vault.

6. Conclusions

This work presents an analysis of the heat transfer phenomena occurring during the artificial ground freezing (AGF) process on the actual case study of the tunnels between Line 1 and Line 6 of the Underground station in Piazza Municipio in Napoli, southern Italy. An efficient numerical model, based on conductive heat transfer and water phase change, has been developed and validated against the data available in the literature.

The present model has us allowed to simulate, for the first time in the literature, a mixed-method used for the freezing process from the first phase, based on the use of nitrogen, a maintenance phase that allows us to raise the temperature of the probes, and a third phase that involves the use of brine. The numerical model has allowed us to reproduce the evolution of the temperature field during the whole excavation process. The model has been used to illustrate the heat transfer phenomena associated with the phase change and the influence of latent heat, the influence of phase change, and temperature variation.

The present modeling activity allows us to identify the possible solutions for reducing the time required for the completion of the excavation activities and the freezing of the galleries. In addition to properly planning the nitrogen feeding phase, this approach allows the analysis of alternative solutions to accelerate the freezing of the soil, such as increasing the number of probes, or using different configurations.

The results show that the time needed to complete the freezing process and excavation of the two tunnels, consists of 120 days. Instead, 14 days are required to obtain an ice vault with a thickness larger than the minimum value required for safety reasons (1.5 m). In particular, the section that first reaches the formation of the frozen wall is the one at 5 m from the well of Line 6, and from this section, the excavation operations can begin. Moreover, the defrosting phase for the odd tunnel starts after 60 days, and after 90 days, the temperature rises up to 0 °C.

The authors believe that the present model is useful to optimize the AGF technique, and to better understand the heat transfer phenomena occurring between probes and ground.

Author Contributions: Conceptualization, N.M. and A.M.; methodology, Eng. G.N.; software, N.M., A.M., G.N. and P.M.; formal analysis, G.N. and P.M.; resources, N.M.; data curation, G.N.; writing—original draft preparation, G.N.; writing—review and editing, N.M. and A.M.; supervision, N.M. and A.M.; project administration, N.M. and F.C.; funding acquisition, N.M. and F.C. All authors have read and agreed to the published version of the manuscript.

Funding: This research was funded by a research agreement with Ansaldo STS | A Hitachi Group Company.

Acknowledgments: The authors thank the support of Ansaldo STS | A Hitachi Group Company, in particular Eng. Giuseppe Molisso, MetroTec Scarl, and Trevi SpA.

Conflicts of Interest: The authors declare no conflict of interest.

References

1. Russo, G.; Corbo, A.; Cavuoto, F.; Autuori, S. Artificial Ground Freezing to excavate a tunnel in sandy soil. Measurements and back analysis. *Tunn. Undergr. Sp. Technol.* **2015**, *50*, 226–238. [CrossRef]
2. Smith, N. Artificial ground freezing. In *Temporary Works: Principles of Design and Construction*; Grant, M., Pallet, F.P., Eds.; ICE Publishing: London, UK, 2012; pp. 109–116.
3. Rocca, O. *Congelamento Artificiale del Terreno*; Hevelius Edizioni: Napoli, Italy, 2011.
4. Andersland, O.B.; Ladanyi, B. *Frozen Ground Engineering*; Wiley: Hoboken, NJ, USA, 2004.
5. Colombo, G. Il congelamento artificiale del terreno negli scavi della metropolitana di Napoli: Valutazioni teoriche e risultati sperimentali. 2010.
6. Papakonstantinou, S.; Pimentel, E.; Anagnostou, G. Analysis of artificial ground freezing in the Pari-Duomo platform tunnel of the Naples metro. *Numer. Methods Geotech. Eng.* **2000**, 281–284. [CrossRef]
7. Pimentel, E.; Papakonstantinou, S.; Anagnostou, G. Numerical interpretation of temperature distributions from three ground freezing applications in urban tunnelling. *Tunn. Undergr. Space Technol.* **2012**, *28*, 57–69. [CrossRef]
8. Vitel, M.; Rouabhi, A.; Tijani, M.; Guerin, F. Modeling heat transfer between a freeze pipe and the surrounding ground during artificial ground freezing activities. *Comput. Geotech.* **2015**, *63*, 99–111. [CrossRef]
9. Vitel, M.; Rouabhi, A.; Tijani, M.; Guérin, F. Modeling heat and mass transfer during ground freezing subjected to high seepage velocities. *Comput. Geotech.* **2016**, *73*, 1–15. [CrossRef]
10. Marwan, A.; Zhou, M.-M.; Abdelrehim, M.Z.; Meschke, G. Optimization of artificial ground freezing in tunneling in the presence of seepage flow. *Comput. Geotech.* **2016**, *75*, 112–125. [CrossRef]
11. Kang, Y.; Liu, Q.; Cheng, Y.; Liu, X. Combined freeze-sealing and New Tubular Roof construction methods for seaside urban tunnel in soft ground. *Tunn. Undergr. Space Technol.* **2016**, *58*, 1–10. [CrossRef]
12. Panteleev, I.; Kostina, A.; Zhelnin, M.; Plekhov, A.; Levin, L. Intellectual monitoring of artificial ground freezing in the fluid-saturated rock mass. *Procedia Struct. Integr.* **2017**, *5*, 492–499. [CrossRef]
13. Alzoubi, M.A.; Madiseh, A.; Hassani, F.P.; Sasmito, A.P. Heat transfer analysis in artificial ground freezing under high seepage: Validation and heatlines visualization. *Int. J. Therm. Sci.* **2019**, *139*, 232–245. [CrossRef]
14. Fan, W.; Yang, P. Ground temperature characteristics during artificial freezing around a subway cross passage. *Transp. Geotech.* **2019**, *20*, 100250. [CrossRef]
15. Colangelo, F.; De Luca, G.; Ferone, C.; Mauro, A. Experimental and numerical analysis of thermal and hygrometric characteristics of building structures employing recycled plastic aggregates and geopolymer concrete. *Energies* **2013**, *6*, 6077–6101. [CrossRef]
16. Carotenuto, A.; Marotta, P.; Massarotti, N.; Mauro, A.; Normino, G. Energy piles for ground source heat pump applications: Comparison of heat transfer performance for different design and operating parameters. *Appl. Therm. Eng.* **2017**, *124*, 1492–1504. [CrossRef]

17. Colombo, G.; Lunardi, P.; Cavagna, B.; Cassani, G.; Manassero, V. The artificial ground freezing technique application for the Naples underground. In Proceedings of the World Tunnel Congress, Agra, India, 22–24 September 2008; pp. 1–15.
18. Manassero, V.; Di Salvo, G.; Giannelli, F.; Colombo, G. A Combination of artificial ground freezing and grouting for the excavation of a large size tunnel below groundwater. In Proceedings of the International Conference on Case Histories in Geotechnical Engineering, Arlington, VA, USA, 11–16 August 2008; pp. 1–13.

Article

Investigation of Start-Up Characteristics of Thermosyphons Modified with Different Hydrophilic and Hydrophobic Inner Surfaces

Xiaolong Ma, Zhongchao Zhao *, Pengpeng Jiang, Shan Yang, Shilin Li and Xudong Chen

School of Energy and Power, Jiangsu University of Science and Technology, Zhenjiang, Jiangsu 212000, China; marlon@stu.just.edu.cn (X.M.); pengpengjiang@stu.just.edu.cn (P.J.); shanyang33@stu.just.edu.cn (S.Y.); shilinli@stu.just.edu.cn (S.L.); xudongchen@stu.just.edu.cn (X.C.)

* Correspondence: zhongchaozhao@just.edu.cn; Tel.: +86-0511-84493050

Received: 20 December 2019; Accepted: 7 February 2020; Published: 9 February 2020

Abstract: In this paper, the influence of wettability properties on the start-up characteristics of two-phase closed thermosyphons (TPCTs) is investigated. Chemical coating and etching techniques are performed to prepare the surfaces with different wettabilities that is quantified in the form of the contact angle (CA). The 12 TPCTs are processed including the same CA and a different CA combination on the inner surfaces inside both the evaporator and the condenser sections. For TPCTs with the same wettability properties, the introduction of hydrophilic properties inside the evaporator section not only significantly reduces the start-up time but also decreases the start-up temperature. For example, the start-up time of a TPCT with CA = 28° at 40 W, 60 W and 80 W is 46%, 50% and 55% shorter than that of a TPCT with a smooth surface and the wall superheat degrees is 55%, 39% and 28% lower, respectively. For TPCTs with combined hydrophilic and hydrophobic properties, the start-up time spent on the evaporator section with hydrophilic properties is shorter than that of the hydrophobic evaporator section and the smaller CA on the condenser section shows better results. The start-up time of a TPCT with CA = 28° on the evaporator section and CA = 105° on the condenser section has the best start-up process at 40 W, 60 W and 80 W which is 14%, 22% and 26% shorter than that of a TPCT with smooth surface. Thus, the hydrophilic and hydrophobic modifications play a significant role in promoting the start-up process of a TPCT.

Keywords: Thermosyphon; start-up characteristics; hydrophilic and hydrophobic; contact angle

1. Introduction

As a two-phase passive device, the thermosyphon has a wide-range of various industrial applications, for instance, electronic equipment [1], heat-recovery systems [2], solar water heater systems [3] and space applications [4] due to the simple structure, reliability, high efficiency and low cost. The basic concept of heat pipes was first proposed by Gaugler in 1944 [5]. A thermosyphon is a gravity-assisted heat pipe without wicks that depends on phase-change heat transfer in both the evaporator and condenser sections to transfer large amounts of heat with relatively small temperature difference [6] and low thermal resistance [7]. Figure 1 shows the schematic diagram of the cross section and working principle of a two-phase closed thermosyphon (TPCT). A TPCT is composed of evaporator, adiabatic and condenser sections. The operating process begins with a certain volume of working fluid in the evaporator section, which is then heated by a source of heat, such as a heating element or a thermal bath. The heating converts the saturated liquid into vapor that rises to the condenser section. Afterwards, the vapor condenses into liquid, which flows back down to the evaporator section by gravity, in the process transferring heat to the heat sink, such as cold water [8].

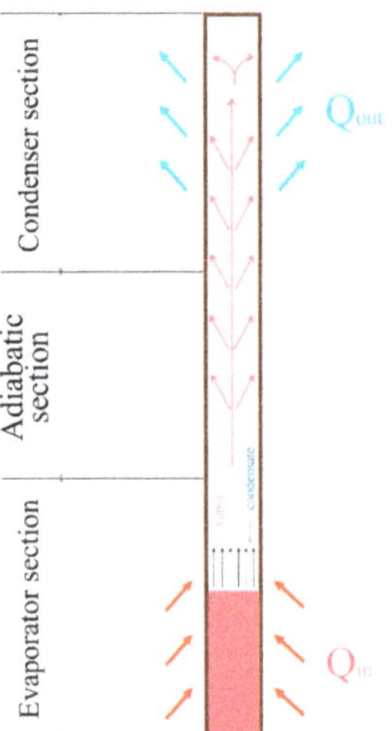

Figure 1. Cross section and working principle of a two-phase closed thermosyphon [9].

At present, the investigation of thermosyphons mainly includes the analysis of the thermal performance and start-up characteristics. In recent years, the thermal performance of thermosyphons in the aspects of filling ratio and surface modifications have been experimentally studied. Lataoui and Jemni [10] conducted an experimental study on a stainless steel thermosyphon to assess the influence of filling ratio, input power and the temperature of the cooling fluid on its thermal performance. Alireza Moradikazerouni et al. [11,12] investigated the effects of surface modifications on the heat sink, and the results found that different structural shapes on the heat surface have various heat transfer mechanisms. Rahimi et al. [13] modified the condenser and evaporator of a thermosyphon and compared the heat performance and resistance of a modified thermosyphon at different input powers with a flat thermosyphon. They found that the average thermal performance at tested heat loads was increased by 15.27% and the average thermal resistance of the thermosyphon was decreased by 2.35 times. Singh et al. [14] investigated the effect of surface modification on the thermal performance in an evaporator and condenser for flat thermosyphons with and without an anodized surface. Solomon et al. [15] studied the heat performance of thermosyphons with surface modifications at diverse inclination angles and input powers. The surface modifications significantly reduce the wall temperature of the evaporator and increase the heat-transfer coefficient. Solomon et al. [16] also analyzed the thermal performance of an anodized surface with a porous structure and observed a 15% reduction in the thermal resistance and 15% increase in the heat-transfer coefficient of the evaporator.

Additionally, Noie [17] investigated the effects of aspect ratio on the thermal performance of a thermosyphon and achieved heat performances of 60%, 90% and 30% for aspect ratios of 11.8, 7.45 and 9.8, respectively. Gedik [18] reported the influence of various operating conditions, such as heat input, inclination angle and the flow rate of cooling water on the heat-transfer characteristics of a TPCT. Moreover, the method whereby a nanofluid was used as working fluid in a thermosyphon has been theoretically and experimentally studied. Ma et al. [19] found that the heat-transfer rate of a nanofluid can rise to 3.11 times in an inclined square enclosure which indicates that the nanofluid is a potential choice as working fluid. Besides, the nanoadditives of various shapes on the fluid flow and heat transfer aspects of a nanofluid have different influences [20]. Hence, the parameters of surface modifications, operating conditions, working fluid and filling ratio have a great effect on the thermal performance of thermosyphons. Similarly, the start-up performance of the thermosyphon will also be affected by these factors.

The reliable operation of thermosyphons requires good start-up performance. The start-up of thermosyphon is a complex, transient process that is affected by several parameters. Sun et al. [21] studied the effects of filling ratio and heat input levels on the start-up characteristics of micro-oscillating heat pipes and observed two different start-up behaviors, start-up processes with and without bubble nucleation, depending principally on the spatial distribution of slugs/plugs in the micro-oscillating heat pipes. Guo et al. [22,23] found that the inclination angle is one of the factors that affect the start-up characteristics of thermosyphon. Then the influence of evaporator length on the start-up performance of a sodium-potassium alloy heat pipe was tested and obtained a uniform temperature distribution by increasing the evaporator section length. The Na-K heat pipe had excellent start-up performance, and the increase of inclination angle raised the temperature of the condenser. Wang et al. [24] analyzed the influence of inclination angle, heat input and flow rate of cooling water on the start-up properties of a thermosyphon with small diameter. Huang et al. [25] introduced the non-condensable gas used for regenerative building heating exchangers in a gravity loop thermosyphon and investigated its effect on the start-up time. They found that the non-condensable gas extended the start-up time of the thermosyphon, with a higher level corresponding to a longer time. Joung et al. [26] observed that a large amount of heat leakage increased the operating temperature and the start-up time of a loop heat pipe. In addition, Singh et al. [27] studied the start-up characteristics of a loop heat pipe and found that the start-up time increased with decreasing the applied heat load. Ji et al. [28] designed a loop heat pipe with composite porous wicks, and studied its heat transfer and start-up characteristics. Huang et al. [29] experimentally and mathematically analyzed the start-up process of a loop heat pipe. They concluded that the start-up process was closely subject to the structural parameters and environment of the loop heat pipe.

Although the thermal performance of a thermosyphon has been well studied from different aspects mentioned above, the start-up characteristics of a thermosyphon have been rarely investigated due to its complex and transient process. The start-up characteristics of a thermosyphon is in an unstable state. How to shorten the start-up time and make the thermosyphon quickly reach a stable state has an important impact on the operation of some equipment. Furthermore, the combination of hydrophilic and hydrophobic properties on the inner surfaces of a thermosyphon is barely investigated. As a high-efficiency heat-transfer device, the start-up characteristics of a thermosyphon is an important index to measure the reliability of the thermosyphon, which must be completed quickly and smoothly. Therefore, it is of great significance to investigate the start-up characteristics of a thermosyphon. In this paper, the effect of wettability properties on the start-up characteristics of TPCTs is fully investigated. Chemical coating and etching techniques are employed to manufacture surface wettability with different contact angles (CAs) at the inner wall of the thermosyphon. The influence of the surface with different CAs on the start-up time and wall superheat degrees of the evaporator section under different input power was compared and analyzed.

2. Methodology

2.1. Experimental System

Figure 2 shows the schematics of the experimental apparatus, while Figures 3 and 4 show the real thermosyphon and the experimental system, respectively. The experimental system is composed of a thermosyphon, a heat supply unit, a cooling unit and a data acquisition unit. The thermosyphon is made of copper with the lengths of evaporator, adiabatic and condenser sections shown in Figure 2 designed to be 100 mm, 50 mm and 100 mm, and the internal and external diameters of 8.32 and 9.52 mm, respectively. Deionized water of 3.2592 g is used as the working fluid and the filling ratio was 24%. The heat supply unit included an electrical resistor, a digital power meter and a voltage-regulating transformer. The evaporator and the adiabatic sections are wrapped with a polytetrafluoroethylene nanoparticle insulation in the inner layer and aluminum foil in the outermost layer for the purpose of reducing the heat loss as shown in Figure 4. The cooling system consisted of a refrigerating unit, a cooling water jacket, a rotameter of 6 ~ 60 L/h and a number of pipelines. The jacket is wrapped with thermal insulation rubber outside. The data acquisition unit is made up of a data logger, a computer and 10 Pt100 thermocouples. The arrangement of 10 thermocouples is shown in Figure 2. In addition, a vacuum pump system consisting of burette, pressure gauge and vacuum pump is used to provide a vacuum in the thermosyphon, and the vacuum degree of each thermosyphon is 10^{-3} Pa. In addition, the boundary conditions of the experiment are shown in Table 1. The 12 TPCTs with different wettability properties on the start-up characteristics are fully investigated under these conditions.

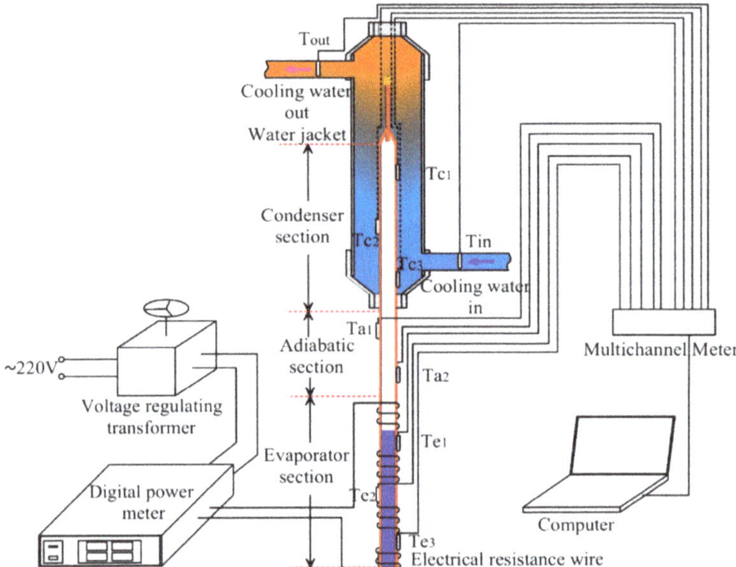

Figure 2. Schematic diagram of experimental apparatus.

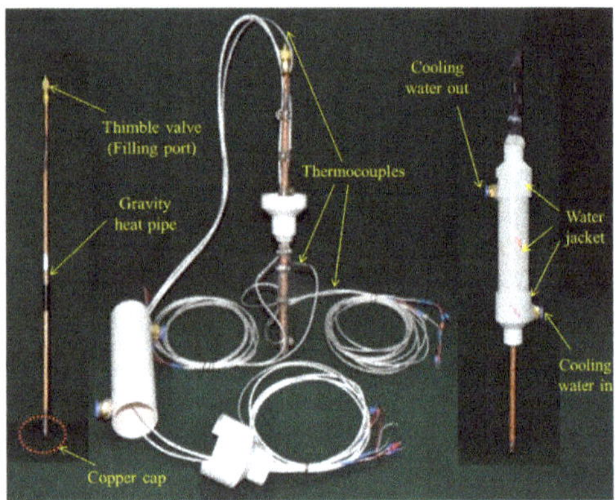

Figure 3. Image of the real thermosyphon used in the experiments.

Table 1. The boundary conditions of the experiment.

Input Power (W)	Temperature of Cooling Water (°C)	Flow Rate of Cooling Water (L/h)
40	18	20
60	18	30
80	18	40

Figure 4. Image of the experimental system.

2.2. Data Reduction and Problem Description

The heat transfer between the condenser section and the cooling water under various operating conditions (e.g., working fluid, applied heating power input and mass flow rate of cooling water) is determined using Equation (1):

$$Q_c = m_w c_p (T_{\mathrm{w},out} - T_{w,in}) \tag{1}$$

where m_w is the mass flow, $T_{w,in}$ is the temperature of cooling water at inlet of the condenser, $T_{\mathrm{w,out}}$ is the temperature at outlet and c_p is the specific heat values of water.

During the experiment, strict insulation measures are taken on the outside of the evaporator section, the adiabatic section and the cooling water jacket in order to ensure minimum heat loss. The heat-balance method [9] is used to determine the heat loss of the system. The power relative error is defined as the ratio of the difference between the input heat Q_e in the evaporator section and the released heat Q_c in the condenser section to the input heat Q_e, and calculated using Equation (2) as:

$$\eta = \frac{Q_e - Q_c}{Q_e} \tag{2}$$

where

$$Q_e = Q_{in} = VI \tag{3}$$

Q_{in} is the heating power input on the evaporator section of the thermosyphon, while V and I are the voltage and the current monitored by the digital power meter. The measurement error of thermocouples was ± 0.5 °C, and the cooling water had a flow rate error of ± 2.5%. In the experiment, the maximum power relative error was 7.3%.

The objective of the paper is to investigate the effect of wettability properties on the start-up characteristics of TPCTs. The measurement includes the average start-up time and wall superheat degree of the evaporator section of the thermosyphon under different input power. The main problem in this study is the machining and preparation of different wettability properties on the inner surface. The values of different CAs need distinct process technology [30,31]. Besides, the cylindrical shape of the inner wall of the thermosyphon leads to a difficulty of processing and long-term stability. For the preparation of inner wettability surfaces, the techniques are chemical etching, electrochemical deposition, composite coating, anodizing, etc. Chemical coating and etching techniques are used to manufacture surface wettability with different CAs at the inner wall of the thermosyphon after a considerable number of experiments.

2.3. Surface Modification

In order to prepare the surfaces with various wettabilities in terms of different CAs, chemical techniques are performed using various materials. NaOH and $(NH_4)_2S_2O_8$ are used to etch the hydrophilic surface with a CA of approximately 28°. By coating various ratios of materials, such as N-butyl, stearic acid, xylenes and acetone, the CAs of surfaces are approximately 61°, 79°, 105°, 117° and 142°, respectively. Detailed surface modification methods have been described in the authors' previous work [9].

The 12 thermosyphons correspondingly produced test samples to verify the coating temperature resistance. The coated samples are put on the thermostatic magnetic stirrer for a high-temperature test. The automatic contact angle meter (Kino-SL150E) with the measuring error of ± 2° is used to measure the CA after high-temperature and long-time testing. It is found that the change in the CA is small and the maximum CA error is 4.2% in the testing temperature from 20 °C to 100 °C.

In order to ensure the accuracy and reproducibility of the CA of the wet surface, 4 μL of deionized water is titrated on each surface 5 times. Finally, the average value of the CA is taken. Figure 5a–f show the low- and high-magnification scanning electron microscopy (SEM) images of the surfaces with CAs of 28°, 61°, 79°, 105°, 117° and 142°, respectively, while Figure 5g shows the SEM image of a smooth surface. Figure 5d–f demonstrate the hydrophobic surfaces on which 4μL of water is dropped, with the static CAs of 105°, 117° and 142°, respectively, illustrating weak interactions between water drops and hydrophobic surfaces. Water drops are dispersed over the surfaces with the CAs of 28°, 61° and 79°, subject to high adhesive force between water and coated copper in the evaporator as shown in Figure 5a–c, respectively. Figure 5g shows that the TPCT7 remained smooth, without any resurface work.

According to different wettability properties, the thermosyphons are classified into 12 different TPCTs from TPCT1 to TPCT12 as listed in Table 2. From TPCT1 to TPCT6, each of their three sections has the same CA on the inner surfaces. A combination of hydrophilic and hydrophobic properties with different CAs is adopted to modify the inner surfaces of the evaporator and condenser of TPCTs from TPCT8 to TPCT12, while the inner wall of the three sections of TPCT7 is smooth surface without fabrication.

Table 2. Different two-phase closed thermosyphons (TPCTs) on inner modified surfaces with hydrophilic and hydrophobic properties.

TPCTs	Evaporator	Adiabatic	Condenser
TPCT1	28°	28°	28°
TPCT2	61°	61°	61°
TPCT3	79°	79°	79°
TPCT4	105°	105°	105°
TPCT5	117°	117°	117°
TPCT6	142°	142°	142°
TPCT7	smooth surface	smooth surface	smooth surface
TPCT8	28°	smooth surface	105°
TPCT9	28°	smooth surface	117°
TPCT10	28°	smooth surface	142°
TPCT11	105°	smooth surface	28°
TPCT12	142°	smooth surface	28°

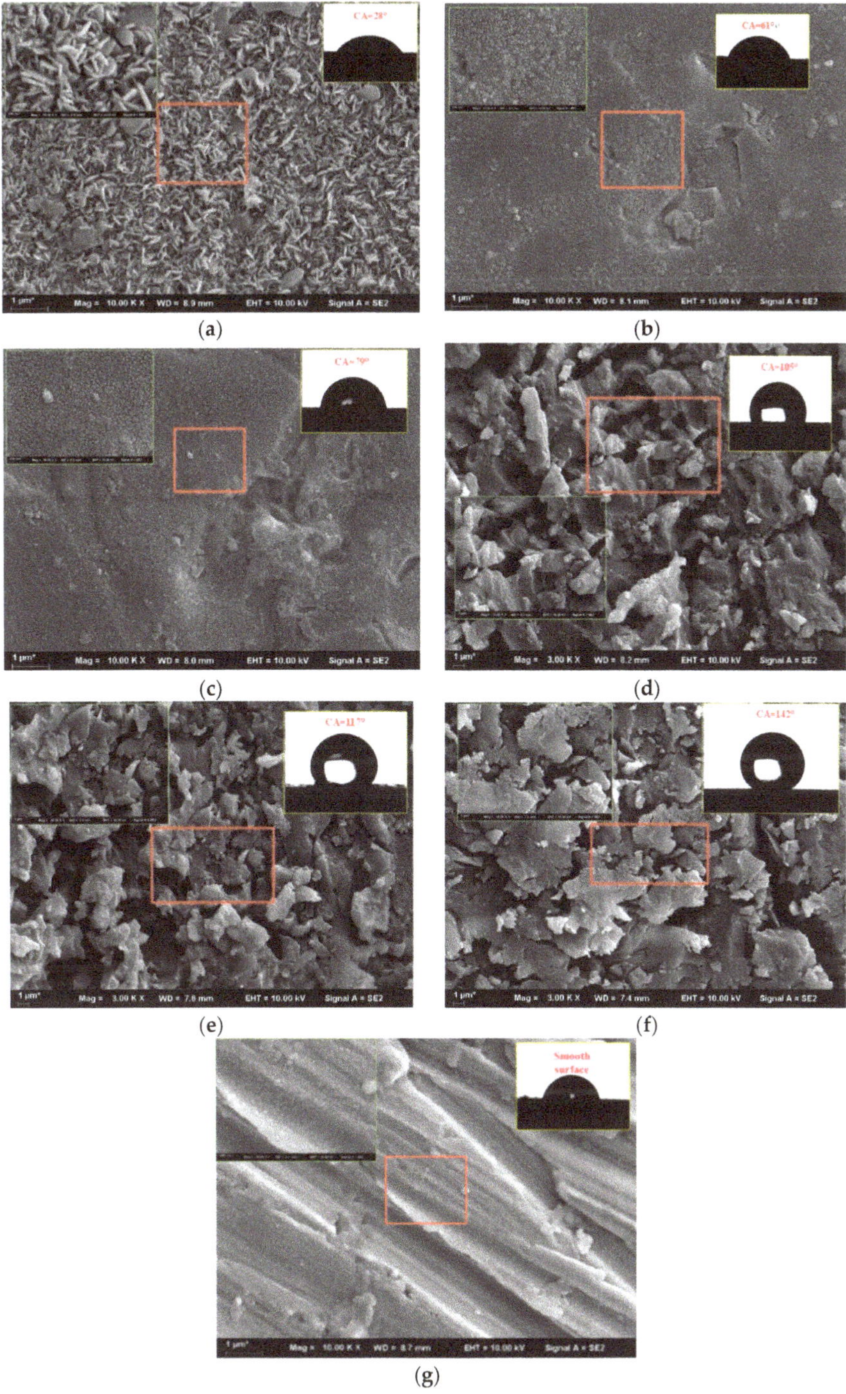

Figure 5. Scanning electron microscopy (SEM) images of: (**a**–**f**) hydrophilic and hydrophobic surfaces with contact angles (CAs) of 28°, 61°, 79°, 105°, 117° and 142°, respectively, and (**g**) smooth surface (inset: 4 μL of water droplet with static CA).

3. Results

The start-up performance is a crucial indicator of the thermosyphon operation. To enhance the overall performance of the thermosyphon, the start-up process in which the evaporator is heated to a steady state must be completed quickly and stably. In order to evaluate the influence of wettability on the start-up performance of the thermosyphon, the start-up times and two typical temperatures, i.e., the average temperatures of evaporator $T_{e.ave}$ (average of points Te1-Te3) and condenser $T_{c.ave}$ (average of points Tc1-Tc3), are selected and measured by thermocouples. In addition, all data are recorded every 5s through a multichannel meter and saved in the computer through Monitor and Control Generated System (MCGS) software.

Figure 6 shows the start-up processes of thermosyphons from TPCT1 to TPCT7 at different input powers: 40 W, 60 W and 80 W. It can be seen from the figure that $T_{e.ave}$ and $T_{c.ave}$ increase first and then maintain two different stable conditions, and all thermosyphons had a successful start-up. After the heat input power is imposed, $T_{e.ave}$ rapidly increases with extended heating process before the start-up is completed, implying that the working fluid is heated at the initial moment, then boiled and evaporated, after which the vapor reaches the condenser section and releases the latent heat. However, $T_{c.ave}$ does not rise at the early stage and the increase in the temperature on the condenser section is slower in the start-up process due to the heat-transfer delay. As $T_{c.ave}$ rises, the process whereby the working fluid is boiled into steam is accelerated by increasing the input power. For example, at the input powers of 40 W, 60 W and 80 W, the $T_{c.ave}$ values of TPCT1–TPCT7 increase sharply at 125 s, 70 s and 45 s, respectively. The generated vapor from the evaporator section is not condensed in the condenser section due to the heat is not taken away rapidly by the external environment and that is why the temperature of the condenser keeps increasing continuously. Once the working fluid cycle is completed, the system enters a steady state.

Figure 7 shows the start-up time taken by the average evaporator temperatures of TPCT1–TPCT7 to stabilize at different input powers with different CAs on the evaporator sections. At the same input power, the TPCT with hydrophobic properties need longer start-up time than those with hydrophilic properties do at different input powers. The start-up times of the evaporator sections of TPCT4–TPCT6 with hydrophobic properties are all longer than those of TPCT1-TPCT3 with hydrophilic properties at 40 W, 60 W and 80 W. The bubbles on both hydrophilic and hydrophobic surfaces go through the entire process of bubble formation, growth, coalescence and separation which reflects the complete cycle inside the thermosyphons. The hydrophilic surface has a more complex microstructure than the hydrophobic surface as shown in Figure 5. The smaller bubble diameters are generated on the hydrophilic surface due to the rapid replenishment of fresh liquid backflow on the hydrophilic surface being faster than that of the hydrophobic surface, which further promotes the heat transfer of the thermosyphon. In contrast, the bubbles produced by the hydrophobic surface could form the gas film with the other bubbles before leaving the hydrophobic surface. Thus, the evaporator with the hydrophobic surface takes longer to start up.

The TPCT1 responds more quickly than TPCT2–TPCT7 do when the thermosyphons enter the steady state at different input powers. Thus, TPCT1 has the best start-up for thermosyphons with the same wettabilities on the evaporator and the condenser sections among TPCT1–TPCT6, while the start-up time of TPCT1 at 40 W, 60 W and 80 W is 46%, 50% and 55% faster than that of TPCT7 (smooth surface). The reason is that the hydrophilic surfaces have more compact structures with stronger tension forces between pore and water, which is conducive to a smaller bubble diameter speeding up the departure frequency of the bubbles [32] and promotes the heat transfer of nucleate pool boiling. Conversely, the bubbles join together to form an air blanket in a very limited time before fleeing the hydrophobic surface [33,34]. The SEM images with different CAs presented in Figure 5 indicate that the pore diameter enlarges with increasing CA. Moreover, the relationship between the CA and the pore diameter is the same as the previous research result [35]. The short knife-like nanostructure coated on the surface with CA of 28° (Figure 5a) increase the heat-transfer area of phase transition and nucleation sites. Meanwhile, the nanostructure above enhances the hydrophilicity characteristics and improved the heat transfer of pool boiling [36], producing smaller bubbles. In consequence, the process of the heat transfer of pool boiling is significantly disturbed prompting more vapor to reach the condenser section to release more latent heat. Thus, the start-up performance of the thermosyphon is enhanced.

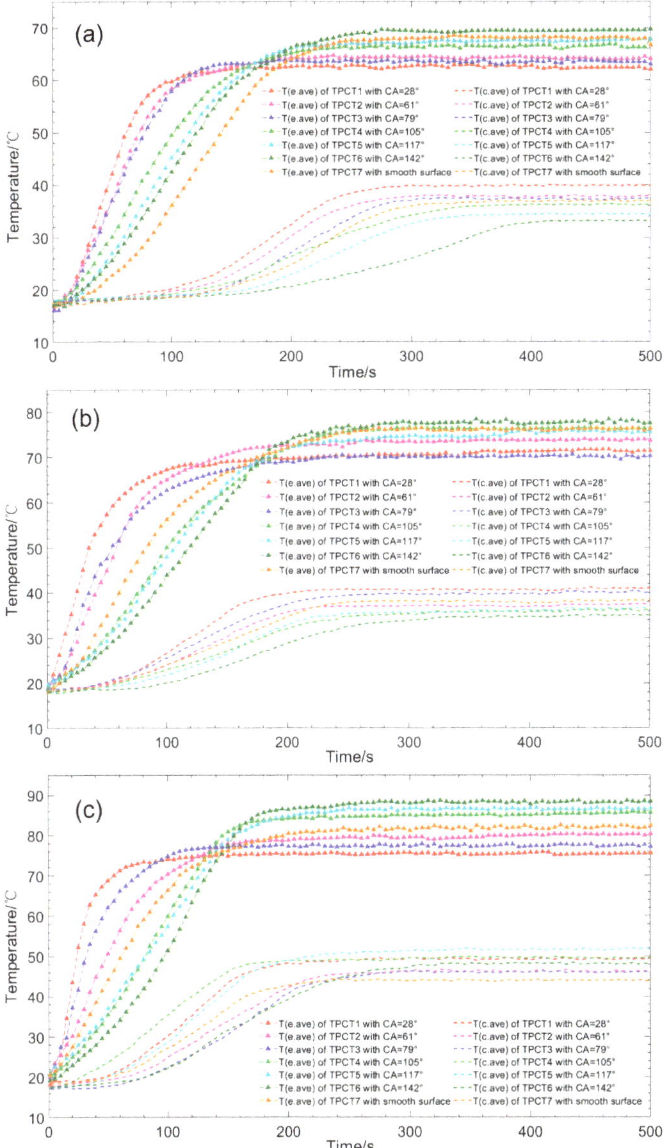

Figure 6. Start-up performances of TPCT1–TPCT7 at different input powers: (**a**) Q_{in} = 40 W; (**b**) Q_{in} = 60 W; (**c**) Q_{in} = 80 W

Generally, the start-up time is prolonged as the CA on the evaporator sections of TPCT1–TPCT6 increase at the same input power. The time of TPCT2 is shorter than that of TPCT3 at the input power of 40 W. However, the time of TPCT2 takes longer at the input powers of 60 W and 80 W. One of the reasons is that the accuracy of CA can affect the experimental results. The CA is the average value that may slightly change in the process of heat transfer. The calculation results show that the CA error of samples is within 4.2%, causing the start-up time of TPCT3 to be shorter than that of TPCT2. In addition, both the wettability's and the roughness of the thermosyphon surface are different, which could lead to different processes of bubble generation, growth and departure. It can be observed from Figure 5b,c that the surface structure is similar, but the grooves and gaps where it is easy to generate nucleation sites density on the surface of TPCT3 are lower than TPCT2. Therefore, the sub-cooled water in the thermosyphon is in full contact with the surface of TPCT3. There are more nucleation sites on the surface of TPCT2, resulting in more bubbles. Adjacent bubbles tend to merge and form large bubbles, which are trapped on the surface of the evaporator section. Thus, the start-up speed of TPCT2 is slower. At low heat flux, the bubble number is less and could not lead to merging of a large number of bubbles. At high heat flux, the surface of TPCT2 has more bubbles that makes merging easier for bubbles. The bubble departure diameter increases and the bubble departure frequencies decrease, so the start-up time of TPCT3 is less than that of TPCT2 at the input power of 60 W and 80 W.

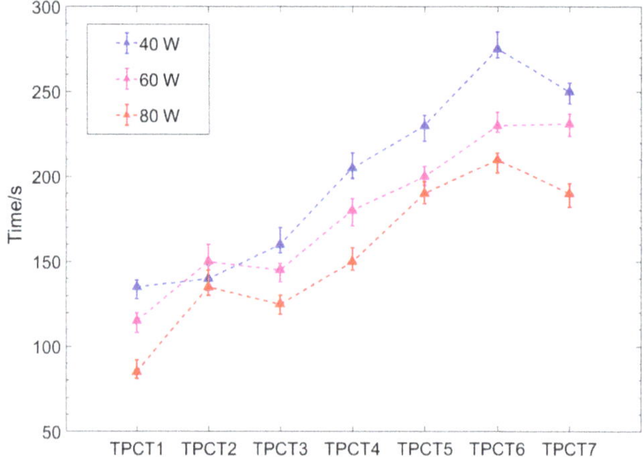

Figure 7. Times taken by the average evaporator temperatures of TPCT1–TPCT7 to stabilize at different input powers with various wettabilities on evaporator sections.

Figure 8 shows the plots of TPCTs with various wettabilities on evaporator sections versus the wall superheat degree at 40 W, 60 W and 80 W. The thermosyphons with hydrophilic surfaces (CA $< 90^{\circ}$) have not only a short start-up time, but also a low wall superheat degree. The wall superheat degree ($\Delta t = t_w - t_{sat}$, t_w: wall temperature, t_{sat}: saturated temperature) of evaporator sections for thermosyphons is an important factor to quantify the characteristics of thermosyphons. Since the heat exchange mechanism of the evaporator section is pool boiling heat transfer, the most important parameter to evaluate the heat-transfer characteristics of pool boiling is the wall superheat degree. The lower wall superheat degree at the same heat power means higher heat-exchange efficiency. As a result, the wall superheat degrees Δt of evaporator sections for TPCT1–TPCT7 were compared. The saturated pressure of each thermosyphon is 0.017212 MPa, and the corresponding saturated temperature is about 57 °C. The TPCT1 not only significantly reduces the start-up time, but also decreases the evaporator wall superheat degrees Δt compared with those of other TPCTs at the same input power. The wall superheat degrees Δt of TPCT1 at 40 W, 60 W and 80 W are 55%, 39% and 28% lower than that of TPCT7. Similarly, it has previously been reported that the modified surfaces with different CAs influence the pool boiling in the thermosyphon [37]. In the process of boiling heat transfer, the surfaces with hydrophobic properties produce a large bubble at low heat flux [38]. There are larger and deeper pores on the hydrophobic surfaces (Figure 5d–f) than those on the hydrophilic surfaces (Figure 5a–c). Thus, some air and gas film exist, which increase the thermal resistance and inhibit the heat transfer. Besides, the hydrophobic surface is close to the gas and then the bubble is easy to polymerize into a gas film, which prevents the liquid from replenishing to the heating wall. Therefore, the heat transfer is impeded and the boiling heat transfer performance begins to deteriorate. As a result, the superheat degrees of hydrophobic surfaces (CAs: 105°, 117° and 142°) surpass those of hydrophilic surfaces (CAs: 28°, 61°and 79°) and a smooth surface.

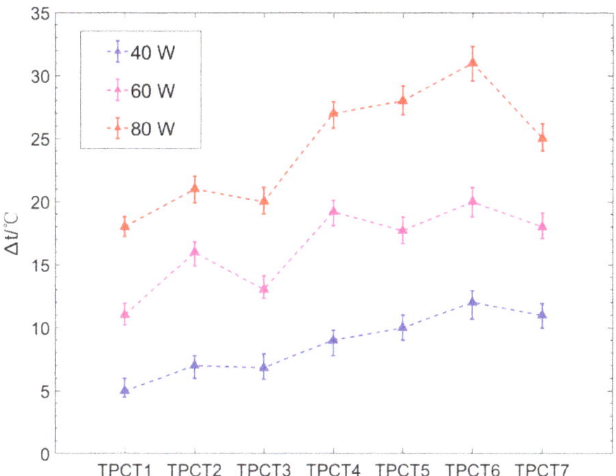

Figure 8. Relationships between wall superheat degree $\Delta t/°C$ on evaporator section and different TPCTs with various wettabilities.

Furthermore, Figure 8 demonstrates that the wall superheat degree Δt gradually increases with rising CA under the same power. As the input power increased, Δt of the hydrophobic evaporator section rises more evidently than that of the hydrophilic evaporator section. For example, the wall superheat degree of TCPT1 increases 13 °C from the input power of 40 W to 80 W while TCPT6 increases 19 °C. Since the hydrophobic surface produces more gas film as the power increases, the resulting large bubbles limit the heat transfer to the working fluid [39]. Thus, the superheat degree is augmented for hydrophobic surface. In contrast, the bubbles on the hydrophilic surface are smaller and quickly departs from the heating surface. Once the bubble departed, the surrounding working fluid quickly fills the remaining area and prevents the formation of large gas films with high thermal resistance. However, Δt of the evaporator of TPCT2 is higher than that of the evaporator of TPCT3 which is consistent with the start-up time due to the inhomogeneous structures as shown in Figure 5b,c. This gives a higher wall superheat degree to the evaporator of TPCT2 than that of the evaporator of TPCT3.

Figure 9 illustrates the start-up processes of TPCT7-TPCT12 modified with a combination of hydrophilic and hydrophobic properties at the input powers of 40 W, 60 W and 80 W. For TPCT8, TPCT9 and TPCT10 with hydrophilic inner surfaces on the evaporator sections, the start-up time is shorter than that of TPCT11 and TPCT12 with hydrophobic inner surfaces, which is consistent with the above discussion. With rising CA for TPCT7–TPCT9 on condenser sections, the start-up process on the evaporator section of TPCT8 is faster than those of TPCT9 and TPCT10. For TPCT8, the difference between the start-up time of the evaporator and the time reaching the stable state of the condenser is gradually decreased from 200 s to 150 s with the increase from 40 W to 80 W, which enhances the evaporator-condenser phenomenon induced by combined hydrophilic and hydrophilic properties. The addition of a hydrophilic surface to the surface of the evaporator section can quickly induce the sub-cooled water after the formation of bubbles, which can quickly take away the heat of the surface of the evaporator section. Thus, the stable temperature reduces. The addition of a hydrophobic surface on the surface of the condenser section rapidly reduces the water droplets in the condenser section, which indirectly speeds up the recycle of working fluid inside the thermosyphon. As the input power increased, TPCT8 and TPCT9 both have almost the same start-up times of 180 s and 200 s at input power of 60 W and 80 W, respectively. However, TPCT8 has a higher average temperature of 2 °C on the evaporator than that of TPCT9. Furthermore, the average temperature on the condenser section of TPCT9 (CA: 117°) gradually approaches that on the condenser section of TPCT8 (CA: 105 °C) with increasing input power at an average temperature of 54 °C.

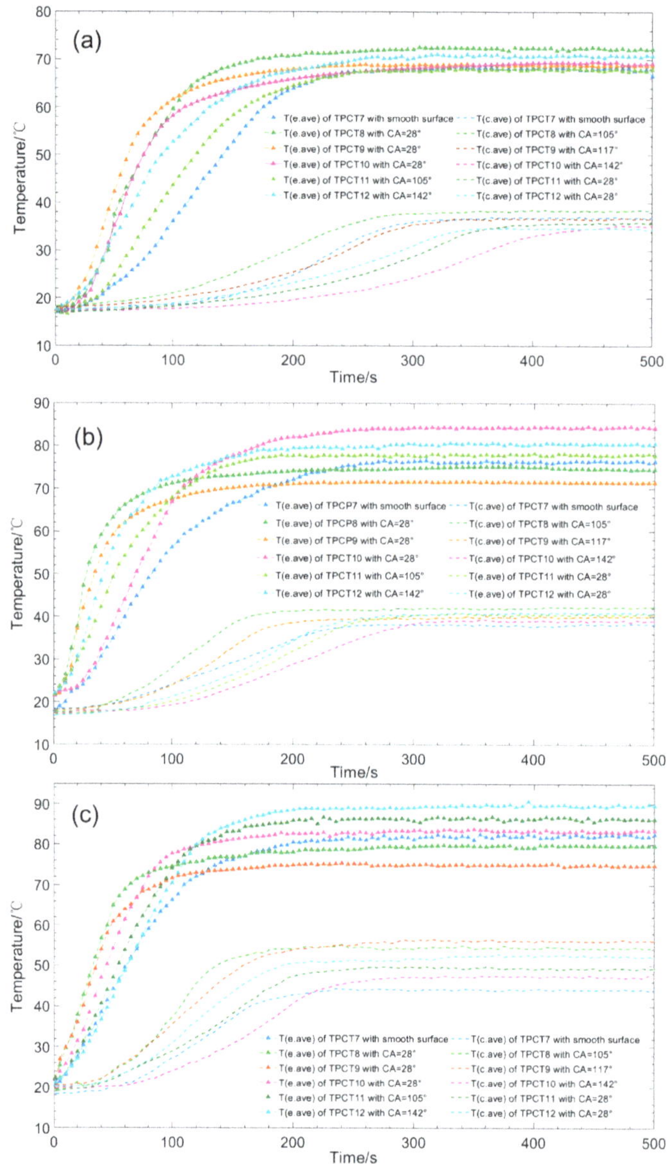

Figure 9. Start-up performances of TPCT7 (smooth surface) and TPCT8–TPCT12 with combined hydrophilic and hydrophobic properties at different heat input powers: (**a**) Q_{in} = 40 W;(**b**) Q_{in} = 60 W; (**c**) Q_{in} = 80 W.

For TPCT11 and TPCT12 with hydrophilic inner surfaces on the condenser sections (both with CA 28°), the condenser sections require shorter times to stabilize than that of TPCT7 with a smooth surface with increasing input power. Although the evaporator sections are hydrophobically modified, the steam still sufficiently releases heat on the inner surface of hydrophilically modified condenser sections, and the condensate fully contacts with the hydrophilic surface to release latent and sensible heats. Furthermore, it is found that the temperature difference between the evaporator (CA: 28°) and condenser (CA: 105 °C) of TPCT8 gradually decreases with increase in the input power, thus surpassing those of TPCT9–TPCT12 concerning isothermal properties. As a result, TPCT8 has the best start-up characteristics among TPCT8–TPCT12 under the same input power. It is postulated that the interaction between the liquid drops and the wall is attenuated with an increase in the CA, and the condensate falls down more easily as smaller drops and fast reflux to the evaporator section, which may suppress the release of latent and sensible heats in the condenser.

4. Conclusions

In this study, the influence of wettability properties inside the inner surface of thermosyphons on the start-up characteristics is fully investigated under different input powers. Chemical techniques are performed to fabricate the surfaces with different wettabilities that is quantified in the form of the CA inside the evaporator and the condenser sections. The experimental results demonstrate that different CAs not only significantly affect the start-up time but also influence the temperature variations and distributions of the outer walls of the evaporator and the condenser sections. Detailed conclusions can be drawn as follows:

(1) For thermosyphons with the same wettabilities on the evaporator and the condenser sections among TPCT1–TPCT6, the introduction of hydrophilic properties inside the evaporator section not only significantly shortens the start-up time but also decreases the start-up temperature. At the same input power, the start-up time of a thermosyphon with CA < 90° is shorter than that with CA > 90°. The start-up time of TPCT with CA = 28° has the shortest start-up time, while under the input powers of 40 W, 60 W and 80 W it is 46%, 50% and 55% shorter than that of TPCT with a smooth surface, respectively. The start-up time becomes longer with the increase of CA of the evaporator sections.

(2) As the CAs on the evaporator sections of TPCT1–TPCT6 increase, the wall superheat degree gradually increases. The TPCT with CA = 28° has a minimum superheat degree at input powers of 40 W, 60 W and 80 W, and the wall superheat degree is 55%, 39% and 28% lower than that of TPCT with smooth surfaces, respectively. In addition, the superheat degree of the hydrophobic evaporator section increases more obviously than that of the hydrophilic with increasing input power. In the experiment, the TPCT with CA = 142° has the highest superheat degree among TPCT1–TPCT6.

(3) For thermosyphons with combined hydrophilic and hydrophobic properties, the start-up time of the evaporator section with CA < 90° and the condenser section with CA > 90° is less than the evaporator section with CA > 90° and the condenser section with CA < 90°. With the increase of CA on the condenser sections of TPCT8–TPCT10, the start-up process of TPCT8 is faster than those of TPCT9 and TPCT10 with the same CA = 28° on the evaporator section. The experimental results and data analysis demonstrate that the start-up time of the TPCT8 with CA = 28° on the evaporator section and CA = 105° on the condenser section is the shortest among TPCT7–TPCT12.

(4) In this paper, the surfaces with CAs of 28°, 61°, 79°, 105°, 117° and 142° are fabricated inside the evaporator and the condenser sections of the thermosyphons. The experimental results show that the TPCT with CA = 28° on both the evaporator and the condenser section has the best start-up characteristics considering start-up time and wall superheat degree, which reflect the optimal wettabilitiy inside the inner surface of thermosyphons for industrial reference. Further research directions should extend to the preparation and investigation of superhydrophobic and superhydrophilic surfaces in the analysis of the thermal performance and start-up characteristics.

Author Contributions: Project administration, supervision, writing-review and editing, Z.Z.; methodology and writing-original draft preparation, X.M. and P.J.; investigation, data curation, formal analysis and validation, X.M., P.J., S.Y., S.L. and X.C. All authors have read and agreed to the published version of the manuscript.

Funding: The authors gratefully acknowledge that this work was supported by Jiangsu marine and fishery science and technology innovation and extension project (HY2017-8) and Zhenjiang funds for the key research and development project (GY2016002-1).

Conflicts of Interest: The authors declare no conflict of interest.

Nomenclature

c_p	Specific heat values (J/kg·K)
m	Mass flow(kg/s)
Q	Heat load (W)
T	Temperature (°C)
I	Current (A)
V	Voltage (V)

Greek symbols

Δt	Wall superheat degree (°C)
η	Efficiency of the thermosyphon

Subscripts

a	Adiabatic section
ave	Average
c	Condenser section
e	Evaporator section
in	Cooling water inlet/ Input power
out	Cooling water outlet
sat	Saturated
w	Wall/water

Acronyms

CA	Contact angle
MCGS	Monitor and Control Generated System
SEM	Scanning electron microscopy
TPCP	Two-phase closed thermosyphon

References

1. Chang, Y.; Cheng, C.; Wang, J.; Chen, S. Heat pipe for cooling of electronic equipment. *Energy Convers. Manag.* **2008**, *49*, 3398–3404. [CrossRef]
2. Jafari, D.; Franco, A.; Filippeschi, S.; Di Marco, P. Two-phase closed thermosyphons: A review of studies and solar applications. *Renew. Sustain. Energy Rev.* **2016**, *53*, 575–593. [CrossRef]
3. Ziapour, B.M.; Khalili, M.B. PVT type of the two-phase loop mini tube thermosyphon solar water heater. *Energy Convers. Manag.* **2016**, *129*, 54–61. [CrossRef]
4. Jouhara, H.; Merchant, H. Experimental investigation of a thermosyphon based heat exchanger used in energy efficient air handling units. *Energy* **2012**, *39*, 82–89. [CrossRef]
5. Gaugler, R.S. Heat Transfer Device. U.S. Patent 2,350,348, 6 June 1994.
6. Faghri, A. *Heat Pipe Science and Technology*; Taylor & Francis: Philadelphia, PA, USA, 1995.
7. Shabgard, H.; Allen, M.J.; Sharifi, N.; Benn, S.P.; Faghri, A.; Bergman, T.L. Heat pipe heat exchangers and heat sinks: Opportunities, challenges, applications, analysis, and state of the art. *Int. J. Heat Mass Transf.* **2015**, *89*, 138–158. [CrossRef]

8. Amatachaya, P.; Srimuang, W. Comparative heat transfer characteristics of a flat two-phase closed thermosyphon (FTPCT) and a conventional two-phase closed thermosyphon (CTPCT). *Int. Commun. Heat Mass Transfer.* **2010**, *37*, 293–298. [CrossRef]
9. Zhao, Z.; Jiang, P.; Zhou, Y.; Zhang, Y.; Zhang, Y. Heat transfer characteristics of two-phase closed thermosyphons modified with inner surfaces of various wettabilities. *Int. Commun. Heat Mass Transfer.* **2019**, *103*, 100–109. [CrossRef]
10. Lataoui, Z.; Jemni, A. Experimental investigation of a stainless steel two-phase closed thermosyphon. *Appl. Therm. Eng.* **2017**, *121*, 721–727. [CrossRef]
11. Moradikazerouni, A.; Afrand, M.; Alsarraf, J.; Mahian, O.; Wongwises, S.; Tran, M.-D. Comparison of the effect of five different entrance channel shapes of a micro-channel heat sink in forced convection with application to cooling a supercomputer circuit board. *Appl. Therm. Eng.* **2019**, *150*, 1078–1089. [CrossRef]
12. Moradikazerouni, A.; Afrand, M.; Alsarraf, J.; Wongwises, S.; Asadi, A.; Nguyen, T.K. Investigation of a computer CPU heat sink under laminar forced convection using a structural stability method. *Int. J. Heat Mass Transf.* **2019**, *134*, 1218–1226. [CrossRef]
13. Rahimi, M.; Asgary, K.; Jesri, S. Thermal characteristics of a resurfaced condenser and evaporator closed two-phase thermosyphon. *Int. Commun. Heat Mass Transf.* **2010**, *37*, 703–710. [CrossRef]
14. Singh, R.R.; Selladurai, V.; Ponkarthik, P.; Solomon, A.B. Effect of anodization on the heat transfer performance of flat thermosyphon. *Exp. Therm. Fluid Sci.* **2015**, *68*, 574–581. [CrossRef]
15. Solomon, A.B.; Daniel, V.A.; Ramachandran, K.; Pillai, B.; Singh, R.R.; Sharifpur, M.; Meyer, J. Performance enhancement of a two-phase closed thermosiphon with a thin porous copper coating. *Int. Commun. Heat Mass Transf.* **2017**, *82*, 9–19. [CrossRef]
16. Solomon, A.B.; Mathew, A.; Ramachandran, K.; Pillai, B.; Karthikeyan, V. Thermal performance of anodized two phase closed thermosyphon (TPCT). *Exp. Therm. Fluid Sci.* **2013**, *48*, 49–57. [CrossRef]
17. Noie, S. Heat transfer characteristics of a two-phase closed thermosyphon. *Appl. Therm. Eng.* **2005**, *25*, 495–506. [CrossRef]
18. Gedik, E. Experimental investigation of the thermal performance of a two-phase closed thermosyphon at different operating conditions. *Energy Build.* **2016**, *127*, 1096–1107. [CrossRef]
19. Ma, Y.; Shahsavar, A.; Moradi, I.; Rostami, S.; Moradikazerouni, A.; Yarmand, H.; Zulkifli, N.W.B.M. Using finite volume method for simulating the natural convective heat transfer of nano-fluid flow inside an inclined enclosure with conductive walls in the presence of a constant temperature heat source. *Phys. A Stat. Mech. Appl.* **2019**, 123035. [CrossRef]
20. Vo, D.D.; Alsarraf, J.; Moradikazerouni, A.; Afrand, M.; Salehipour, H.; Qi, C. Numerical investigation of γ-AlOOH nano-fluid convection performance in a wavy channel considering various shapes of nanoadditives. *Powder Technol.* **2019**, *345*, 649–657. [CrossRef]
21. Sun, Q.; Qu, J.; Yuan, J.; Wang, H. Start-up characteristics of MEMS-based micro oscillating heat pipe with and without bubble nucleation. *Int. J. Heat Mass Transf.* **2018**, *122*, 515–528. [CrossRef]
22. Guo, Q.; Guo, H.; Yan, X.; Wang, X.; Ye, F.; Ma, C. Experimental study of start-up performance of sodium-potassium heat pipe. *J. Eng. Thermophys.* **2014**, *35*, 2508–2512.
23. Guo, Q.; Guo, H.; Yan, X.; Wang, X.; Ye, F.; Ma, C. Effect of the evaporator length on start-up performance fo sodium-potassium alloy heat pipe. *J. Eng. Thermophys.* **2016**, *37*, 1717–1720.
24. Wang, X.; Xin, G.; Tian, F.; Cheng, L. Start-up behavior of gravity heat pipe with small diameter. *CIESC J.* **2012**, *63*, 94–98.
25. Huang, J.; Wang, L.; Shen, J.; Liu, C. Effect of non-condensable gas on the start-up of a gravity loop thermosyphon with gas–liquid separator. *Exp. Therm. Fluid Sci.* **2016**, *72*, 161–170. [CrossRef]
26. Joung, W.; Yu, T.; Lee, J. Experimental study on the loop heat pipe with a planar bifacial wick structure. *Int. J. Heat Mass Transf.* **2008**, *51*, 1573–1581. [CrossRef]
27. Singh, R.; Akbarzadeh, A.; Mochizuki, M. Operational characteristics of a miniature loop heat pipe with flat evaporator. *Int. J. Therm. Sci.* **2008**, *47*, 1504–1515. [CrossRef]
28. Ji, X.; Wang, Y.; Xu, J.; Huang, Y. Experimental study of heat transfer and start-up of loop heat pipe with multiscale porous wicks. *Appl. Therm. Eng.* **2017**, *117*, 782–798. [CrossRef]
29. Huang, B.; Huang, H.; Liang, T. System dynamics model and startup behavior of loop heat pipe. *Appl. Therm. Eng.* **2009**, *29*, 2999–3005. [CrossRef]

30. Dai, C.; Yang, L.; Xie, J.; Wang, T. Nutrient diffusion control of fertilizer granules coated with a gradient hydrophobic film. *Colloids Surf. A Physicochem. Eng. Asp.* **2020**, *588*, 124361. [CrossRef]
31. Anosov, A.; Smirnova, E.Y.; Sharakshane, A.; Nikolayeva, E.; Zhdankina, Y.S. Increase in the current variance in bilayer lipid membranes near phase transition as a result of the occurrence of hydrophobic defects. *Biochim. Biophys. Acta (BBA)-Biomembr.* **2020**, *1862*, 183147. [CrossRef]
32. Li, Y.; Liu, Z.; Wang, G. A predictive model of nucleate pool boiling on heated hydrophilic surfaces. *Int. J. Heat Mass Transfer.* **2013**, *65*, 789–797. [CrossRef]
33. Phan, H.T.; Caney, N.; Marty, P.; Colasson, S.; Gavillet, J. Surface wettability control by nanocoating: The effects on pool boiling heat transfer and nucleation mechanism. *Int. J. Heat Mass Transfer.* **2009**, *52*, 5459–5471. [CrossRef]
34. Hsu, C.; Chen, P. Surface wettability effects on critical heat flux of boiling heat transfer using nanopartical coatings. *Int. J. Heat Mass Transfer.* **2012**, *55*, 3713–3719. [CrossRef]
35. Leese, H.; Bhurtun, V.; Lee, K.P.; Mattia, D. Wetting behavior of hydrophilic and hydrophobic nanostructured porous anodic alumina. *Colloids Surf. A Physicochem. Eng. Asp.* **2013**, *420*, 53–58. [CrossRef]
36. Shi, B.; Wang, Y.; Chen, K. Pool boiling heat transfer enhancement with copper nanowire arrays. *Appl. Therm. Eng.* **2015**, *75*, 115–121. [CrossRef]
37. Bankoff, S. Ebullition from solid surfaces in the absence of a pre-existing gaseous phase. *Trans. Am. Mech. Eng.* **1957**, *79*, 735–740.
38. Zheng, X.; Ji, X.; Wang, Y.; Xu, J. Study of the pool boiling heat transfer on the superhydrophilic and Superhydrophobic surface. *Prog. Chem. Ind.* **2016**, *12*, 3793–3798.
39. Thiagarajan, S.J.; Yang, R.; King, C.; Narumanchi, S. Bubble dynamics and nucleate pool boiling heat transfer on microporous copper surfaces. *Int. J. Heat Mass Transfer* **2015**, *89*, 1297–1315. [CrossRef]

Article

Design and Comparison of Resonant and Non-Resonant Single-Layer Microwave Heaters for Continuous Flow Microfluidics in Silicon-Glass Technology

Tomislav Markovic [1,2,*], Ilja Ocket [1,2], Adrijan Baric [3] and Bart Nauwelaers [2]

1 IOT Group, imec, Kapeldreef 75, 3001 Heverlee, Belgium; ilja.ocket@imec.be
2 Division TELEMIC, Department of Electrical Engineering (ESAT), KU Leuven, Kasteelpark Arenberg 10, 3001 Leuven, Belgium; bart.nauwelaers@kuleuven.be
3 Department of Electronics, Microelectronics, Computer and Intelligent Systems, Faculty of Electrical Engineering and Computing, University of Zagreb, Unska 3, 10000 Zagreb, Croatia; adrijan.baric@fer.hr
* Correspondence: tomislav.markovic@kuleuven.be; Tel.: +32-16-32-96-89

Received: 27 April 2020; Accepted: 19 May 2020; Published: 21 May 2020

Abstract: This paper presents a novel concept for the co-design of microwave heaters and microfluidic channels for sub-microliter volumes in continuous flow microfluidics. Based on the novel co-design concept, two types of heaters are presented, co-designed and manufactured in high-resistivity silicon-glass technology, resulting in a building block for consumable and mass-producible micro total analysis systems. Resonant and non-resonant co-planar waveguide transmission line heaters are investigated for heating of sub-micro-liter liquid volumes in a channel section at 25 GHz. The heating rates of 16 and 24 °C/s are obtained with power levels of 32 dBm for the through line and the open-ended line microwave heater, respectively. The heating uniformity of developed devices is evaluated with a Rhodamine B and deionized water mixture on a micrometer scale using the microwave-optical measurement setup. Measurement results showed a good agreement with simulations and demonstrated the potential of microwave heating for microfluidics.

Keywords: microwave heating; microfluidics; silicon; chip integration

---

1. Introduction

Microfluidic devices enable various applications through incorporated technological functions on a chip, as illustrated in Figure 1. In lab-on-a-chip systems, microfluidic mixers are often achieved using meandered channels [1] or ridges in a channel [1]. Liquid filtering and particle separation are achieved using microfluidic filters with an array of pillars [2], while signal detection to evaluate reactions is done using optical [3] or electrical devices [4–6]. Temperature control is achieved utilizing contact-based heating devices [7]. The traditional heaters perform adequately if thermal runaway to the whole fluidic device is not a limiting factor, and the substrate material of the fluidic chip can conduct heat sufficiently well. If one of these two factors cannot be respected, a different heating technique is required.

An alternative solution for temperature control is microwave heating as it does not rely its operation on heat conduction—heat is generated by the liquid itself once it is exposed to an alternating electric field. Several research groups have reported microwave heating devices [8–18] for digital and continuous microfluidics in the past decade. For continuous flow microfluidics, the focus has been on heating of individual pico- and nano-liter droplets flowing through fluidic channels [8–12], microliter volumes in wells placed along fluidic channels [13,14], or on heating of liquid located over the complete fluidic chip [15,16]. Although all reported devices present breakthroughs in their fluidic subdomains, they cannot be entirely applied and up- or down-scaled to heat the liquid located in a section of a channel section on a planar fluidic chip, as illustrated in Figure 1. In summary, the reported investigations dealt with two extreme points concerning the liquid volume—individual droplets or relatively large individual reservoirs. In other words, a microwave heater investigation for sub-microliter volumes in continuous microfluidics has been lagging so far.

Thus, this work investigates the possibilities of microwave heaters for sub-microliter volumes located in a channel section in a continuous flow microfluidic chip, as illustrated in Figure 1, and the novelty of this work lies in a co-design of microwave heaters and microfluidic channels on a high-thermally conductive substrate, demonstrated by the design and evaluation of two types of microwave heaters. The investigated microwave devices provide uniform, contactless and localized microwave heating without restrictions on existing optical and fluidic techniques. All of this makes the novel microwave heating devices a basic building block that can be up- or down-scaled for the lab-on-a-chip systems. In our investigation, devices are realized in a single metal layer and integrated with a microfluidic chip in silicon-glass technology [19]. Alternatively, the proposed approach for the design of microwave heaters can be implemented in the standard printed circuit board (PCB) technology. Furthermore, the manufactured heater on a PCB can be further integrated with numerous fluidic devices manufactured using the established [20] or novel [21] fabrication processes and materials in microfluidics, such as glass, poly(methyl methacrylate) (PMMA) or polydimethylsiloxane (PDMS). Finally, complete versatility required nowadays in the lab-on-a-chip device development is fully achievable, given that the proposed microwave devices can be stacked up with existing microfluidic devices.

Previously, we reported the microwave characterization of manufactured devices on a die using the probe station and microwave ground-signal-ground (GSG) probes in conference proceedings [22,23]. Here, we present in detail the design approach of two types of devices, their comparison and integration with the supporting microwave circuitry, and finally, evaluation of heating performance on deionized water samples.

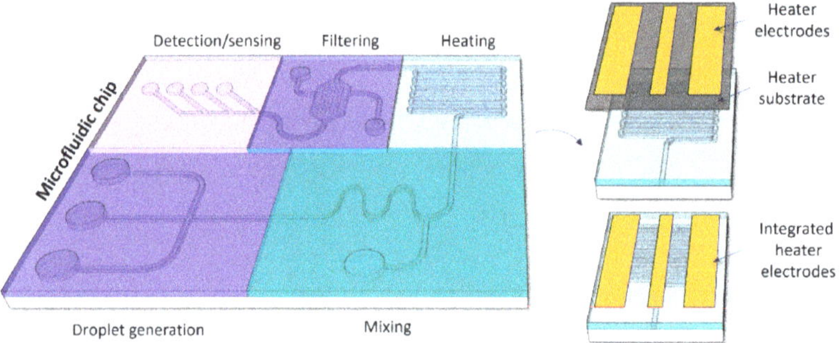

Figure 1. Microfluidic chip having several zones on the chip, incorporating functions such as droplet generation, liquid mixing, heating and filtering, and signal sensing. Microwave heaters can be implemented directly on top of the microfluidic device or can be implemented using the printed circuit board (PCB) technology and stacked-up with the microfluidic chip.

2. Materials and Methods

2.1. Device and System Fabrication and Integration

The microfluidic channels are micro machined in an 8 ″ high-resistivity (bulk = 10^{-5} S/cm) silicon (HR-Si) wafer (Topsil Semiconductor Materials A/S, Frederikssund, Denmark, part number 334-11000-0525-012), which is then bonded to a borosilicate glass wafer and subsequently singulated [19]. The height and width of the microfluidic channels are 100 and 240 μm, respectively, while the total thickness of the HR-Si is 400 μm, as shown in Figure 2a. The cover glass thickness is 300 μm. The buried fluidic channels comprise a reaction chamber, as shown in Figure 2a. The purpose of trenches in the silicon substrate is to thermally isolate the reaction chamber from the rest of the device [19] due to the high thermal conductivity of silicon. Aluminum metal layer is on top of the HR-Si, in which heaters are realized.

To integrate the manufactured devices on HR-Si dies with the microwave laboratory equipment for evaluation, a Rogers PCB (Rogers Corporation, Evergem, Belgium, part number RO4350B) was used. The 254 μm thick substrate was selected together with the copper thickness of 35 μm, on top where a 25 μm thick gold/nickel metal layer was electroplated to enable further wire bonding of microwave heaters on HR-Si with feeding transmission lines on the PCB. For feeding transmission lines on the PCB, a grounded co-planar waveguide (GCPW) topology was selected. The vias connecting the top and bottom metal layers of the PCB are spaced 500 μm apart to avoid substrate mode losses at high microwave frequencies. Additionally, the 10-mil thick substrate was chosen as it allows 50 Ohm transmission line dimensions similar to dimensions of microwave heaters on HR-Si. The metal layers on HR-Si dies and the PCB were wire bonded using a HB16 wire bonder (TPT, Munich, Germany). To interconnect the wire bonded dies on the PCB with the microwave equipment, a high frequency end launch connectors (Johnson Cinch Connectivity-Bel Stewart GmbH, Friedrichsdorf, Germany, part number 142-0761-861) were soldered to the PCB. Finally, the obtained stack-up shown in Figure 2b was fixed to an acrylic glass holder—a 4 mm thick acrylic sheet was sized using a speedy 100R laser cutter (Trotec Laser, Haaksbergen, The Netherlands).

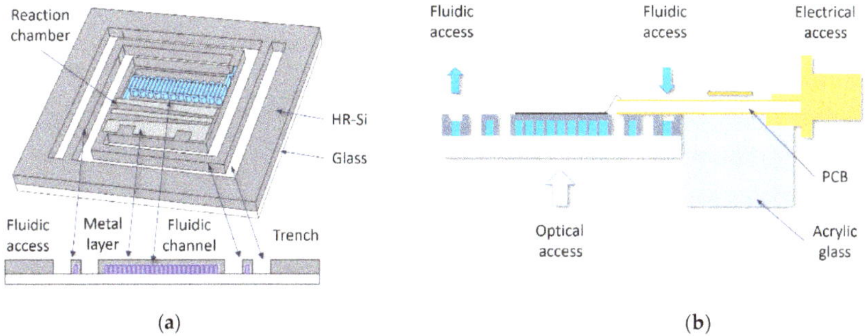

Figure 2. Technology and stack-up overview: (**a**) The 3D and cross section view of the reaction chamber on a high-resistivity silicon (HR-Si) die with the aluminum metal layer on top and buried microfluidic channels covered with glass; (**b**) cross-section of the stack-up consisting of the HR-Si die, PCB and acrylic glass holder.

2.2. Heater Design and Integration

Microwave heaters are designed to heat the liquid located in a channel section, as illustrated in Figure 1. In our investigation, this channel section is located in a reaction chamber, as presented in Figure 2a. The reaction chamber with its fluidic inlet and outlet is a basic building block that can be used in any lab-on-a-chip design case, as simple as a fluidic mixer or a filter would be used. The size of the reaction chamber is fixed to 3.6 × 3.6 mm, and the fluidic channel dimensions of 100 × 240 μm are kept constant in the chamber. Microwave heaters can be realized in only one metal layer that comes on top of the fluidic substrate, as presented in a device cross section in Figure 2a. By maintaining this technological restriction in terms of the single metal layer, design versatility from an integration perspective is retained because the single metal layer can be placed on top of the fluidic substrate in multiple ways. Moreover, when the spacing between the buried channel in the fluidic chip and the single metal layer is present, contactless microwave heating is achieved. Additionally, placing microwave heaters only over the reaction chamber results in localized microwave heating that does not heat the surrounding liquid on the fluidic chip.

To achieve uniform microwave heating, equal amounts of power should be dissipated over the liquid volume across the complete reaction chamber. To achieve this requirement, lossy microwave transmission lines are chosen because their physical implementation and corresponding dimensions can be designed in such a way that microwaves traveling along the line dissipate an equal amount of power in the liquid located in the reaction chamber. In this investigation, lossy transmission lines are realized using co-planar waveguides that consist of three electrodes in the single metal layer, as presented in Figure 2a. One electrode represents a central conductor of the co-planar waveguide line while the remaining two electrodes represent ground conductors of the transmission line. Other transmission line topologies such as microstrip or grounded co-planar waveguide could also be used at the cost of several metal layers, which does not comply with our initial requirement. Additionally, these transmission line technologies would restrict the optical access to the liquid located in the reaction chamber, which is not desired for the novel microwave-optical-fluidic platform.

Lossy co-planar waveguide transmission lines offer numerous possibilities in design and flexibility with respect to their size that is a limiting parameter in this investigation—a meticulous theoretical analysis of transmission lines is presented in textbook of D. Pozar [24]. In addition to the design possibilities, lossy microwave transmission lines allow two modes of heating—a traveling wave mode and a standing wave mode. In the traveling wave mode, microwaves travel along the line without any local reflections and exit the line at the output port that is terminated with 50 Ohm. In the standing wave mode, microwaves travel along the line without any local reflections but encounter a reflection at the end of the line because the output port is left open—from the transmission line theory, the open output port causes a reflection that is responsible for creating a standing wave along the line. In both modes of operation, the dissipated microwave power is given by Equation [24]:

$$P(z) = |V(z)|^2/Z_0, \tag{1}$$

in which *P(z)* represents the amount of dissipated power along the line, *V(z)* represents the voltage along the line and Z_0 represents the characteristic impedance of the line. Furthermore, the voltage along the line *V(z)* can be expressed using the Equation [24]:

$$V(z) = V^+ \cdot e^{-\gamma z} \cdot (1 + \Gamma_L \cdot e^{2\gamma z}), \tag{2}$$

in which V^+ represents an incident wave entering the line, γ represents a propagation constant of the line, z represents the position along the line, and Γ_L represents the reflection coefficient of the load at the end of the line, given by:

$$\Gamma_L = (Z_L - Z_0)/(Z_L + Z_0), \tag{3}$$

In other words, the voltage (i.e., electric field) along the line responsible for microwave dielectric heating is a sum of two waves on the line, among which the reflected wave is caused by the mismatch between the characteristic impedance of the line (i.e., Z_0) and the load impedance (i.e. Z_L). Based on Equation (3), Γ_L is 0 in the traveling wave mode as the load impedance is chosen to be 50 Ohm and the characteristic impedance of the line is designed to be 50 Ohm. Moreover, this consequently results in a single term in Equation (2). In the standing wave mode, Γ_L is different from zero and preferably designed to be 1 or −1 to obtain large wave reflections. To obtain Γ_L of −1 or 1, we should keep the output port shorted or open, which consequently results in a very small or a very large impedance (i.e., 0 and ∞ Ohm, respectively). For the detailed analysis of microwave transmission line theory, we refer the reader to the textbook of D. Pozar [24].

In this investigation, we have chosen modes of heating having Γ_L close to 0 and 1. A heater having Γ_L close to 0 (the traveling wave mode) is a through line microwave heater (TLMH), while a heater having Γ_L close to 1 (the standing wave mode) is an open-ended line microwave heater (OELMH) because an open circuit is used at the end of the transmission line. Both microwave heaters are designed at an operating frequency of 25 GHz. This choice is made to maximize the amount of dissipated power in the liquid in the fluidic channel caused by the presence of an electric field of microwave signals traveling along the line. The amount of dissipated power due to the presence of the electric field in the liquid without ions is given by the Equation [24]:

$$P(x, y, z, f, T) = 2 \cdot \pi \cdot f \cdot \varepsilon'' \cdot (x, y, z, f, T) \cdot |E(x, y, z, f, T)|^2, \quad (4)$$

in which f is the operating frequency, ε'' is the imaginary part of the complex permittivity of liquid, and $|E(x, y, z)|$ is the magnitude of the electric field strength within the liquid volume. To elaborate more on the operating frequency choice, it is necessary to look into the imaginary part of the complex permittivity of the liquid ε'' and the amount of dissipated power in the liquid volume $P(x, y, z)$, which are presented in Figure 3a,b for the deionized (DI) water case—we are conducting the analysis for DI water as liquid mixtures in biology often have high water content, which has the most dominant influence on the permittivity of mixtures at high frequencies. If another liquid would be chosen (e.g., a solvent in chemistry), the analysis should be conducted starting from the complex permittivity data for the chosen liquid. In the case when the permittivity data is not available, the liquid should be characterized done based on dielectric spectroscopy using commercially available solutions such as coaxial dielectric probes, or in-house developed spectroscopy chips and protocols [25–27].

The imaginary part of the complex permittivity ε'' sets the operating frequency around 20 GHz at room temperature to use the peak of ε'' and have rapid heating at the onset of the heating process. Nevertheless, the product governing heat losses steers the frequency choice in a different direction. If we assume that enough power is available for heating at a frequency range of interest, the operating frequency should be set to the highest possible value. This is concluded by calculating the amount of the dissipated power in a liquid volume having the same electric field strength of 1 V/m at all frequencies—results shown in Figure 3b. The upper limit on the frequency choice in this work is set by available microwave laboratory equipment, and therefore, a frequency of 25 GHz is chosen.

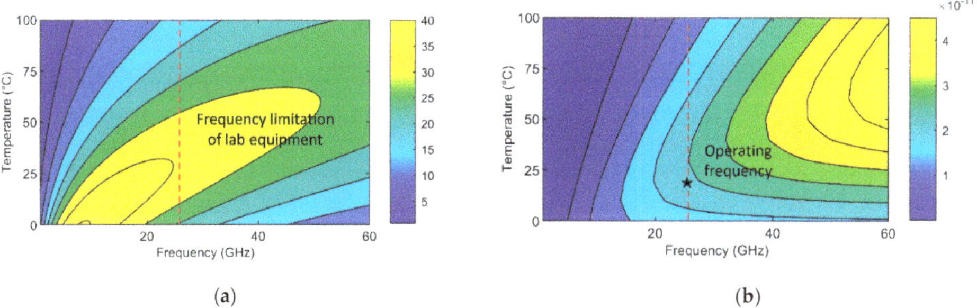

Figure 3. Data used to select operating frequency: (**a**) The imaginary part of complex permittivity of deionized water from 1 to 60 GHz and from 0 to 100 degrees Celsius [28]; (**b**) power losses in deionized water calculated using Equation (4) and the data in Figure 3a for an electric field of 1 V/m at all frequencies.

3. Results

3.1. Heater Design

In the heater design process, it is possible to design the microwave and the fluidic part. Designing the microwave part means sizing the width of the central conductor and the spacing between the central and ground conductors of the co-planar waveguide transmission line. Designing the fluidic part means arranging the layout of the channel in the reaction chamber to further help the heating uniformity. In the through line design case, the size of fluidic channel sections is kept constant and perpendicular to the transmission line length, as shown in Figure 4a,b. By keeping them perpendicular, the interaction between the electric field and the liquid in the channel is maximized, as previously reported by others [13]. Because the size of the channel sections is kept constant in the through line microwave heater (TLMH) design case, only the microwave transmission line is sized to dissipate equal amounts of power in the liquid located in all fluidic channel sections. This is done in COMSOL Multiphysics (COMSOL AB, Stockholm, Sweden) for 20 channel sections that result in a volume of 465 nL. The total amount of dissipated power is 295 mW for an excitation signal of 1000 mW at 25 GHz—the microwave theory was previously reported [22]. The final design dimensions of the TLMH obtained using COMSOL are depicted in Figure 4c. The TLMH dissipates an average amount of power of 0.63 mW/nL with a variation of 0.066 mW/nL between different sections. All microwave power not transformed into heat goes out at the output port of the TLMH. The electric field strength in the channel shown in Figure 4c is presented for the cross-section A in Figure 4g, 200 µm away from the metal layer of the TLMH, and calculated using an excitation signal of 1000 mW at 25 GHz in COMSOL Multiphysics. This top-view 2D electric field plot provides more insight into the heating uniformity caused by the electric field in the liquid. It shows that the electric field is present in the complete liquid volume with limited gradients along the complete channel length. The side-view plot in Figure 4e for the cross section B, 100 µm away from the edge of the central conductor of the co-planar waveguide transmission line, also shows the 2D spatial distribution of the electric field in the liquid located in the channel and indicates that most of the electric field is contained around the transmission line.

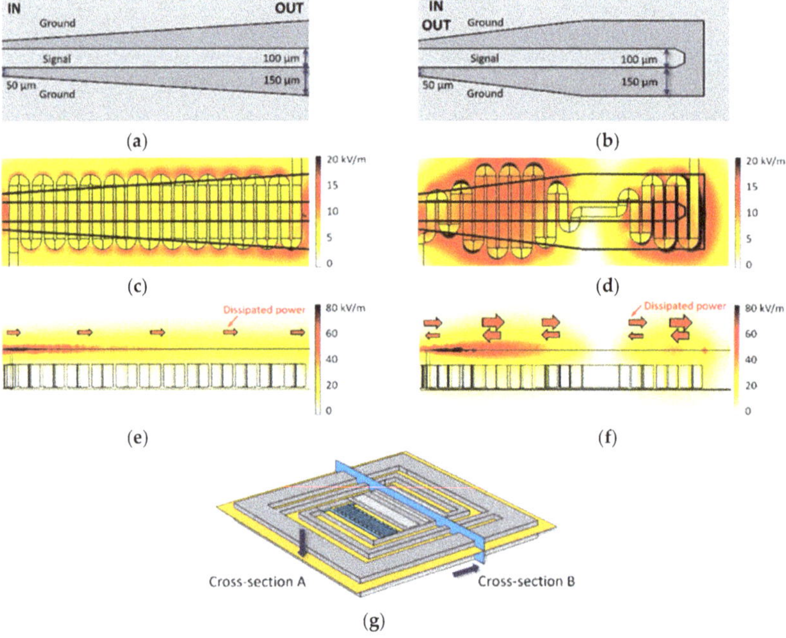

Figure 4. Design of microwave heaters: (**a**) The through line microwave heater (TLMH) with final optimized dimensions; (**b**) the open-ended-line microwave heater (OELMH) with final optimized dimensions; (**c**) the electric field strength for the TLMH at the plane of the cross-section A; (**d**) the electric field strength for the OELMH at the plane of the cross-section A; (**e**) the electric field strength for the TLMH at the plane of the cross-section B; (**f**) the electric field strength for the OELMH at the plane of the cross-section B; (**g**) illustration of cross-sections used for electric field plots.

The open-ended-line microwave heater (OELMH) employs a resonant structure for heating because a standing wave pattern is created. Consequently, the resonant heating structure creates zones without the electric field in which dielectric heating does not occur, while in other zones where the electric field is very strong, fast dielectric heating occurs. Therefore, microwave and microfluidic parts are co-designed together for heating uniformity across the complete liquid volume. The co-design is done by sizing the length of channel sections and spacing between signal-ground conductors to obtain the same normalized dissipated power per unit volume. This can be seen in Figure 4d,f for the final design dimensions of the OELMH—a large volume is present in zones of strong electric fields, while a small volume is present in zones of weak electric fields. Additionally, the spacing between single-ground conductors is smaller in zones of strong electric field than in zones of the weaker electric field. The total OELMH length is chosen to be 3/4 of the wavelength that fits into the reaction chamber. This topology design is also carried out in COMSOL Multiphysics and the obtained OELMH device design results in power delivery of 487 mW to a volume of 315 nL over the transmission line length for an excitation signal of 1000 mW at 25 GHz. The final design dimensions of the OELMH depicted in Figure 4b dissipate an average amount of power of 1.55 mW/nL with a variation between channel sections of 0.46 mW/nL. Once volumes around the minima of the standing wave pattern are excluded from the analysis, 419 mW is dissipated in 241 nL (1.74 mW/nL) with a variation between channel sections of 0.28 mW/nL showing satisfying uniformity of power delivery. The electric field strength in the channel resulting from the OELMH is obtained from COMSOL Multiphysics using an excitation signal of 1000 mW at 25 GHz—it is presented in Figure 4d,f for the cross sections A and B in Figure 4g. The plots presented in Figure 4d,f confirm larger dissipated power than in the case of TLMH due to the presence of stronger electric fields. Additionally, the same plots confirm the larger variation in power uniformity due to the presence of larger electric field gradients than in the case of the through line heater.

3.2. Platform Design

Designed microwave heaters in the transmission line topology are placed on individual reaction chambers depicted in Figure 5a,b, top and bottom view, respectively. Four chambers are put on a single high resistivity-silicon-glass (HR-Si-glass) die that can be integrated with a microwave feeding network, as depicted in Figure 5c, top view. The feeding network is designed with grounded co-planar waveguides to connect the HR-Si die to standard coaxial connectors, as depicted in Figure 5c. Additionally, the ends of transmission lines are intentionally designed to be capacitive to compensate for the additional inductance of the bond wires and minimize reflections. Finally, the designed microwave-microfluidic platform is presented in Figure 5c,d—the microwave and microfluidic part are interconnected from the top side, while the optical access is provided from below. The dimensions of the platform holder in acrylic glass are sized to fit the microscope top stage.

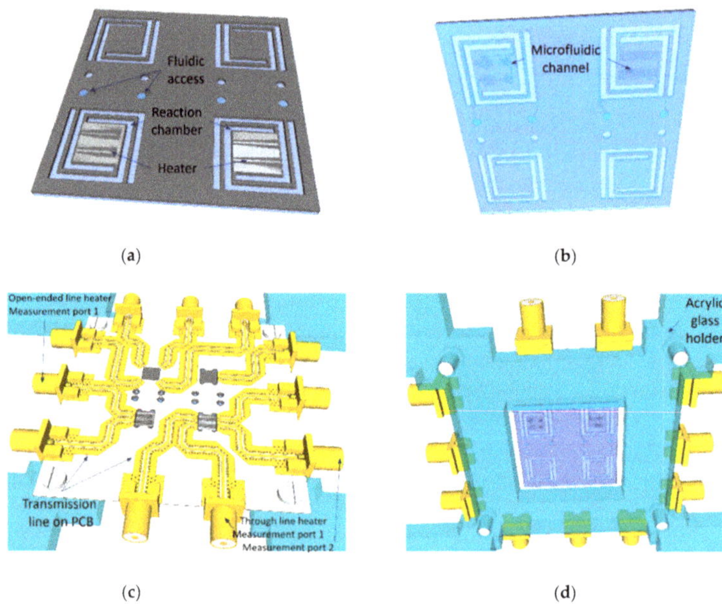

Figure 5. Integrated heaters on a die with the PCB and the acrylic glass holder: (**a**) The top view of the HR-Si-glass die with four reaction chambers showing microwave heaters; (**b**) the bottom view of the HR-Si-glass die showing microfluidic channels; (**c**) the top view of the integrated HR-Si-glass dies with the PCB and the acrylic glass holder showing feeding transmission lines; (**d**) the bottom view of the integrated HR-Si-glass dies with the PCB and the acrylic glass holder showing the optical access to the microfluidic channels.

3.3. Microwave Measurements

Microwave measurements are often carried out by measuring S-parameters, which represent the ratio of incident and reflected waves at ports of interest [24]. For example, S_{11} is a ratio of a reflected wave at the port 1 and an incident wave at the same port 1, while S_{21} is a ratio of a reflected wave at the port 2 and an incident wave at the port 1. In other words, S_{11} is used to calculate how much power is returned to the source, and S_{21} is used to calculate how much power is transmitted from the port 1 to the port 2 of a device. S_{11} of −10 dB means 10% of the input power is reflected by a device and 90% is delivered to the device. S_{21} of 3 dB means 50% of the incident power passed from port 1 to port 2.

In our case, we measured parameters S_{11} and S_{21}, for ports indicated in Figure 5c. Microwave measurement results of the heaters on an HR-Si-glass die and integrated heaters with the PCB, for the air and deionized (DI) water filled channels are presented in Table 1. The first investigated results are S_{11} parameters of the integrated silicon die with microwave heaters and the printed circuit board (PCB) for the case when the channels are filled with DI water—measurement ports of the die integrated with the PCB are indicated in Figure 5c. In both cases S_{11} is low, which means most of the incident microwave power is delivered to the heating system. Afterward, this amount of the delivered power travels to the heater on HR-Si. While traveling over the feeding transmission line, the microwave signal gets attenuated due to the dielectric and conductive losses of the transmission line on the PCB—this effect can be seen by comparing S_{21} of the TLMH (at the level of the die) and S_{21} of the TLMH integrated with the PCB for the air-filled channel. Additionally, for the OELMH, this can be seen by comparing S_{11} of the heater on the HR-Si die and S_{11} of the integrated die for the air-filled channel. Finally, the power reaching the heaters on HR-Si heats liquid in the reaction chamber, as demonstrated by the heater measurements at the silicon die level for the water-filled channels.

Table 1. Microwave measurement results.

Frequency 25 GHz		TLMH		OELMH
		S_{11}	S_{21}	S_{11}
HR-Si die	Air	−17.74	−0.93	−0.92
	DI Water	−23.57	−2.63	−4.45
Integrated die and PCB	Air	−11.62	−5.26	−4.82
	Di Water	−10.39	−8.55	−16.92

3.4. Heating Experiments

Heating experiments were carried out with 1.58 W (32 dBm) at 25 GHz at the coaxial connector plane on the PCB using the microwave-optical measurement setup [29]—the measurement port 1 is indicated in Figure 5c. Images of the Rhodamine B and deionized water liquid mixture loaded into different reaction chambers before heating are presented in Figure 6a,b. Temperature profiles of the liquid in the channel, calculated based on these images, are presented in Figure 6c,d. The obtained temperature profiles show a spatial temperature variation of ±1 °C, which is the limitation of this measurement technique [30]. Figure 6e–j shows calculated temperature profiles of heated liquid for the first 10 s of heating. It is possible to observe that the liquid mixture is heated uniformly in the first stage of the heating when the highest heating rates are measured. The heating rate and the average temperature of the liquid mixture for two heaters are shown in Figure 6k,l. More specifically, the highest recorded heating rate one second after the heating has started is 16 °C/s for the TLMH and 24 °C/s for the OELMH. A few seconds after the heating has started, the warm liquid acts as a heat source for the HR-Si reaction chamber and the bottom glass substrate, resulting in lower heating rates than in the beginning, as presented in Figure 6k,l. At this point of time, the heat transfer from the heated liquid to the silicon and glass surrounding is highest as the temperature gradient is the largest. The dominant path for the heat losses responsible for the temperature saturation is through the HR-Si substrate due to its thermal conductivity, which is 187.5 times larger than the glass thermal conductivity (150 to 0.8 W/mK). In the end, the system enters the steady state and the liquid does not significantly heat any further, as indicated in Figure 6k,l. The highest measured average temperature is 53 °C for the through line heater and 65 °C for the open-ended line heater.

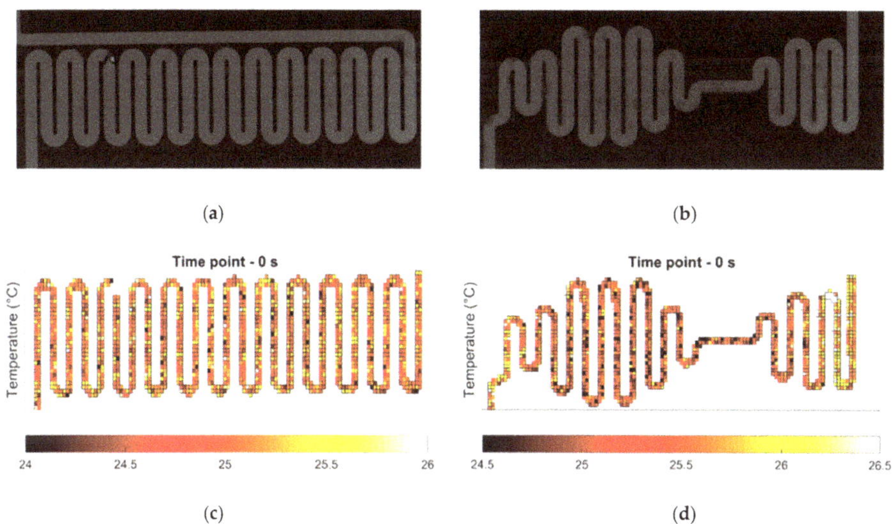

(**a**) (**b**)

(**c**) (**d**)

Figure 6. *Cont.*

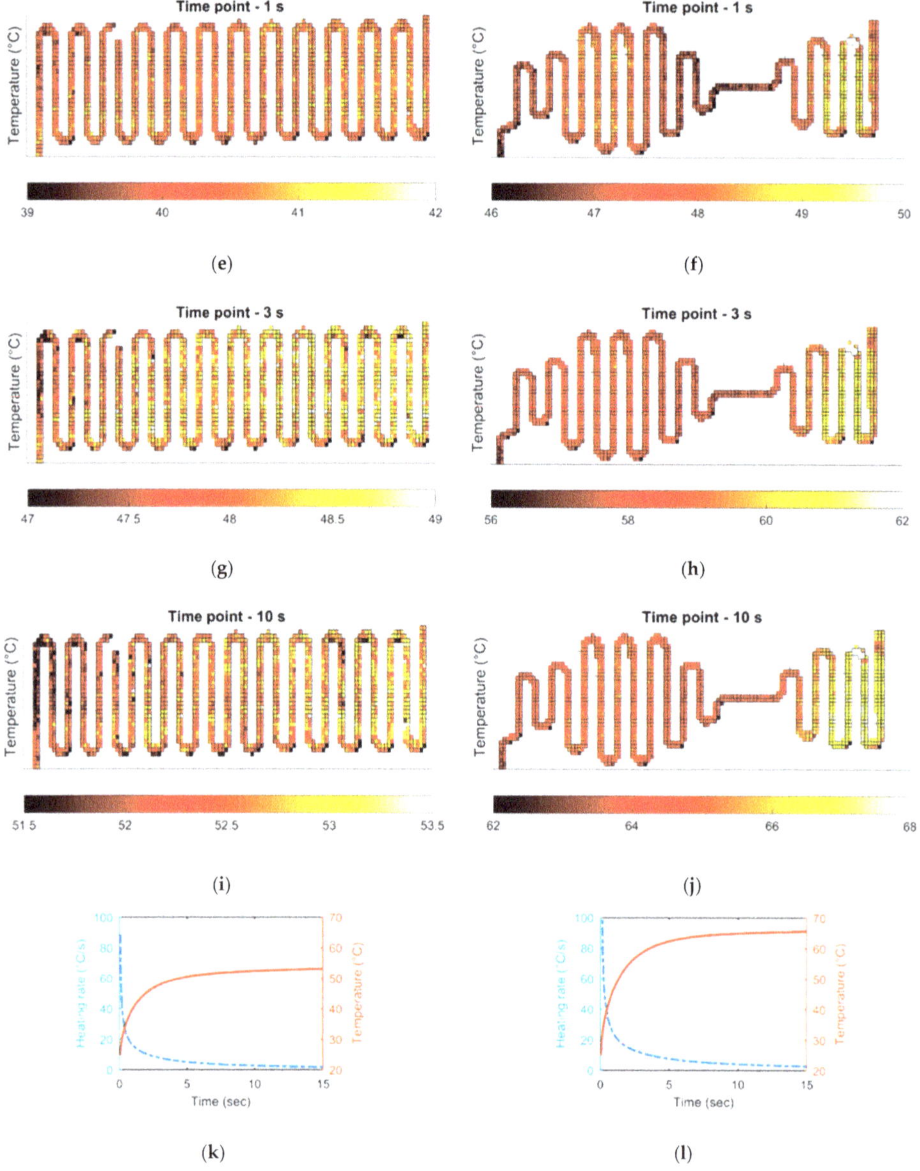

Figure 6. Microwave heating results: (**a**) Image of the fluorescent liquid mixture located in the channel of the TLMH before heating; (**b**) image of the fluorescent liquid mixture located in the channel of the OELMH before heating; (**c**) the calculated temperature profile of the liquid mixture prior to heating for the TLMH; (**d**) the calculated temperature profile of the liquid mixture prior to heating for the OELMH; (**e**–**j**) the calculated temperature profiles of the liquid mixture during heating for the TLMH and OELMH; (**k**) the calculated average temperature of the liquid mixture and corresponding rate for the TLMH; (**l**) the calculated average temperature of the liquid mixture and corresponding rate for the OELMH.

4. Discussion and Conclusions

The first comparison between the TLMH and the OELMH is based on the highest achieved temperature. According to the microwave design in COMSOL Multiphysics, the OELMH dissipates more power in the microfluidic channel than the TLMH, which is confirmed by microwave measurements—the TLMH dissipates 37% of the input power, while the OELMH dissipates 51% of the input power in the liquid located in the reaction chamber. Therefore, the liquid temperature obtained using the OELMH recorded in heating measurements agrees well with our design results and expectations. The second comparison between the heaters is based on the temperature uniformity of heating. This comparison is possible thanks to the calculated spatial temperature distribution of the liquid in the channel from measurement data. Results in Figure 6c–j show that the TLMH achieves better temperature uniformity at different time points than the OELMH across the whole liquid volume—the temperature uniformity of the TLMH is within ±1.5 °C, while the OELMH shows gradients of ±3 °C. Additionally, the OELMH creates two different temperature zones in its two liquid sections. All these observations in measurement results can be correlated to the 2D electric field distribution presented in Figure 4c,d—the electric field distribution of the TLMH is more uniform once compared to the OELMH. Because of these electric field gradients created by the resonant device, a larger temperature non-uniformity occurs, as reported in Figure 6. Although electric field gradients created by the resonant device result in less uniform heating, the HR-Si substrate helps to locally uniformize the temperature of the loaded liquid due to its high thermal conductivity. In addition, thermally induced flow could also be a factor in uniformizing the temperature. At the end, each heater has its advantages and should be selected accordingly to meet desired specifications.

In summary, this work demonstrates a new class of microwave heaters for sub-microliter volumes in continuous microfluidics for micro total analysis systems. Both types of microwave heaters achieve uniform, contact-less and localized microwave heating without the need for additional microwave structures. This is achieved thanks to the design method that employs a novel co-design of microwave topologies and microfluidic channels, as demonstrated on the design of two different devices. Among many possible modifications of the presented versatile heaters, it is important to stress within the two heating concepts it is possible to adjust channel dimensions to obtain a desired liquid volume, an amount of the dissipated power in liquid to achieve different heating rates over time and the heater-microfluidic system size to fit miniature micro total analysis systems—all parameters often found in specifications of heating systems. Finally, the heating measurement results confirmed the designed behavior of heaters and demonstrated that microwave heating for lab-on-a-chip devices has a large potential in heating applications in cases when a heater cannot be put in the immediate proximity of the liquid and contact between the heater and the liquid should be avoided.

Author Contributions: Conceptualization, T.M. and I.O.; methodology, T.M. and B.N.; investigation, T.M.; resources, I.O., A.B. and B.N.; writing—original draft preparation, T.M.; writing—review and editing, I.O. and A.B.; visualization, T.M.; supervision, I.O., A.B. and B.N.; funding acquisition, I.O. and B.N. All authors have read and agreed to the published version of the manuscript.

Funding: This research was funded by KU Leuven C2 project microwave microbiology, and FWO projects G0B6713N and G0A1220N. This publication was made possible through funding support of the KU Leuven Fund for Fair Open Access.

Acknowledgments: The authors would like to acknowledge the Life Science Technologies department design team at imec for the realization of the microfluidic wafer, and M. Magerl for wire bonding of dies to PCBs.

Conflicts of Interest: The authors declare no conflict of interest.

References

1. Ward, K.; Hugh Fan, Z. Mixing in microfluidic devices and enhancement methods. *J. Micromech. Microeng.* **2015**, *25*, 094001. [CrossRef] [PubMed]
2. Yoon, Y.; Kim, S.; Lee, J.; Choi, J.; Kim, R.K.; Lee, S.J.; Sul, O.; Lee, S.B. Clogging-free microfluidics for continuous size-based separation of microparticles. *Sci. Rep.* **2016**, *6*, 26531. [CrossRef]

3. Alves, P.U.; Vinhas, R.; Fernandes, A.R.; Birol, S.Z.; Trabzon, L.; Bernacka-Wojcik, I.; Igreja, R.; Lopes, P.; Baptista, P.V.; Águas, H.; et al. Multifunctional microfluidic chip for optical nanoprobe based RNA detection—Application to Chronic Myeloid Leukemia. *Sci. Rep.* **2018**, *8*, 381. [CrossRef]
4. Curto, V.F.; Marchiori, B.; Hama, A.; Pappa, A.M.; Ferro, M.P.; Braendlein, M.; Rivnay, J.; Fiocchi, M.; Malliaras, G.G.; Ramuz, M.; et al. Organic transistor platform with integrated microfluidics for in-line multiparametric in vitro cell monitoring. *Microsyst. Nanoeng.* **2017**, *3*, 17028. [CrossRef]
5. Bao, J.; Yan, S.; Markovic, T.; Ocket, I.; Kil, D.; Brancato, L.; Puers, L.; Nauwelaers, B. A 20-GHz microwave miniaturized ring resonator for nL microfluidic sensing applications. *IEEE Sens. Lett.* **2019**, *3*, 4500604. [CrossRef]
6. Pan, Y.; Jiang, D.; Gu, C.; Qiu, Y.; Wan, H.; Wang, P. 3D microgroove electrical impedance sensing to examine 3D cell cultures for antineoplastic drug assessment. *Microsyst. Nanoeng.* **2020**, *23*, 6. [CrossRef]
7. Miralles, V.; Huerre, A.; Malloggi, F.; Jullien, M.C. A review of heating and temperature control in microfluidic systems: Techniques and applications. *Diagnostics* **2013**, *3*, 33–67. [CrossRef] [PubMed]
8. Markovic, T.; Bao, J.; Maenhout, G.; Ocket, I.; Nauwelaers, B. An interdigital capacitor for microwave heating at 25 GHz and wideband dielectric sensing of volumes in continuous microfluidics. *Sensors* **2019**, *19*, 715. [CrossRef]
9. Abduljabar, A.A.; Choi, H.; Barrow, D.A.; Porch, A. Adaptive coupling of resonators for efficient microwave heating of microfluidics systems. *IEEE Trans. Microw. Theory Tech.* **2015**, *63*, 3681–3690. [CrossRef]
10. Boybay, M.S.; Jiao, A.; Glawdel, T.; Ren, C.L. Microwave sensing and heating of individual droplets in microfluidic devices. *Lab Chip* **2013**, *13*, 3840–3846. [CrossRef]
11. Issadore, D.; Humphry, K.J.; Brown, K.A.; Sandberg, L.; Weitz, D.A.; Westervelt, R.M. Microwave dielectric heating of drops in microfluidic devices. *Lab Chip* **2009**, *9*, 1701–1706. [CrossRef] [PubMed]
12. Shah, J.J.; Sundaresan, S.G.; Geist, J.; Reyes, D.R.; Booth, J.C.; Rao, M.V.; Gaitan, M. Microwave dielectric heating of fluids in an integrated microfluidic device. *J. Micromech. Microeng.* **2007**, *17*, 2224–2230. [CrossRef]
13. Marchiarullo, D.J.; Sklavounos, A.H.; Oh, K.; Poe, B.L.; Barker, N.S.; Landers, J.P. Low-power microwave-mediated heating for microchip-based PCR. *Lab chip* **2013**, *13*, 3417–3425. [CrossRef] [PubMed]
14. Kempitiya, A.; Borca-Tasciuc, D.A.; Mohamed, H.S.; Hella, M.M. Localized microwave heating in microwells for parallel DNA amplification applications. *Appl. Phys. Lett.* **2009**, *94*, 064106. [CrossRef]
15. Morgan, A.J.L.; Naylon, J.; Gooding, S.; John, C.; Squires, O.; Lees, J.; Barrow, D.A.; Porch, A. Efficient microwave heating of microfluidic systems. *Sens. Actuators B Chem.* **2013**, *181*, 904–909. [CrossRef]
16. Shaw, K.J.; Docker, P.T.; Yelland, J.V.; Dyer, C.E.; Greenman, J.; Greenway, G.M.; Haswell, S.J. Rapid PCR amplification using a microfluidic device with integrated microwave heating and impingement cooling. *Lab Chip* **2010**, *10*, 1725–1728. [CrossRef]
17. Markovic, T.; Liu, S.; Ocket, I.; Nauwelaers, B. A 20 GHz microwave heater for digital microfluidics. *Int. J. Microw. Wirel. Technol.* **2017**, *9*, 1591–1596. [CrossRef]
18. Markovic, T.; Bao, J.; Ocket, I.; Kil, D.; Brancato, L.; Puers, R.; Nauwelaers, B. Uniplanar microwave heater for digital microfluidics. In Proceedings of the 2017 First IEEE MTT-S International Microwave Bio Conference (IMBioC), Gothenburg, Sweden, 15–17 May 2017.
19. Majeed, B.; Jones, B.; Tezcan, D.S.; Tutunjyan, N.; Haspeslagh, L.; Peeters, S.; Fiorini, P.; de Beeck, M.O.; van Hoof, C.; Hiraoka, M.; et al. Silicon based system for single-nucleotide-polymorphism detection: Chip fabrication and thermal characterization of polymerase chain reaction. *Jpn. J. Appl. Phys.* **2012**, *51*, 04DL01. [CrossRef]
20. Duffy, D.C.; McDonald, J.C.; Schueller, O.J.A.; Whitesides, G.M. Rapid prototyping of microfluidic systems in poly(dimethylsiloxane). *Anal. Chem.* **1998**, *70*, 4974–4984. [CrossRef]
21. Bao, J.; Markovic, T.; Brancato, L.; Kil, D.; Ocket, I.; Puers, R.; Nauwelaers, B. Novel fabrication process for integration of microwave sensors in microfluidic channels. *Micromachines* **2020**, *11*, 320. [CrossRef]
22. Markovic, T.; Ocket, I.; Jones, B.; Nauwelaers, B. Characterization of a novel microwave heater for continuous flow microfluidics fabricated on high-resistivity silicon. In Proceedings of the 2016 IEEE MTT-S International Microwave Symposium (IMS), San Francisco, CA, USA, 22–27 May 2016.
23. Markovic, T.; Ocket, I.; Jones, B.; Nauwelaers, B. Contactless microwave heating of continuous flow microfluidics on silicon. In Proceedings of the 20th International Conference on Miniaturized Systems for Chemistry and Life Sciences, Dublin, Ireland, 9–13 October 2016.
24. Pozar, D.M. *Microwave Engineering*, 4th ed.; Wiley: Hoboken, NJ, USA, 2011.

25. Liu, S.; Ocket, I.; Barmuta, P.; Markovic, T.; Lewandowski, A.; Schreurs, D.; Nauwelaers, B. Broadband dielectric spectroscopy calibration using calibration liquids with unknown permittivity. In Proceedings of the 84th ARFTG Microwave Measurement Conference, Boulder, CO, USA, 4–5 December 2014.
26. Barmuta, P.; Bao, J.; Zhang, M.; Marković, T.; Nauwelaers, B.; Schreurs, D.; Ocket, I. Broadband electrical determination of liquid mixing ratios for microfluidics. In Proceedings of the 2019 European Microwave Conference in Central Europe (EuMCE), Prague, Czech Republic, 13–15 May 2019.
27. Liu, S.; Orloff, N.D.; Little, C.A.; Zhao, W.; Booth, J.C.; Williams, D.F.; Nauwelaers, B. Hybrid characterization of nanolitre dielectric fluids in a single microfluidic channel up to 110 GHz. *IEEE Trans. Microw. Theory Tech.* **2017**, *65*, 5063–5073. [CrossRef]
28. Ellison, W.J. Permittivity of pure water, at standard atmospheric pressure, over the frequency range 0–25 THz and the temperature range 0–100 °C. *J. Phys. Ref. Data* **2007**, *36*. [CrossRef]
29. Markovic, T.; Jones, B.; Barmuta, P.; Ocket, I.; Nauwelaers, B. A transmission line based microwave heater at 25 GHz for continuous flow microfluidics fabricated on silicon. In Proceedings of the 2019 49th European Microwave Conference (EuMC), Paris, France, 1–3 October 2019.
30. Shah, J.J.; Gaitan, M.; Geist, J. Generalized temperature measurement equations for Rhodamine B dye solution and its application to microfluidics. *Anal. Chem.* **2009**, *81*, 8260–8263. [CrossRef] [PubMed]

Article

Off-Design Exergy Analysis of Convective Drying Using a Two-Phase Multispecies Model

Andrea Aquino * and Pietro Poesio

Department of Mechanical and Industrial Engineering, University of Brescia, 25123 Brescia, Italy; pietro.poesio@unibs.it
* Correspondence: andrea.aquino@unibs.it; Tel.: +39-030-3715646

Received: 9 October 2020; Accepted: 31 December 2020; Published: 4 January 2021

Abstract: The design of a convective drying cycle could be challenging because its thermodynamic performance depends on a wide range of operating parameters. Further, the initial product properties and environmental conditions fluctuate during the production, affecting the final product quality, environmental impact, and energy usage. An off-design analysis distinguishes the effects of different parameters defining the setup with the best and more stable performance. This study analyzes a reference scenario configured as an existing system and three system upgrades to recover the supplied energy and avoid heat and air dumping in the atmosphere. We calculate their performance for different seasons, initial product moisture, input/output rate, and two products. The analysis comprises 16 simulation cases, the solutions of a two-phase multispecies Euler–Euler model that simulates the thermodynamic equilibrium in all components. Results discuss the combination of parameters that maximizes the evaporation rate and produces the highest benefits on global performance up to doubling the reference levels. The advantages of heat recovery vary by the amount of wasted energy, increasing the exergy efficiency by a maximum of 17%. Energy needs for air recirculation cut the performance at least by 50%. Concluding remarks present the technical guidelines to reduce energy use and optimize production.

Keywords: drying; energy analysis; exergy analysis; multiphase model; multispecies model; thermodynamics

---

1. Introduction

Thermal drying plays a crucial role in several industries, such as chemical, pharmaceutical, agricultural, and food production. It involves heating a wet product to evaporate its liquid fraction and generating a thermally induced mass flux [1]. According to the dominant energy transfer mechanism, thermal drying can be categorized into convective [2], conductive [3], and radiative drying [4]. The present study focuses on convective drying operated as a continuous process on a horizontal fluidized bed (Figure 1).

The fluidized bed systems present several advantages as a good mixing quality (homogeneous temperature distribution along the bed) and high heat and mass transfer rate caused by the extended contact surface between the solid particles, and the gaseous phase [5,6]. However, reaching the optimal thermodynamic performance of such systems needs careful tuning of numerous operating parameters that influence the final energy consumption, product quality, and environmental effects.

For example, the finest particles (powder) tend to agglomerate, affecting the evaporation rate and quality of the final product [7]. This effect increases the bed pressure drop; thus, a faster airflow becomes necessary to preserve product quality, but it increases the final energy consumption, and costs [8]. Further limitations of the global performance derive from the unavoidable inefficiencies of the process: the sensible heating of the dry fraction, the heat losses of the drying chamber, and the low thermal conductivity of the heat transfer media (air), which demands high operative temperatures to drive adequate heat flux [9]. Another issue regards the effect of the environmental conditions on the final energy use: the analysis of Reference [10] reveal the drying chamber as particularly susceptible to the external temperature; its performance deteriorate with temperature fluctuations of approx. 5 °C. Finally, some specific applications such as the convective drying of biomass and hazardous materials, release several pollutants into the atmosphere; this deteriorates local air quality and contributes to greenhouses gas emissions [11–13]. For such reasons, stringent environmental regulations limit their functioning.

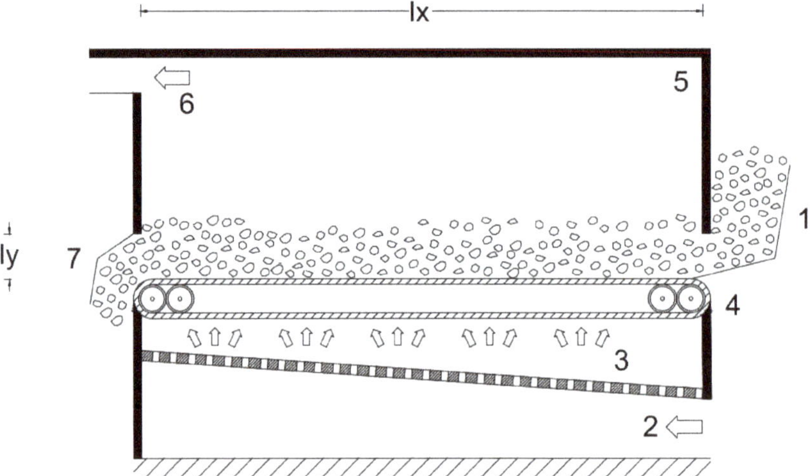

Figure 1. Fluidized bed drying chamber: (1) product inlet; (2) air inlet; (3) air diffuser; (4) conveyor belt; (5) insulated wall; (6) air outlet; and (7) product outlet.

In the design and optimization practice, predicting all the effects mentioned above can be challenging because of the relations among the drying parameters and the mutual dependencies among the system components [14]. The standard approach studies the relation between the drying conditions and final performance by the energy analysis of the drying chamber. Dincer et al. [15] calculated the efficiency of the cycle (i.e., drying efficiency) as the ratio between the energy invested in the evaporation process over the total energy entering the drying chamber by hot airflow. Further studies [16] included the specific energy consumption, calculated as the amount of consumed energy per unit mass of the evaporated moisture.

Energy analysis by itself cannot calculate the operative costs and any form of environmental effect because it does not distinguish the primary energy source; such an approach does not provide any information regarding the effects of the climate on system productivity. To overcome these limitations, several authors included the exergy analysis in the evaluation of system performances.

The first exergy analysis of convective drying is presented in Reference [17], wherein the authors calculate the exergy efficiency of the drying chamber as the ratio between the exergy invested in the evaporation process to the total exergy of the entering airflow. Assuming the entering product is at a dead-state, the exergy investment is the exergy of the total evaporated moisture leaving the drying chamber. Several works follow the approaches presented above. Akpinar et al. [18] related the energy utilization of the drying chamber with an evaporation rate and initial product moisture; they measured an increase in exergy losses using the drying air temperature and velocity. Yogendrasasidhar et al. [19] analyzed the effect of the several drying parameters on the energy utilization, exergy losses, and exergy efficiency of the system: by increasing the wall temperature and air velocity, the energy usage augments but the exergy losses decreases with benefits on the final exergy efficiency; the prolonging of drying time produces a two-fold benefit by reducing energy usage and exergy losses and by increasing exergy efficiency. Aviara et al. [20] showed a linear dependence between energy efficiency and the drying air temperature.

More recent research on convective drying oriented the exergy analysis toward optimization by comparing different system configurations and operative conditions. Icier et al. [21] compared the exergy performance of two open drying cycles and a closed cycle, where drying air was recovered from the drying chamber and heated by a heat pump. Xiang et al. [22] analyzed a drying system coupled to a heat pump; in particular, they investigated how the system performance changes under different operative conditions, varying the amount of recirculating air (from an open to a fully-closed cycle). Cay et al. [23,24] analyzed and compared two open cycles, in which internal and external combustion chambers heat the drying air, respectively. Erbay [10], and Gungor [25] investigated how dead-state temperature affects the exergy performance of a drying system fed by a ground source and a gas heat pump.

All studies presented above refer to the given system configurations; their results are valid only under the working conditions observed for the analyzed system and can be used for designing the drying cycle of a specific material. Furthermore, limiting the performance analysis to the drying chamber misses the most recent and accurate optimization techniques of energy systems that extend the exergy analysis to a multicomponent level (they calculate the exergy destruction in a single component by considering interdependencies with all other components) [26,27]. Thus, the research gap in the current literature is based on the design approach specifically formulated for convective drying, which considers the effects of changing the system configuration and operative conditions (e.g., dry another product, adding another component, and varying the climate) for the optimization of thermodynamics and costs [28].

The design of a drying system needs a theoretical model to calculate the state of working flows in the different components (i.e., thermodynamic equilibrium model). The Euler–Euler description is a common approach to simulate the thermodynamic equilibrium of a drying process due to its low computational costs and the capability to simulate fluids with a high concentration of the dispersed phase [29–31]. Assari et al. [32,33] studied the drying of wheat grain: first, they formulated a two-fluid model to investigate the effects of varying the operative conditions on the main operative parameters of the drying bed (e.g., void percentage and air humidity); in the following study, they analyzed the exergy performance. Li et al. [34] modeled two fluids, the bubbly and emulsion phases, a mixture of an interstitial gas, and a solid phase; using this formulation, they performed a sensitivity analysis of the drying performance with respect to the state of inlet air, particle diameter, and wall temperature of the drying chamber. Ranjbaran et al. [35] developed a two-fluid model for paddy drying; they investigated the temporal variation of the energy and exergy efficiency during the drying process with effects of air temperature and flow rate; a source term in transport equations is used to model the evaporation of moisture. Rosli et al. [36] simulate the drying of sago waste by a two-fluid model of the drying column; they investigate by CFD the effects of different drying conditions (e.g., the air velocity, temperature, and particle size) on the fluidization of the bed. Jang et al. [37] simulate a fluidized bed dryer by an Euler-Euler model coupled to empirical correlations representing the inter-phase exchanges; the authors investigate the advantages of such a model in the design and scaling-up of pharmaceutical applications.

The above studies simulated the drying process by two homogeneous phases—the drying air and the wet solid—without focusing on their chemical composition. A multispecies approach enhances the versatility and accuracy of the two-fluids theory—it comprehends the effects of each species on the mass and energy fluxes [38]. Furthermore, this approach can simulate applications of reactive flows according to the stoichiometry of the ongoing chemical reactions (e.g., combustion of a hydrocarbon for air heating). The multicomponent theory generally finds applications in petroleum distillation [39,40], while the available literature lacks references for applications to convective drying.

In addition to the thermodynamic equilibrium model, the design of a drying system needs characteristic equations describing the heat and mass transfer phenomena that occur in each component; as an example, the characteristic equation of the drying chamber is the evaporation model. Defraeye et al. [41] derive the characteristic equation of drying by a theoretical approach—the authors analyze the convective drying of a porous flat plate by solving the transport phenomena at the interface between the porous media and the airflow, explicitly. Such an approach, known as conjugate modeling, describes in detail the physics of the heat and mass transfer; however, despite its accuracy, just a few academic applications use this technique because of the high complexity, and computational costs [30]. Quite the opposite, the empirical or non-conjugated models derive from experimental observations and describe the heat and mass transfer by constant coefficients, with a limited understanding of the involved physics. Some well-known examples of empirical models are Newton's law of cooling, and the evaporation model of Page [42].

Our study investigates the thermodynamic performance of convective drying under different operating conditions and system configurations; the final aim is distinguishing the parameters with the most significant effects on the energy use and product quality to define the setup with the best performance. We follow a theoretical approach, named off-design analysis applied successfully for the optimization and control strategies of various energy systems [43–45]. The novelty of our work is the nature of the studied application: in the literature, there is not any off-design analysis of convective drying; in particular, the current design practice lacks a theoretical formulation for modeling the state of working flows in the drying chamber and all system components, including the devices for heat and mass recovery. The level of detail of our theoretical approach is a further innovation in the field of drying modeling: we formulate the two-fluids theory describing the thermodynamic equilibrium of the single chemical species to simulate all components of the drying system. Finally, due to the adaptability of the solving algorithm, we present an innovative tool for both design new drying cycles and verify the states of working flows in existing systems.

The presented equilibrium model includes the multispecies approach in the two-fluid theory to simulate all the components of a drying system; in particular, we follow a two-fluid multispecies Euler–Euler (TFMM) approach to simulate the following processes.

- *Mass and energy exchange between nonreactive phases:* Convective drying of a wet solid and air-water counter-current mixing;
- *Mass and energy exchange between reactive phases:* combustion of airflow by a hydrocarbon jet;
- *Energy exchange between nonreactive phases:* Air-to-air and air-to-water heat exchange within a tube bank.

Our analysis aims to support the design practice of a drying system rather than investigate drying physics in detail. Thus, we simulate heat and mass exchanges by empirical equations because of their appropriate accuracy for the scale of simulated processes (macro-scale between 0.01–10 m [30]), as well as the advantages of a simple mathematical formulation and low computational costs.

The analysis starts with the exergy analysis of a reference drying cycle, named the baseline scenario, based on an existing industrial system; this scenario mounts the essential components of convective drying: fan, combustion chamber, and drying chamber (Figure 1). We calculate the baseline performance at the operative conditions of the existing system and validate the results against experimental data. Later, we change the drying conditions, dead-state conditions, and dried material to investigate the effects on energy consumption and exergy efficiency. By the last set of operative conditions, we assess the drying of municipal sewage sludge, thereby aiming to contribute to this crucial but scarcely investigated sector. Finally, we modified the reference system to investigate the effects on the global performance of heat and mass recovery by three different layouts.

2. Research and Method

2.1. Governing Equations

We assume that all thermodynamic processes of a convective drying cycle occur in a one-dimensional open system of infinitesimal dx length (Figure 2). To support the assumption of a one-dimensional system, we calculate the heat and mass exchange rates by the lumped-system model (e.g., the Page model for drying and Newton's cooling law for heat exchangers; see more details in the Appendix A).

The TFMM describes the mass and energy exchanges, entropy generation, and exergy destruction: Phase 1 is drying air, which is a semi-perfect gas mixture; Phase 2 is the substance that exchanges mass and heat with drying air (its state and composition vary by the nature of the simulated process).

Assuming equilibrium between the *j*-species dispersed in the *i*-phase, we track them by their respective mass fractions

$$\alpha_i = \frac{m_i}{m_{os}} \tag{1}$$

$$\epsilon_{ij} = \frac{m_{ij}}{m_i} \tag{2}$$

where

$$m_{os} = \sum_i m_i \tag{3}$$

$$\sum_i \alpha_i = \sum_j \epsilon_{ij} = 1. \tag{4}$$

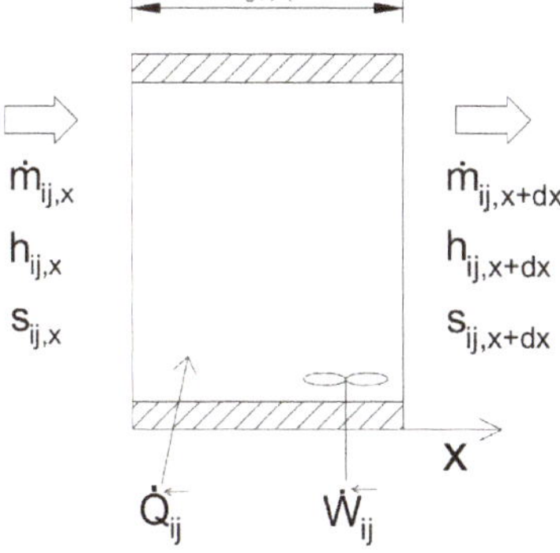

Figure 2. The two-fluid multispecies model of the i-phase formed by the dispersed j-species.

- Mass balance of *j*-species per unit length is

$$\frac{\partial m'_{ij}}{\partial t} + \frac{\partial}{\partial x}(\dot{m}_{ij}) = (\Delta\dot{m}_{12,j})'. \tag{5}$$

The mass balance of the i-phase is obtained by summing the j-species as

$$\frac{\partial m_i'}{\partial t} + \frac{\partial}{\partial x}(\dot{m}_i) = (\Delta \dot{m}_{12})'. \tag{6}$$

The term

$$\Delta \dot{m}_{12,j} = -\Delta \dot{m}_{21,j} \tag{7}$$

represents the mass exchange rate between two phases.

- Energy balance is

$$\frac{\partial E_{ij}'}{\partial t} + \frac{\partial}{\partial x}(\dot{m}_{ij} h_{ij}) = (\dot{W}_{ij}^{\leftarrow})' + (\dot{Q}_{ij}^{\leftarrow})' + (\Delta \dot{H}_{12,j})' \tag{8}$$

and that for the whole i-phase is

$$\frac{\partial E_i'}{\partial t} + \frac{\partial}{\partial x}(\dot{m}_i h_i) = (\dot{W}_i^{\leftarrow})' + (\dot{Q}_i^{\leftarrow})' + (\Delta \dot{H}_{12})'. \tag{9}$$

The terms $\dot{W}_{ij}^{\leftarrow}$ and $\dot{Q}_{ij}^{\leftarrow}$ represent the shaft work and the heat absorbed by the j-species (from the environment or other species). The term $\Delta \dot{H}_{12,j}$ is the heat exchanged between two phases by any phase transition and/or chemical reaction

$$\Delta \dot{H}_{12,j} = -\Delta \dot{H}_{21,j}. \tag{10}$$

- Entropy balance is given by

$$\frac{\partial S_{ij}'}{\partial t} + \frac{\partial}{\partial x}(\dot{m}_{ij} s_{ij}) = \left(\frac{\dot{Q}_{ij}^{\leftarrow}}{T_b}\right)' + (\Delta \dot{S}_{12,j})' + (\dot{S}_{irr,ij})' \tag{11}$$

and that for the whole i-phase is

$$\frac{\partial S_i'}{\partial t} + \frac{\partial}{\partial x}(\dot{m}_i s_i) = \left(\frac{\dot{Q}_i^{\leftarrow}}{T_b}\right)' + (\Delta \dot{S}_{12})' + (\dot{S}_{irr,i})'. \tag{12}$$

The entropy generated by $\dot{Q}_{ij}^{\leftarrow}$ depends on the system boundary temperature T_b, whereas $\dot{S}_{irr,ij}$ is the entropy generated by process irreversibilities; the term $\Delta \dot{S}_{12,j}$ represents the entropy related to any phase change of the j-species:

$$\Delta \dot{S}_{12,j} = -\Delta \dot{S}_{21,j}. \tag{13}$$

- Exergy balance is given by

$$\frac{\partial Ex_{ij}'}{\partial t} + \frac{\partial}{\partial x}(\dot{m}_{ij} ex_{ij}) = (\dot{W}_{ij}^{\leftarrow})' + (\dot{Q}_{ij}^{\leftarrow})'\left(1 - \frac{T_0}{T_b}\right) + (\Delta \dot{E}x_{12,j})' - (\dot{E}x_{d,ij})' \tag{14}$$

and the contribution for the whole i-phase is

$$\frac{\partial Ex_i'}{\partial t} + \frac{\partial}{\partial x}(\dot{m}_i ex_i) = (\dot{W}_i^{\leftarrow})' + (\dot{Q}_i^{\leftarrow})'\left(1 - \frac{T_0}{T_b}\right) + (\Delta \dot{E}x_{12})' - (\dot{E}x_{d,i})', \tag{15}$$

where T_0 is the dead-state temperature. The exergy exchange $\Delta\dot{E}x_{12,j}$ and the exergy destroyed by process irreversibility $\dot{E}x_{d,ij}$ are

$$\Delta\dot{E}x_{12,j} = \Delta\dot{H}_{12,j} - T_0\Delta\dot{S}_{12,j} = -\Delta\dot{E}x_{21,j} \tag{16}$$

$$\dot{E}x_{d,ij} = T_0\dot{S}_{irr,ij}. \tag{17}$$

In the following sections, we assume steady-state conditions, and therefore, all time derivatives $\frac{\delta}{\delta t}$ are equal to zero.

2.2. Solving Algorithm

A custom-developed code written in C language is used to solve the TFMM balance equations. The algorithm (Figure 3) consists of modules connected by one or more dataflows (representing matter and energy streams). There is a node between two connected modules.

A module is a function block that uses the upstream node values as input and calculates the state of the *i*-phases at the downstream node as a result; each module includes the following parts:

1. TFMM governing equations;
2. State equation of the i-phase;
3. Component characteristic equations;
4. End-of-file (EOF) values.

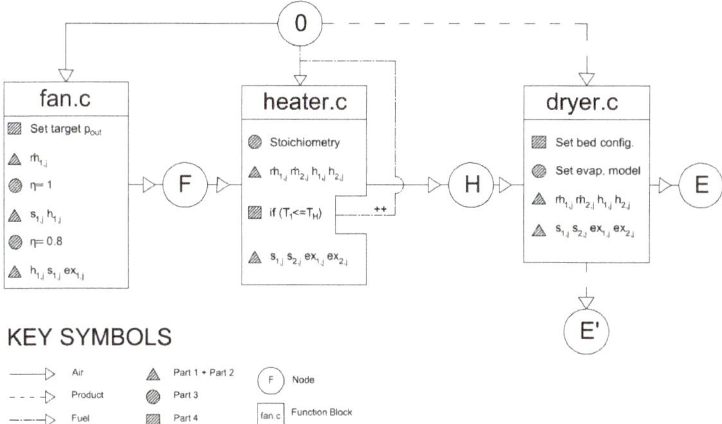

Figure 3. Solving algorithm applied to the baseline scenario.

The Part 1 is the general equilibrium model and it is unvaried in all modules; Part 2 depends on the nature of the *i*-phases entering/leaving the module, and it defines the coefficients of TFMM differential equations (e.g., perfect gas or incompressible fluid, constant or polynomial specific heat); Parts 3–4 are strictly related to the features of the simulated component, and they are different for each module.

The advantage of such a structure is the high adaptability of its running logic, which is suitable for both design and off-design analysis of energy systems. Following the equations of Parts 2–3, the function block implicitly solves Part 1 under a steady state using the Newton–Raphson method [46]. This procedure is an iterative process that produces a sequence of solutions for consecutive *dx* space intervals; it stops when the solution matches the target values (design) or when the total number of solved intervals equals the component dimensions (off-design), both fixed by Part 4. The ideal gas law is the equation of state used in this analysis. State functions are calculated by a polynomial T-dependent specific heat for gaseous species and a constant specific heat for liquid and solid species ([47–49]). Further details about state equations, component-specific equations (Part 3), and design targets (Part 4) are in Appendix A.

3. Case Study: Baseline Scenario

The case study of the current analysis is an industrial drier located at Kedah, Malaysia; this system, designed to dry rice paddy, involves three centrifugal fans (maximum capacity of each: 15 kW) and a furnace to heat the drying air at the operative temperature; the drying chamber is a fluidized bed system (Figure 1).

The bed dimensions l_x, l_y are 4.85 m and 0.97 m; the bed thickness l_z varies based on the operative conditions. However, we assume it is equal to 0.1 m according to most experimental observations [50,51]. The baseline scenario (Figure 4) reproduces the case study. This configuration is the benchmark of our analysis because it runs the fundamental steps of convective drying: a fan flows the external air to the combustion chamber; the combustion chamber heats up the drying air at the set temperature (T_H); and the heated air enters the drying chamber, where the drying process takes place as cross-flow heat and mass exchange.

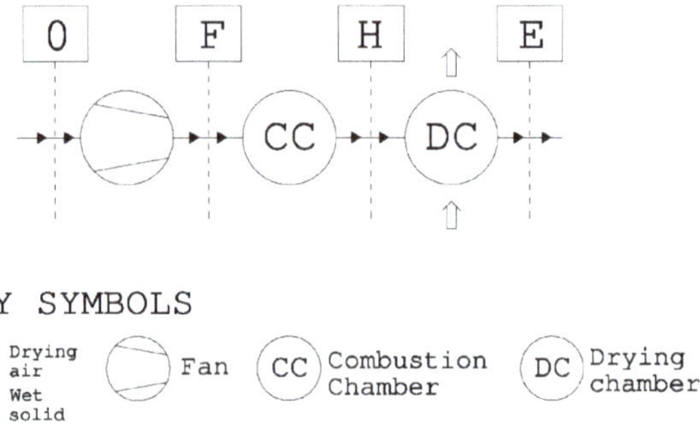

Figure 4. Baseline scenario corresponds to the layout of the case study.

4. Validation of TFMM

The validation estimates the accuracy of TFMM for simulating the case study; it compares the model predictions to the experimental data of two works ([50,51]) that describe the functioning of the reference system at different seasons and under different operating conditions. The numerical comparison focuses on two parameters that strictly depend on the mass and energy balance of the drying process:

1. Final moisture content of the dried product (u_f), which indicates the total evaporated moisture (i.e., evaporation rate);
2. Air temperature at the outlet of the drying chamber (T_E), which measures the energy wasted by the drying chamber.

We run validation cases to increase the heat supply (by increasing T_H) and vary the initial moisture content (u_0) and external temperature (T_0) of the wet solid (see values in Table A3). Compared to the experimental observations (Figure 5), the mean percent error (MPE) of the predicted u_f and T_E are 5.5%

and 7.1%, respectively. These values are consistent with the MPE of numerical simulations available from References [34,35,52].

The second validation part investigates the accuracy of TFMM after changing the operating conditions of the reference cycle. We compared the simulation results with three further experimental datasets obtained from Reference [53] and Reference [54] who studied rice drying at higher T_H and feed rate ($\dot{m}_s$) and at a lower airflow rate ($\dot{m}_a$) than the reference case study, and from the research of Reference [55] who performed an experimental analysis of drying municipal sewage sludge (MSS) (see values in Table A4). These studies measured only u_f values, and the simulation results (Figure 6a,b) showed high accuracy of the TFMM when it runs off-design conditions; for rice drying, the u_f values presented a mean percent error (MPE) of 1.5% and 4.8%; the mean error of MSS drying is approximately 5.1%.

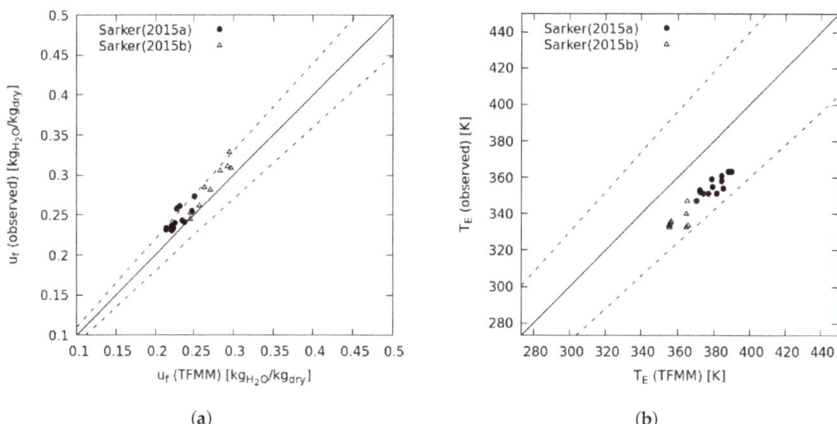

Figure 5. Simulation vs. experimental data: u_f (**a**) and T_E (**b**) of the case study; values on the dot lines represent a prediction error of ±10%.

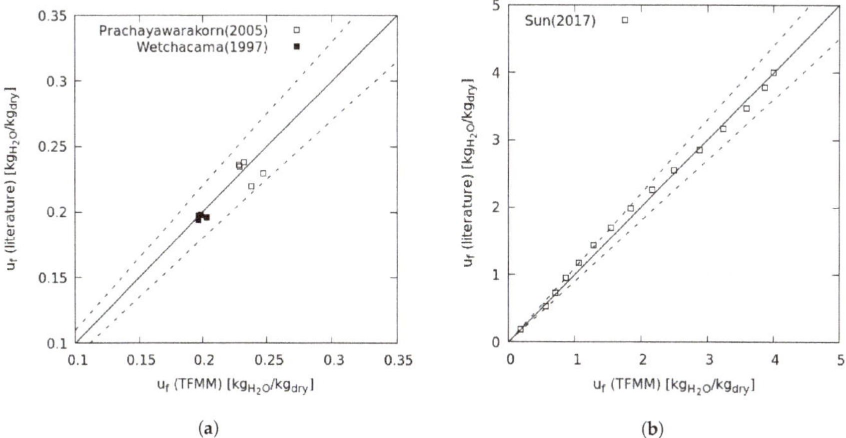

Figure 6. Value of u_f under off-design conditions of rice (**a**) and MSS (**b**); values on the dot lines present the prediction error of ±10%.

5. Off-Design Analysis Setup

Our analysis involves a case study under off-design conditions; we start by simulating the baseline scenario as in the experiments, and then, we vary the operative conditions and cycle configuration to optimize its performance and reduce environmental effects. We upgraded the baseline scenario thrice—each one running 4 different combinations, named *set*, of operative conditions, for a total of 16 different setups—and compared the results.

5.1. Operating Conditions

All sets are listed in Table 1. The set1 reproduces the experimental conditions of the case study; in set2, the system dries a higher amount of rice at a higher drying temperature; thus, we investigated the effects of increasing the total mass and energy supplied to the drying chamber. The drying conditions of set3 are the same as those of set1. However, set3 operates in a colder and more humid season. Finally, we investigated the drying of a different product, the MSS, by running set4. The k, n parameters (i.e., Page's model constants) are specific for each material and vary by the drying conditions; further details about the calculation of these parameters in each set are given in the Appendix A.

Table 1. The drying conditions of each set: T_0 and T_H are the external and drying air temperature; $\dot{m}_a$ and $\dot{m}_s$ are the mass flow rates of the drying air and product; the terms u_0 and ω_0 measure the initial moisture content within the product and drying air and are expressed in kg of water on kg of dry matter.

SET	T_0 $[K]$	u_0 -	ω_0 -	$\dot{m}_a$ $[\frac{kg}{s}]$	$\dot{m}_s$ $[\frac{kg}{s}]$	T_H $[K]$	Material	k,n
1	300	0.3	0.011	10.98	2.36	363	Rice paddy	0.103 (k) 0.771 (n)
2	300	0.3	0.011	12.86	4.17	388	Rice paddy	0.166 (k) 0.695 (n)
3	288	0.5	0.008	10.98	2.36	363	Rice paddy	0.103 (k) 0.771 (n)
4	300	0.3	0.011	10.98	4×10^{-4}	363	MSS	1.72×10^{-6} (k) 1.487 (n)

5.2. Heat Recovery by Scenario 1

The baseline scenario is an open cycle. The exhaust air is sensibly far from the dew point (experimental observations show a minimum $T_E \approx 330$ K and maximum relative humidity of approximately 6%); thus, the system expels evaporated moisture and a sensible heat fraction to the environment, which is unexploited by the drying process.

Waste heat is a crucial indicator of the energy performance of the system because it measures the energy productively invested in the evaporation processes (i.e., drying efficiency); moreover, the heat released in the environment is one of the leading environmental affects a thermal system [56]. To decrease heat wastage, we include a heat-recovery unit in the baseline scenario; this component exploits the wasted heat to reduce the thermal load of the combustion chamber.

The result is Scenario 1 (Figure 7), in which the airflow from the drying chamber recirculates to preheat the fresh air intake by HE1; the latter is a cross-flow and air-to-air heat exchanger, configured as a tube bank and fabricated by aluminum tubes with an inner diameter of 7.5×10^{-3} m and a shell thickness of 2×10^{-3} m (Figure 8a,b).

5.3. Heat and Mass Recovery by Scenarios 2 and 3

A form of environmental effect of a drying system is the emission of pollutants in the atmosphere caused by processing hazardous materials. For example, References [12,57,58] showed that drying MSS, agricultural wastes, and biomass emits VOC, NH_3, and CO, thereby affecting the local air quality and contributing to greenhouses gas emissions.

Scenarios 2 and 3 reduce the emission of air pollutants and mitigate the wastage of heat. After they have recovered most of the waste heat by HE1, the systems restore the drying airflow at the initial conditions and reuse it in a new cycle instead of dumping it in the atmosphere. Both systems include two separate loops, one circulating the heating and one the drying air (Figures 8b and 9a); these loops exchange heat by the air-to-air heat exchangers HE0 and HE3, configured as HE1, with tubes of the same diameter and wall thickness. An external combustion chamber generates the driving heat. Its combustion exhausts are at a fixed temperature ($T_{H'}$) of 500 K, and they feed HE0 to heat the drying air to the target temperature (T_H). When the drying process is completed and HE1 has recovered most of the wasted heat, the systems restore the initial conditions of drying air using the following steps.

- Air cooling (restoration of the absolute humidity by dehumidification)
- Air post-heating (restoration of the temperature and relative humidity).

As shown in the analysis of other air-handling systems (e.g., desalination [59]) and as confirmed by our results, the dehumidification of humid air represents an intensive energy use and entropy source. The rate of the entropy generation of a dehumidifier varies between the saturation and condensation steps because of the effect of the preponderant heat and mass transfer mechanism. As claimed by the theorem of the equipartition of entropy production [60], it is minimal when its distribution in space approaches uniformity.

To simulate air dehumidification, the TFMM solves the saturation and the condensation steps separately by calculating the state variables of the humid air along the length of the dehumidifier (x-direction). Thus, by comparing two cooling systems characterized by different cooling rates, costs, and heat\mass transfer mechanisms, we can choose the best technique that minimizes entropy generation.

- Scenario 2 uses HE2, a serpentine tube bank made of aluminum tubes with the same dimensions (tube diameter and thickness) as those of the air-to-air units (Figure 8c,d); this unit employs direct air cooling because a thin nonpermeable layer (tube shell) separates water from the drying air;
- Scenario 3 provides a direct-cooling system: An evaporative cooling tower which mixes the nebulized cooling water and the airstream in a counter-current flow; the tower diameter is 1 m. Heat exchange occurs through a porous packed bed with a specific surface ratio of 300 m^2/m^3 [61].

A chiller produces the cooling water for both units, and the heat subtracted by air cooling could feed low-temperature thermal processes (e.g., Reference [62]).

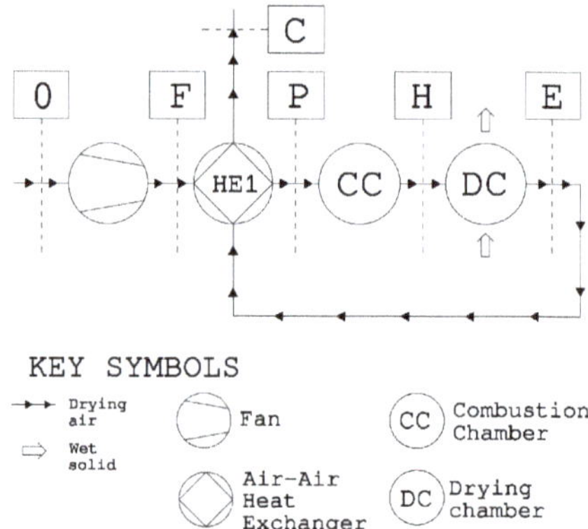

Figure 7. Scenario 1 is the first upgrade of the baseline scenario: An open cycle installed with a heat recovery unit.

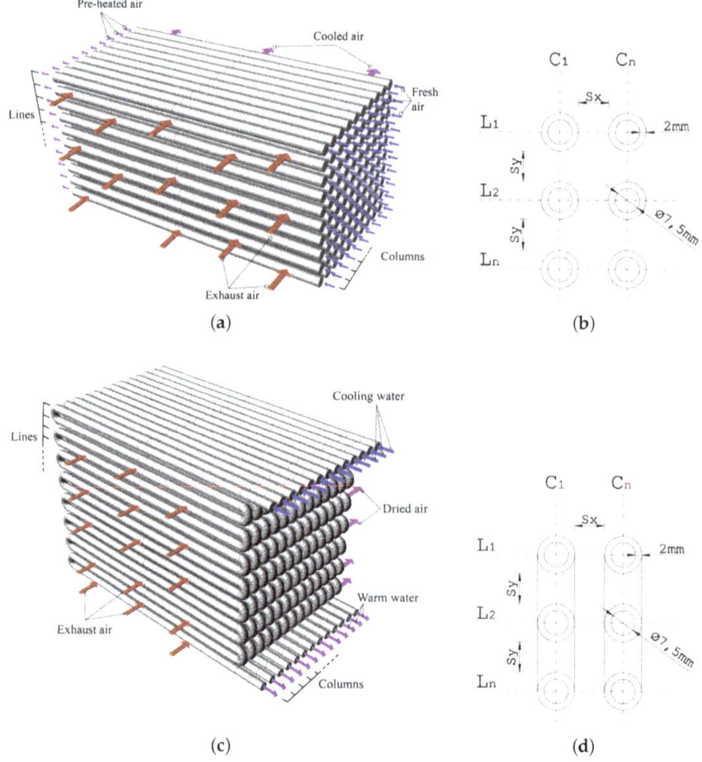

Figure 8. The spacing between the lines (L) and columns (C) of tube banks is equal in the horizontal and vertical direction s_x, s_y; in air-to-air units (**a**,**b**) tubes are free and in air-to-water units (**c**,**d**) they are connected in a serpentine.

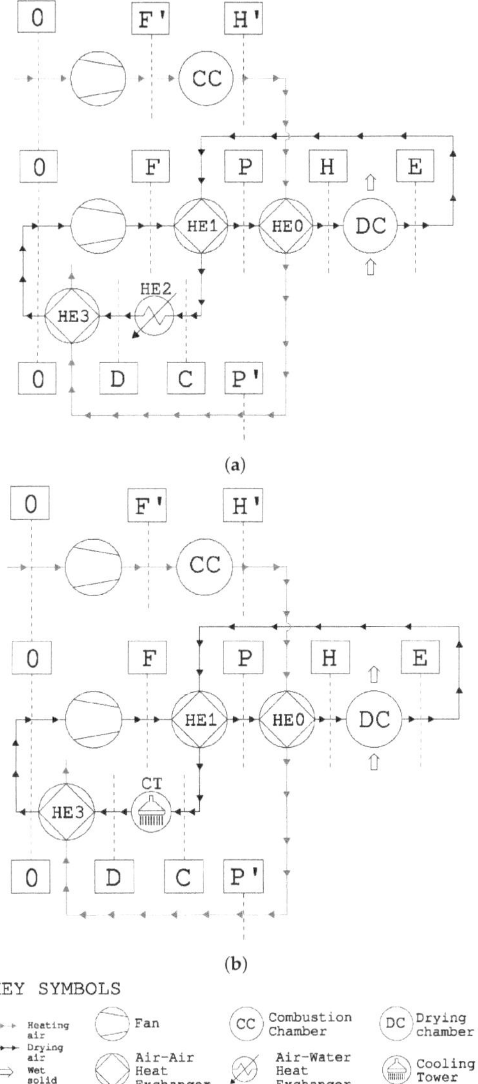

Figure 9. Closed cycles are Scenarios 2 (**a**) and 3 (**b**): An external combustion system keeps the chemical composition of the drying air unaltered; the latter is regenerated at the initial conditions by a cooling and a post-heating unit.

5.4. Performance Indicators

As energy efficiency indicators, we use the drying efficiency η_{dc}, which relates the heat fraction effectively used by the evaporation process to the total heat supplied to the drying chamber; also, we use the specific thermal energy consumption (STEC) and specific electric energy consumption (SEEC) to obtain the first distinction of the nature of input energies.

$$\eta_{dc} = \frac{q_{ev}}{\dot{E}_{th}} \quad (18)$$

$$SEEC = \frac{\dot{E}_{el}}{\dot{m}_{ev}} \quad (19)$$

$$STEC = \frac{\dot{E}_{th}}{\dot{m}_{ev}}. \quad (20)$$

The exergy efficiency η_{ex} depends on the total exergy inputs of the running cycle (i.e., the exergy of fuel $\dot{E}x_f$) and the exergy effectively invested in the drying process ($\dot{E}x_{ev}$):

$$\eta_{ex} = \frac{\dot{E}x_{ev}}{\dot{E}x_f}. \quad (21)$$

The fuel exergy $\dot{E}x_f$ includes the fan power, enthalpy of the reaction of the fuel, and chiller power inputs. The term $\dot{E}x_{ev}$ measures the exergy change of the amount of water that passes from the initial liquid state to the dispersed vapor state [17]. Assuming the wet product entering into the drying chamber at the dead state ($ex_{2,0}$), the term $\dot{E}x_{ev}$ depends on the exergy of the exhaust air collected at the outlet of the drying chamber ($ex_{1,E}$).

$$\dot{E}x_f = \dot{m}_{CH_4} \Delta H^0_{f,CH_4} + \dot{E}_{el} \quad (22)$$

$$\dot{E}x_{ev} = \dot{m}_{ev}(ex_{1,E} - ex_{2,0}). \quad (23)$$

The exergy efficiency as defined in the Equation (21) measures the effects of different drying conditions and system setup on the drying process; further, it compares the exergy costs for recovering the residual exergy of the drying exhausts (i.e., the air outflow of the drying chamber).

Finally, the exergy destruction ratio y_k measures the weight of the exergy destruction by the irreversibilities of the k-component ($\dot{S}_{irr,k}$) in a system formed by a total of n components:

$$\dot{E}x_{d,k} = T_0 \dot{S}_{irr,k} \quad (24)$$

$$y_k = \frac{\dot{E}x_{d,k}}{\sum_n \dot{E}x_{d,k}}. \quad (25)$$

The parameter y_k identifies the k-component affected by the largest irreversibility, and it can address the designer toward the most effective design strategies for enhancing the system's performance by replacing single components.

6. Results and Discussion

This section discusses the results of the analysis by comparing the 16 simulation cases. Each case involves a specific configuration of the drying cycle running an operative *set*.

6.1. Heating and Flowing Loads

Loads of fan(s) and combustion chamber are shown in Figure 10.

Fans balance the system's pressure losses, and therefore, their power loads increase by adding a heat exchanger to the baseline layout. The average power load of the baseline scenario augments by 3 times in Scenario 1 and 7 and 6 times in Scenarios 2 and 3. The airflow rate ($\dot{m}_a$) affects the fan power load: from *set*1 to *set*2, the average load increases by 31%; the other *sets*, where $\dot{m}_a$ is unvaried, presents slight differences caused by different dimensions (i.e., pressure drop) of the heat exchangers.

In the baseline scenario, the thermal loads are equal to the total heat supplied to the drying chamber: from *set*1, it increases by 47% in *set*2 because of the higher drying temperature and mass flow rate ($T_H, \dot{m}_a$) and by 7% in *set*3 because of the colder external air. Comparing the baseline scenario to Scenario 1, we observe that the heat-recovery successfully reduces thermal loads by 14%, below the operating conditions of the *set*2, up to 27% in *set*1. Closed cycles need more thermal energy than the baseline scenario because of the external combustion chamber and the post-heating process; their thermal load is three times higher than that of the baseline scenario running *sets*1, 2, *and* 4; the *set*3 produces a further increase (+6%) because of the lower temperature of feeding air (T_0).

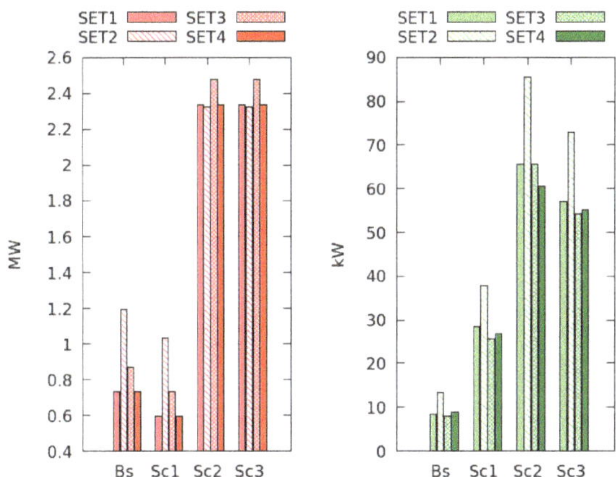

Figure 10. The heating (**red**) and flowing loads (**green**).

6.2. Drying

Figure 11 shows a plot of the spatial distribution along the bed length of the most relevant operative parameters of the drying chamber; these results show the influence of the operating conditions on the drying process and product quality. In all simulations, the evaporation rate ($\dot{m}_{ev}$) gradually reduces to zero because the moisture content of the wet product tends to equilibrium levels (u_{eq}). Furthermore, $\dot{m}_{ev}$ is inversely proportional to the material-specific resistance, which is represented by the exponential term in Page's evaporation model [63].

The results of *set*2 show that the evaporation rate is augmented with the mass and energy supplied to the drying chamber: $\dot{m}_{ev}$ of *set*2 is 86% higher than that of *set*1, although the final product moisture is almost the same ($u_f = 0.25$). Thus, a hotter airflow reduces u_{eq} and evaporates a more in-depth moisture layer, enhancing the drying efficiency η_{dc} (+1% as shown in Figure 12).

As shown by *set*3, the initial product moisture has a critical effect on the evaporation rate and drying performance. A more humid product presents a lower resistance to the evaporation; thus, when more humid rice is fed into the drying chamber, the thermal load of *set*1 produces 2.5 times higher $\dot{m}_{ev}$ and the η_{dc} increases up to 18%.

Finally, *set*4 showed that the nature of the processed material is the most significant parameter in the design of a drying system. The mass flow rate $\dot{m}_s$ must decrease by 10^3 times to reduce the moisture content of municipal sewage sludge (MSS) at the same level of rice; this inevitably reduces the $\dot{m}_{ev}$ (10^{-2} times lower), and the drying efficiency falls below 0.3%, which indicates the current dimensions of the drying chamber are inadequate for drying MSS.

The results above are solutions of the baseline scenario, where the specific humidity of drying air at the inlet of the drying chamber is higher than the outdoor level because of the water vapor generated by methane combustion ($\omega_H > \omega_0$). This is an adverse effect of open cycles that can reduce the evaporation rate (i.e., the air humidity is closer to the equilibrium level). As shown in Figure 13, ω_H is augmented with the thermal load: comparing the Baseline systems, it increases from the external level by 27% in *set*1 and by 38% in *set*2. These proportions decrease in Scenario 1 to 22% and 33%, respectively, because of the heat recovery. In general, ω_H varies between the limit calculated for the baseline scenario and those of the closed cycles, where the external combustion maintains $\omega_H = \omega_0$ (within the colored fields of Figure 13). Although the effects of ω_H on product moisture are negligible in the analyzed systems (see u in Figure 13), these could become significant in larger-scale systems, where the designer must adjust the drying temperature considering airflow rate.

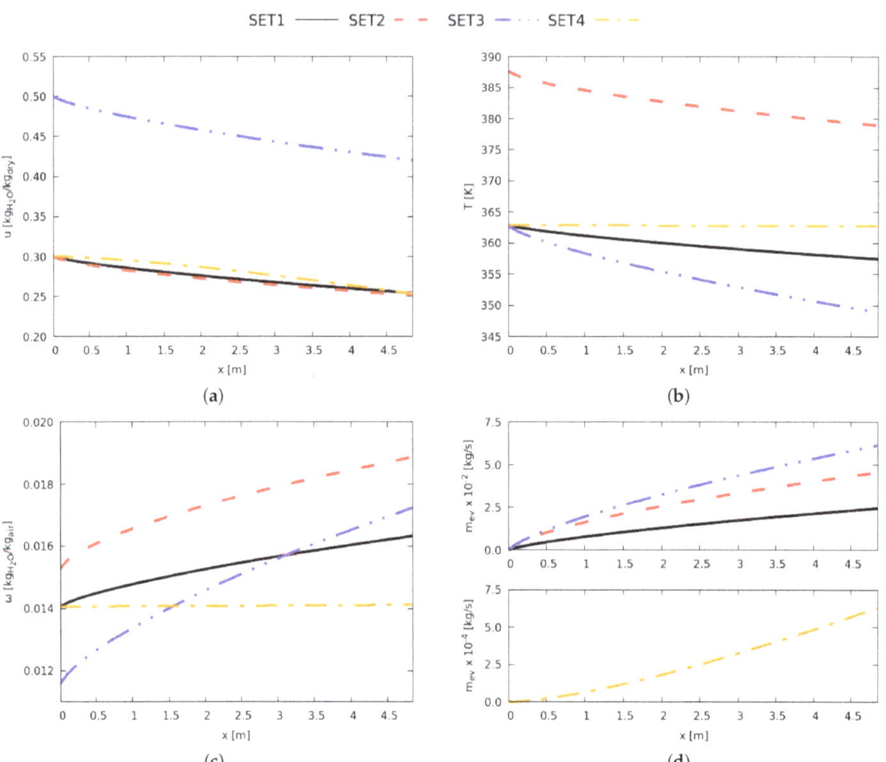

Figure 11. Moisture content of the product (**a**), air temperature (**b**), absolute humidity (**c**), and cumulative evaporation rate (**d**) along the bed length.

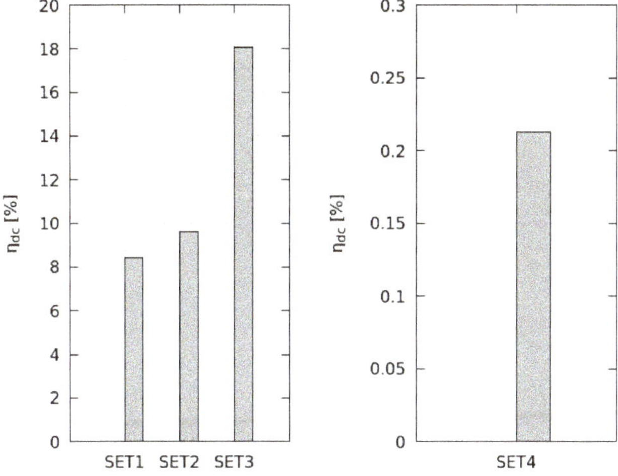

Figure 12. Drying efficiency η_{dc}.

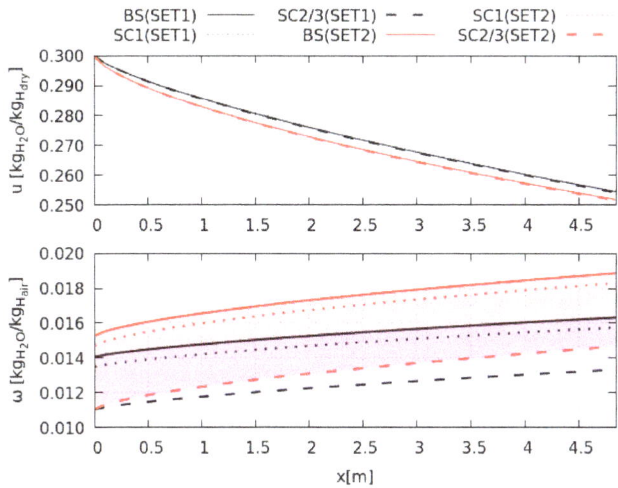

Figure 13. Effects of combustion on the u and ω values compared among different sets and scenarios.

6.3. Heat Exchangers

This section discusses how the heat and mass recovery units change their size based on cycle configuration and operative conditions. The total heat exchange surfaces ($\sum \Delta A_{he}$) are shown in Figure 14; these surfaces vary as a function of the design targets, initial conditions of entering flows (temperature regimes in Table 2), and the mass and energy exchange technique (transfer coefficients in Table 3).

6.3.1. Heat Recovery

HE1 (Figure 7) pre-heats the external air intake by 10K (design target), and its total surface depends on the initial temperature of the entering streams: the airflow from the drying chamber (T_E) and that compressed by the fan (T_F). Further, the HE1 dimensions change with $\Delta T = T_E - T_0$, which measures the total heat wasted by the cycle. At a given dead-state temperature, a higher T_E reduces HE1: from *set*1, its average surface decreases by 26% in *set*2 and by 13% in *set*4 (T_E values in Figure 11). When running *set*3, ΔT remains higher than that in *set*1, and the average surface of HE1 decreases by 8%.

The effects of T_F become evident when different scenarios that run the same *sets* (the T_E remains constant) are compared: when T_F increases because of a larger energy dissipation from the fan, the HE1 area is diminished (by approximately 1.8% per unit temperature). Based on these results, the sizes (i.e., economic costs) of heat-recovery are strictly related to the cycle performance. When the drying efficiency η_{dc} decreases, the system wastes more energy, and a smaller HE1 fulfills the design targets.

6.3.2. Mass Recovery

HE0 (Figure 9) heats the drying air from T_P to the target value T_H by combustion exhausts at 500 K. Hence, its size depends on the target heat-transfer rate, given by $\Delta T = T_H - T_P$. A higher ΔT entails a larger HE0. The average surface of *set*1 increases by 2.5 times in *set*2, when T_H is augmented, and by 24% in *set*3, when the external air gets colder. No differences were observed by running *set*4. As observed for HE1, the drying systems need a smaller unit to fulfill the target T_H when T_P increases because of the effect of more powerful compression (the HE0 area reduces by approximately 2% per unit temperature).

The cooling units are the largest ones. Below the same dead-state conditions, dimensions increase with the heat supply (i.e., heat to dissipate)—from *set*1 to *set*2, the HE2 area increases by 41%, and the tower area increases by 37%. When it is difficult to dry the product and the $\dot{m}_{ev}$ decreases, air-cooling is less demanding, and the HE2 reduces from *set*1 to *set*4 by 16%, while the tower dimensions do not change. The dead-state conditions produce the most significant variations on the exchange surfaces: From *set*1 to *set*3, the area of the HE2 and tower increases by 85% and 48%, respectively.

Based on the results above, the dimensions of the cooling units depend on the heat supply, dead-state conditions, and nature of dried materials; these operative conditions can be defined as follows:

1. Sensible heat to dissipate in the saturation step (i.e., wasted heat unrecovered by HE1);
2. Amount of moisture to condense, which is set equal to $\dot{m}_{ev}$ (i.e., target heat\mass-transfer rate).

Although these quantities are interdependent (e.g., a lower $\dot{m}_{ev}$ corresponds to higher wasted heat), we can recognize which parameter has the largest influence because the TFMM calculates the cooling-rate of the saturation and the condensation step separately (see Figure 15):

- Air-saturation takes most of HE2, but its incidence on the total exchange area is more sensible to $\dot{m}_{ev}$ than to the wasted heat: from 70% of *set*1, it decreases to 64% in *set*2, although the highest heat waste occurs, and to 36% in *set*3, where the wasted heat is only 2% less than that in *set*1 but $\dot{m}_{ev}$ is more than double; in contrast, the air-saturation needs almost the whole HE2 in *set*4, where $\dot{m}_{ev} \approx 0$;
- The cooling tower is more efficient than HE2 because it fulfills the design targets by a smaller surface ($\sim$10 times). The air saturation is almost instantaneous, and as in HE2, the $\dot{m}_{ev}$ has the largest effect on its incidence on the total tower surface. From 11% of *set*1, it decreases to 7.6% in *sets*2 and 7.9% in *set*3.

The results above show that the target $\dot{m}_{ev}$ is the most influencing parameter on the dimensions of cooling units. This is because the cooling rate is generally higher in the saturation than in the condensation step (Slopes of curves in Figure 15); therefore, an increase of $\dot{m}_{ev}$ augments the surface needs, while the wasted heat has a secondary effect in indirect cooling and becomes irrelevant in direct cooling.

An indicator that includes both $\dot{m}_{ev}$ and the wasted heat is the drying efficiency η_{dc}. When it is augmented, systems need a large cooling unit; thus, it is similar to the observation for HE1, and air recirculation becomes more convenient when the product is difficult to dry and the drying efficiency is low.

HE3 is the smallest unit, and its dimensions depend on the target heat-transfer rate, given by the temperature difference $\Delta T = T_D - T_0$ as well as the inlet temperature $T_{P'}$ of the combustion exhausts. In particular, the HE3 area decreases when the dead-state\target conditions (T_0, ω_0) are reduced and it is augmented with the heat demand because $T_{P'}$ reduces when more heat is taken from the combustion exhausts: from *set*1, the exchange area increases by 46% in *set*2, decreases by 69% in *set*3, and remains constant in *set*4.

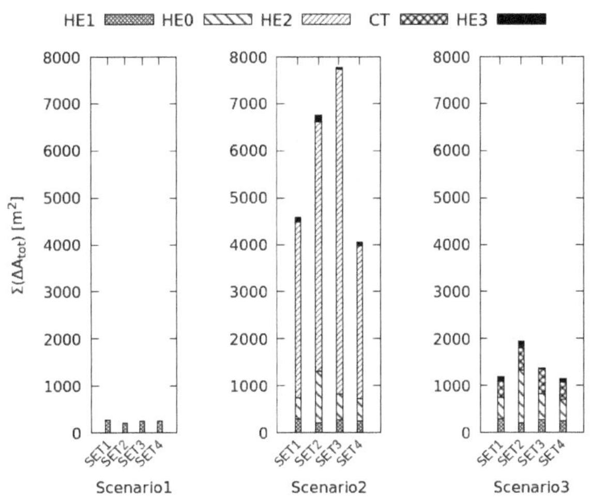

Figure 14. Total exchange surfaces varying among the different sets and scenarios.

Table 2. The inlet and outlet temperatures of the heat exchangers.

SET	Temperature [K]	Scenario 1	Scenario 2				Scenario 3			
		HE1	HE0	HE1	HE2	HE3	HE0	HE1	CT	HE3
SET1	T1,in	357.50	314.70	357.60	347.60	288.75	313.90	357.60	347.60	288.70
	T1,out	347.50	363.15	347.60	288.75	300	363.15	347.60	288.70	300
	T2,in	302.50	500	304.70	280.15	451.90	500	303.90	280.15	451.12
	T2,out	312.50	451.90	314.70	284.87	440.65	451.12	313.90	284.85	439.83
SET2	T1,in	378.81	314.43	379.12	369.20	288.76	313.50	379.12	369.20	288.70
	T1,out	368.90	388.15	369.20	288.76	300	388.15	369.20	288.70	300
	T2,in	302.77	500	304.43	280.15	414.25	500	303.49	280.15	413.16
	T2,out	312.77	414.25	314.43	287.72	403	413.15	313.49	287.53	401.86
SET3	T1,in	348.73	302.81	348.88	339.05	283.87	301.82	348.87	338.87	283.84
	T1,out	338.85	363.15	338.98	283.86	288	363.15	338.97	283.83	288
	T2,in	290.21	500	292.81	280.15	440.10	500	291.82	280.15	439.12
	T2,out	300.21	440.10	302.81	284.58	435.98	439.12	301.82	284.57	434.96
SET4	T1,in	362.79	314.27	363.01	353.01	289.35	313.77	363	352.74	288.70
	T1,out	352.81	363.15	353.01	289.35	300	363.15	353	288.70	300
	T2,in	302.30	500	304.26	280.20	451.46	500	303.77	280.15	450.96
	T2,out	312.30	451.45	314.26	285.27	440.81	450.96	313.77	285.25	439.67

Table 3. The heat and mass transfer coefficients of the heat exchangers.

SET	Coefficients	Scenario 1	Scenario 2				Scenario 3			
		HE1	HE0	HE1	HE2	HE3	HE0	HE1	CT	HE3
SET1	HTC $[W/m^2K]$	9.30	9.01	9.30	8.34	9.26	9.01	9.30	126.80	9.26
	MTC $[m/s]$	-	-	-	0.0063	-	-	-	0.096	-
SET2	HTC $[W/m^2K]$	9.73	8.56	9.73	8.35	9.69	8.56	9.73	136.19	9.69
	MTC $[m/s]$	-	-	-	0.0063	-	-	-	0.103	-
SET3	HTC $[W/m^2K]$	9.31	9.01	9.30	7.97	9.26	9.01	9.30	127.05	9.26
	MTC $[m/s]$	-	-	-	0.0060	-	-	-	0.096	-
SET4	HTC $[W/m^2K]$	9.29	9.01	9.29	8.02	9.26	9.01	9.29	126.75	9.26
	MTC $[m/s]$	-	-	-	0.0060	-	-	-	0.096	-

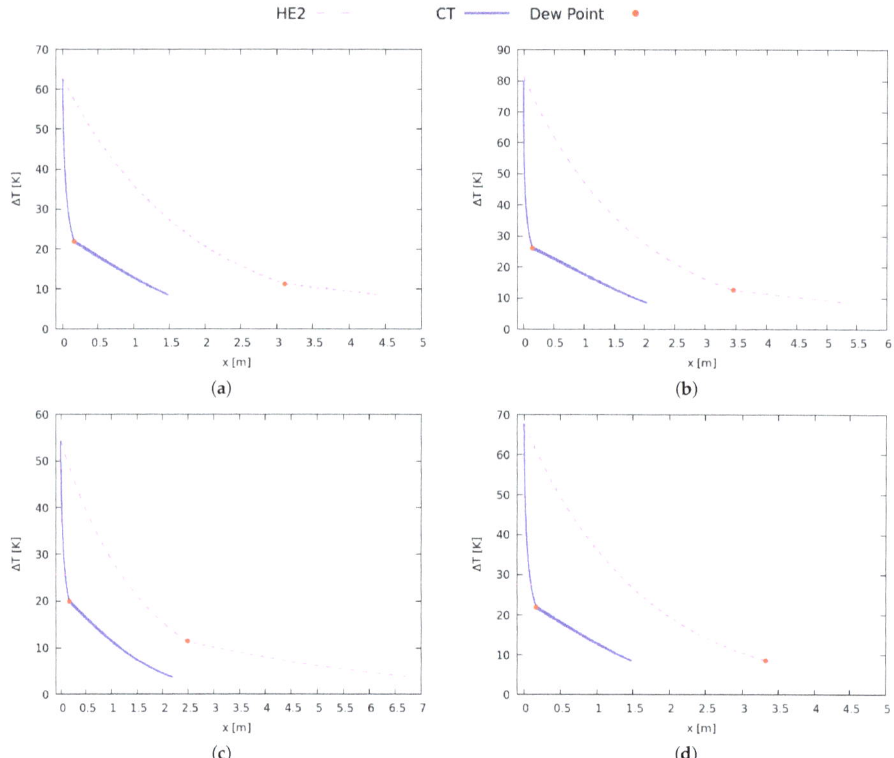

Figure 15. The temperature difference between two phases inside HE2 and the cooling tower (CT); the dew point (red point) separates the saturation from the condensation step: SET1 (**a**), SET2 (**b**), SET3 (**c**), and SET4 (**d**).

6.4. Thermodynamic Performance

We show the variations in the thermodynamic performances of a drying cycle among different *sets* and configurations. The designer can assess the productivity and the convenience of the operating conditions based on specific energy consumptions and second law indicators defined in Section 5.4.

6.4.1. Specific Energy Consumptions

Figure 16 shows the specific electric and thermal energy consumptions. HE1 reduces the heat demand by pre-heating the feeding air of the combustion chamber, but it increases the electrical needs by additional pressure losses. From baseline to Scenario 1, the STEC decreases at least by 14% below the operating conditions of the *set*2 and up to 27% under the *set*1; the average SEEC augments by 3 times.

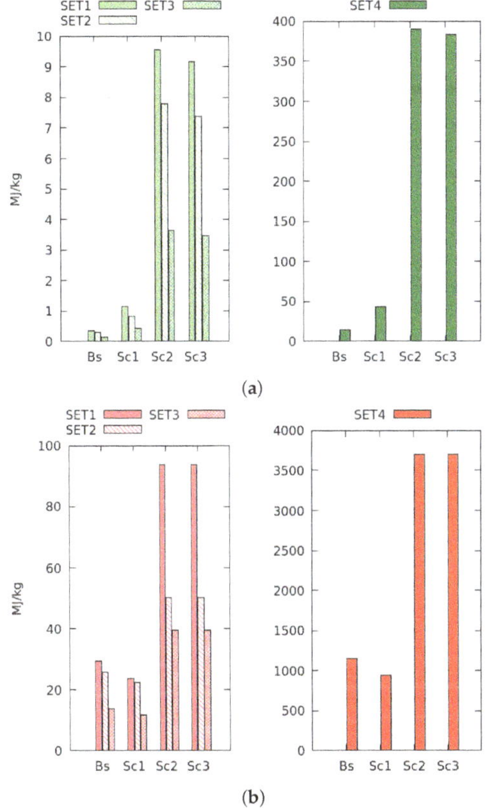

Figure 16. Specific electric energy consumption (SEEC) (**a**) and the specific thermal energy consumption (STEC) (**b**) collected by operative sets.

The energy needs of closed cycles increase because of the effect of the additional pressure drops and the energy demand of the processes for air recirculation. From the values of the baseline scenario, the average STEC increases by 3.2 times because of the external combustion and air post-heating; the average SEEC increases by 27 times owing to the electricity needs of the heat-pump feeding cooling units.

When a *set* improves η_{dc}, it reduces the energy consumptions: from *set*1 to *set*2, the average SEEC and STEC decrease by 19% and 39%, respectively; *set*3 performs the best because it reduces the SEEC by 63% and the STEC by 57%. *Set*4 performs the worst, and it augments both indicators by approximately 40 times.

When comparing the single *set*-Scenario, it becomes clear how the SEEC and STEC describe the effects of the operating conditions on the system configuration and its final energy consumption: for example, from Baseline to Scenario 1, the STEC diminishes by 27% below *set*1 and by 31% below *set*2; the SEEC increases by 3.4 times in *set*1 and by 2.8 times in *set*2. Thus, the benefits of the heat-recovery are more conspicuous in *set*2 than in *set*1 because an increase of the thermal load augments the η_{dc}, and simultaneously, it reduces the dimensions of HE1 with related pressure drops.

6.4.2. Irreversibility and Exergy Efficiency

The exergy efficiencies (η_{ex}) and the exergy destruction rates ($\dot{E}x_d$) with the ratios of system components are shown in Figures 17 and 18.

The effects of methane and electricity consumption on exergy efficiency are evident when comparing the different cycles. From the baseline scenario, the average η_{ex} increases by 15% in Scenario 1 because of heat recovery, whereas it reduces by 64% in closed cycles, characterized by the highest exergy inputs. On the total $\dot{E}x_d$, the incidence of the fan is generally irrelevant ($\approx$1%), whereas the combustion chamber plays the most significant role (90–80%). The primary entropy source of air-heating is the combustion reaction; therefore, $\dot{E}x_{d,CC}$ is augmented with fuel consumption: its average value in the baseline scenario decreases by 20% in Scenario 1, and it is augmented by 2.2 times in closed cycles.

The exergy performances of the drying chamber follow the drying efficiency (see η_{dc} in Figure 12). The average η_{ex} is augmented by 70% from *set*1 to *set*2 and by 1.2 times in *set*3; $\dot{E}x_{d,DC}$ of *set*1 is doubled in *set*2 and is augmented by 2.5 times in *set*3. Both parameters are decreased by more than 10 times when *set*4 is run. When it is difficult to dry the product or the bed dimensions are inappropriate, a shorter amount of exergy is invested in the evaporation process. In contrast, a higher heat supply or more favorable dead-state conditions enhance the exergy efficiency by augmenting $\dot{m}_{ev}$.

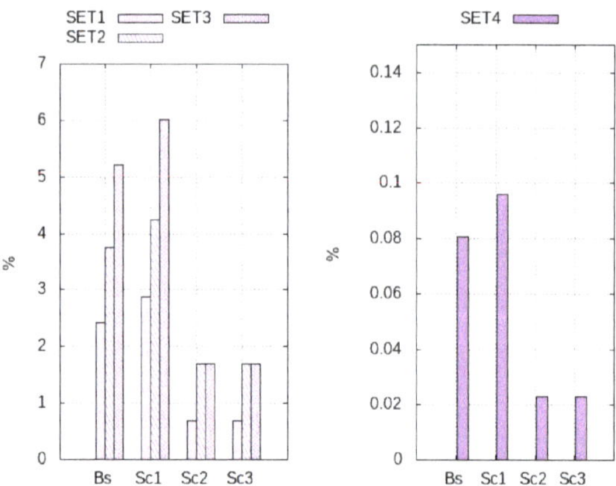

Figure 17. Exergy efficiency among different sets and scenarios.

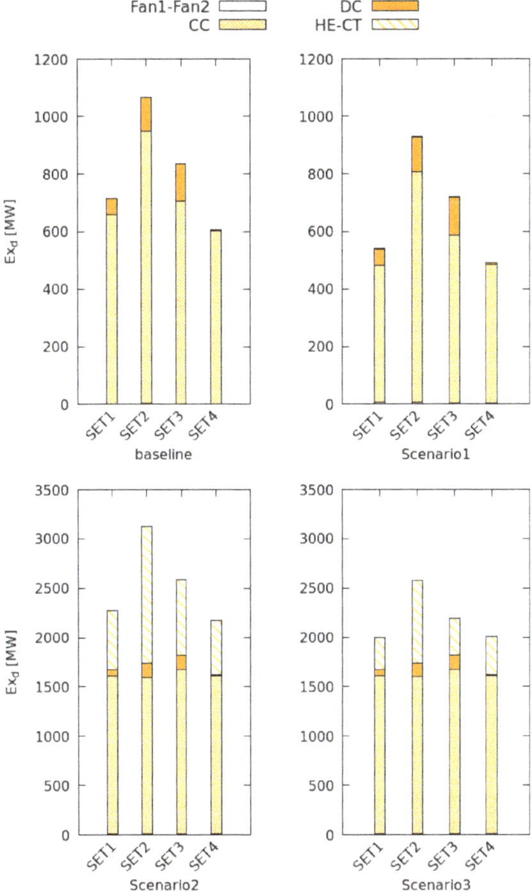

Figure 18. Exergy destroyed in different cycles by flowing (Fan1-Fan2), heating (CC), drying (DC), and heat and mass recovery (HE-CT) processes.

All heat exchangers are additional entropy sources that contribute to the total exergy destruction. The entropy generation of these components depends on their respective heat and mass transfer mechanisms, as well as the specific design targets. No mass transfer occurs within the air-to-air units; therefore, their entropy generation exclusively depends on the inlet temperature difference between the airflows and the target heat-transfer rate:

- HE0 is the air-to-air unit with the highest destruction ratio ($y_{HE0} \approx 5\%$); results show that $\dot{E}x_{d,HE0}$ is augmented with the heat-transfer rate (given by $\Delta T = T_H - T_P$). Its average value below *set*1 increases by 3 times in *set*2 and by 1.5 times in *set*3, and it remains unvaried in *set*4.
- HE1 is the heat exchanger with the lowest exergy destruction rate, and its influence on the total $\dot{E}x_d$ is insignificant ($y_{HE1} < 0.5\%$). $\dot{E}x_{d,HE1}$ increases with energy wastage (derived by temperatures T_E, T_F): its average value is augmented by 50% from *set*1 to *set*2, by 18% in *set*3, and 16% in *set*4.

- The destruction ratio of HE3 is similar to that of HE1. $\dot{E}x_{d,HE3}$ depends on the heat-transfer rate: from *set*1 to *set*2, it is diminished by 15%, and in *set*3 by 63%. Thus, it is diminished when $T_{P'}$ is lowered or the dead-state temperature reduces.

The entropy generation of the cooling units depends on the initial humidity and temperature of the airflow and the target cooling-rate; their respective $\dot{E}x_d$ changes by the cooling technique, and it is different between the saturation and the condensation steps (see Figure 19).

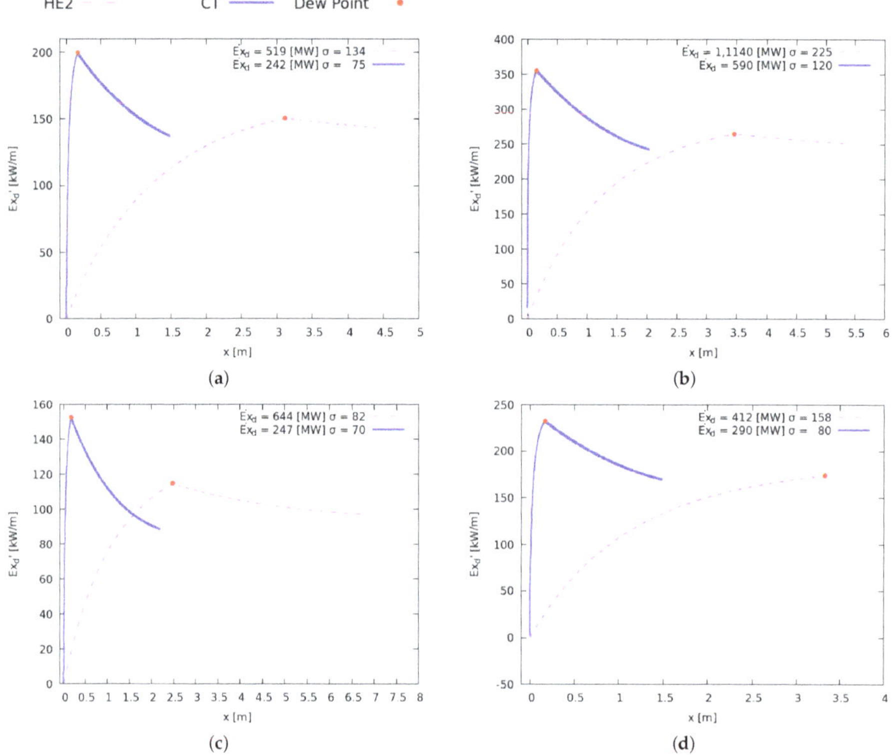

Figure 19. Spatial distribution of exergy destruction rate within HE2 (purple) and cooling tower (blue); the dew point (red point) separates the saturation from the condensation step: SET1 (**a**), SET2 (**b**), SET3 (**c**), and SET4 (**d**).

- HE2 is the heat exchanger characterized by the highest destruction ratio ($y_{HE2} \approx 26\%$). As shown in Figure 19, most of the exergy destruction occurs in the saturation step; $\dot{E}x_{d,HE2}$ doubles from *set*1 to *set*2 and increases by 10% in *set*4 where the condensation is almost instantaneous ($\dot{m}_{ev} \approx 0$).
- The cooling tower reduces the irreversibility of air-cooling: the destruction ratio is $y_{CT} = 15\%$, and with respect to HE2, the exergy destruction generally decreases by half. In terms of exergy destruction, 90% occurs in the condensation step, and the total $\dot{E}x_{d,CT}$ doubles in *set*2 and is augmented by 20% in *set*4, while it shows no changes in *set*3.

Waste heat is the primary entropy source of air-cooling. When the temperature of the inlet air (T_C) increases, it extends the saturation step where the exergy destruction occurs at a faster rate. Finally, for each cooling-unit, we calculate the standard deviation (σ) of its respective exergy destruction rate. This parameter measures the spatial distribution of entropy generation (a lower variance corresponds to a more homogeneous distribution), as shown in Reference [64]. Data reported in Figure 19 show that the unit that destroys more exergy presents the highest variance of exergy generation rate; thus, our results verify the theorem of equipartition of entropy production [60].

6.5. Climate Effects

The dead-state conditions have a significant effect on the evaporation rate and on the performance of the entire drying cycle (see Section 6.2). Therefore, we present a reduced solution of TFMM, focusing on the effects of climate on the exergy efficiency of the baseline scenario. Based on these results, we propose some adjustments to be made to the operative conditions to stabilize the production target (the final product moisture u_f) in terms of yearly climatic variations.

Figure 20 shows the η_{ex} of the baseline scenario, calculated below the Brescia climate year 2018, on a monthly scale. The dried product is rice paddy, and the initial moisture content is 0.7 in season 1 (from November to May) and 0.6 in season 2 (from June to October). The system is initially running *set1* (data label is $E_{th} = 1$ in Figure 20): η_{ex} varies along the year ($\pm 2\%$ on average), and it is the maximum in the colder season. The u_f never falls in the set target region (the green field, where $u_f = 0.5 \pm 0.02$), indicating that the current energy supply is inadequate. As a solution, we double the heat supply when the system is undersized and reduce it by 25% when over-sizing; the results show an enhancement of η_{ex} in all months by +21% on an average in season 1 and +13% in season 2 and stabilization of the final u_f in the target region.

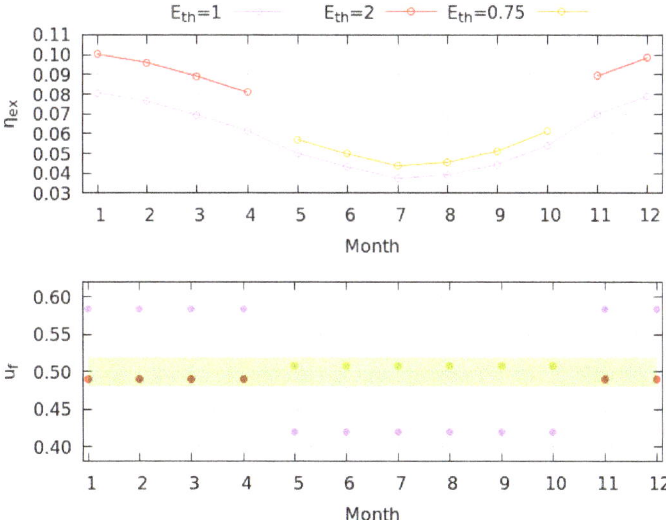

Figure 20. The exergy efficiency and final moisture content of the baseline scenario in the Brescia climate year 2018 at different operating conditions; starting from inputs of set1 (data in purple), we doubled (data in red) and reduced by 25% (data in orange) the thermal loads.

7. Conclusions

The performance of convective drying systems depends on a wide range of interrelated operating parameters that affect the energy use, exergy efficiency, dimensions (i.e., costs) of system components, and product quality. We proposed a methodology that calculates the system performance considering exogenous variables (climatic conditions, initial product moisture, and physical properties of the dried product), energy\mass intake, and cycle configuration.

On comparing the results of different *sets* to the reference *set*1, the evaporation rate showed the most significant effect on the global performance because it determines the energy and exergy productively invested in the drying process. When increasing heating and flowing loads (*set*2), the system evaporates a deeper moisture layer with benefits on the exergy efficiency (+97%). Higher initial product moisture (*set*3) augments the evaporation rate, and the drying cycles presents the best performance in terms of drying efficiency (+114%) and exergy efficiency (+127%). When drying the MSS (*set*4), the evaporation rate reduces by approximately 10^2 times, worsening the exergy performance (−97%); such results reveal the dimensions of the drying bed inadequate to dry that particular product.

Scenario 1 is the optimal system configuration. The heat-recovery increases the baseline electrical consumption and reduces the thermal consumption by maximum +211% and −17%, respectively, with benefits on the exergy efficiency (+17%). The efficiency of heat recovery depends on the wasted energy; at low drying efficiency, or when the fan dissipates more shaft work, the HE1 fulfills the production target reducing the exchange surface by 25%. Thus, the heat-recovery is more convenient in low efficiency processes when the material is difficult to dry.

Closed cycles cut the exergy efficiency by maximum −67%. These systems use four additional heat exchangers increasing the electric consumption by 20 times. Moreover, external combustion and air regeneration processes augment the thermal energy needs (+180%). The performances of Scenario 3 are slightly better than that of Scenario 2 because the tower presents a faster cooling-rate than HE2, especially in the saturation step. This process occurs almost instantaneously in the tower that reaches the target cooling conditions with a 10 times smaller surface than HE2. Furthermore, when the saturation step becomes shorter, the total exergy destruction rate diminishes (−48%), and its spatial distribution tends to become more uniform (−39%). Thus, the tower reduces system irreversibility.

Based on the above results, we derive some technical recommendations that can help the designer optimize the energy and exergy use of a convective drying system.

- The performance of the system varies along the climatic year because of the variations of the initial temperature and humidity of the working flows (air intake and processed product). However, the designer can ensure the production targets are unaltered by adjusting the energy input with benefits on the exergy efficiency. Best performances are observed in the cold season.
- Operating conditions shall be oriented to maximize the evaporation rate; the increasing thermal loads is valid to this purpose; however, it is limited by some adverse effects as the depletion of the product quality and the humidification of drying air by combustion. As an alternative, the designer can augment the dimensions of the drying bed and extend the residence time of the processed product in the drying chamber.
- When the bed cannot be enlarged, the drying efficiency is low, and a heat recovery unit can reuse a fraction of the heat wasted by the drying chamber to preheat the external air intake; this practice is particularly advantageous in high-temperature processes where small units significantly increase energy and exergy efficiency.
- Air recirculation dramatically reduces the performance of the system because of air regeneration processes. More than an optimization practice, the cycle closure can be a safety procedure for drying

hazardous materials and limit the emissions of harmful substances in the environment. The cooling tower is particularly suitable for this purpose; compared to an indirect cooling system, it presents the highest energy and exergy efficiency and is configurable as a wet scrubber to wash away toxic species from the airflow and restore its initial conditions.

Future developments will overcome the limitations of the current work. Using a real gas model (e.g., Van der Walls), the TFMM can simulate the compression and throttle of a refrigerant fluid and predict the performances of a drying cycle driven by the heat pump, thereby promising remarkable enhancement of the exergy efficiency caused by the low-temperature of the heat generation process. Finally, the coupling of the TFMM to an analytical model (e.g., upscaled porosity model) will increase the accuracy of TFMM to describe the drying phenomenology and simulate the process at a higher level of detail.

Author Contributions: Methodology, A.A.; software, A.A.; validation, A.A.; formal analysis, A.A.; investigation, A.A.; resources, P.P.; data curation, A.A.; writing–original draft preparation, A.A.; writing–review and editing, A.A. and P.P; visualization, A.A.; supervision, P.P.; project administration, P.P. All authors have read and agreed to the published version of the manuscript.

Funding: This research received no external funding.

Institutional Review Board Statement: Not applicable.

Informed Consent Statement: Not applicable.

Data Availability Statement: Data sharing not applicable.

Acknowledgments: The authors would like to thank the reviewers for their insightful comments, which have resulted in an improved manuscript.

Conflicts of Interest: The authors declare no conflict of interest.

Appendix A

Appendix A.1. Equations of State

The state functions of Phase 1 are calculated on a molar basis using the molar mass M_{1j} of single j-species:

$$\dot{m}_1(\epsilon_{1j}) = \sum_{j=1}^{r}\left[\frac{\alpha_1\epsilon_{1j}\dot{m}_{os}}{M_{1j}}\right] = \sum_{j=1}^{r}[\dot{n}_{1j}] \tag{A1}$$

$$\dot{H}_1(\epsilon_{1j}, T) = \sum_{j=1}^{r}\left[\frac{\alpha_1\epsilon_{1j}h_{1j}\dot{m}_{os}}{M_{1j}}\right] = \sum_{j=1}^{r}[\dot{n}_{1j}h_{1j}] \tag{A2}$$

$$\dot{S}_1(\epsilon_{1j}, T, P) = \sum_{j=1}^{r}\left[\frac{\alpha_1\epsilon_{1j}s_{1j}\dot{m}_{os}}{M_{1j}}\right] + S_{mix} = \sum_{j=1}^{r}[\dot{n}_{1j}s_{1j}] + S_{mix} \tag{A3}$$

$$S_{mix}(\epsilon_{1j}) = -R\sum_{j=1}^{r}\left[\dot{n}_{1j}ln(\epsilon_{1j})\right] \tag{A4}$$

$$\dot{E}x_1(\epsilon_{1j}, T, P) = \sum_{i=1}^{r}\left[\frac{\alpha_1\epsilon_{1j}ex_{1j}\dot{m}_{os}}{M_i}\right] = \sum_{j=1}^{r}[\dot{n}_{1j}ex_{1j}]. \tag{A5}$$

The term S_{mix} is the entropy of mixing; the specific state functions h_{1j} and s_{1j} are calculated assuming the specific heat of each r-species as a polynomial T-dependent; the values of constants $a_{1j}, b_{1j}, c_{1j}, and\ d_{1j}$ are listed in Tables [49]:

$$c_{p,1j}(T) = a_{1j} + b_{1j}T + c_{1j}T^2 + d_{1j}T^3 \tag{A6}$$

$$h_{1j}(T) = \int_{T_0}^{T}[c_{p,1j}(T)]dT = \left[a_{1j}T + \frac{b_{1j}}{2}T^2 + \frac{c_{1j}}{3}T^3 + \frac{d_{1j}}{4}T^4\right]_{T_0}^{T}, \tag{A7}$$

$$s_{1j}(T,P) = \int_{T_0}^{T} \left[\frac{c_{p,1j}(T)}{T}\right] dT - \int_{P_0}^{P} \left[\frac{R}{P}\right] dP = \left[a_{1j} ln(T) + b_{1j} T + \frac{c_{1j}}{2} T^2 + \frac{d_{1j}}{3} T^3\right]_{T_0}^{T} - \left[R ln(P)\right]_{P_0}^{P}. \quad \text{(A8)}$$

Within the air-to-air heat exchangers, Phase 2 is a mixture of gaseous species, and its state functions are derived by the equations above. In all other components, Phase 2 is a mixture of solid and/or liquid species, and the state functions are calculated as follows:

$$\dot{m}_2(\epsilon_{2j}, T) = \sum_{j=1}^{k} [\alpha_2 \epsilon_{2j} \dot{m}_{os}] = \sum_{i=1}^{k} [\dot{m}_{2j}] \quad \text{(A9)}$$

$$\dot{H}_2(\epsilon_{2j}, T) = \sum_{j=1}^{k} [\alpha_2 \epsilon_{2j} \dot{m}_{os} h_{2j}] = \sum_{i=1}^{k} [\dot{m}_{2j} h_{2j}] \quad \text{(A10)}$$

$$\dot{S}_2(\epsilon_{2j}, T) = \sum_{j=1}^{k} [\alpha_2 \epsilon_{2j} \dot{m}_{os} s_{2j}] = \sum_{i=1}^{k} [\dot{m}_{2j} s_{2j}] \quad \text{(A11)}$$

$$\dot{E}x_2(\epsilon_{2j}, T) = \sum_{j=1}^{k} [\alpha_2 \epsilon_{2,i} \dot{m}_o sex_{2j}] = \sum_{i=1}^{k} [\dot{m}_{2j} ex_{2j}], \quad \text{(A12)}$$

where h_{2j} and s_{2j} are calculated assuming a constant specific heat of each k-component (values are found in References [47–49]):

$$c_{p,2j}(T) = a_{2j} \quad \text{(A13)}$$

$$h_{2j}(T) = \int_{T_0}^{T} [c_{p,2j}(T)] dT = a_{2j} \left[T\right]_{T_0}^{T} \quad \text{(A14)}$$

$$s_{2j}(T) = \int_{T_0}^{T} \left[\frac{c_{p,2j}(T)}{T}\right] dT = a_{2j} \left[ln(T)\right]_{T_0}^{T}. \quad \text{(A15)}$$

Finally, the specific exergy of the j-component is calculated by this general expression for both i-phases:

$$ex_{ij}(T,P) = [h_{ij}(T) - h_{ij}(T_0)] - T_0[s_{ij}(T,P) - s_{ij}(T_0, P_0)]. \quad \text{(A16)}$$

Appendix A.2. Fan Equations

In this section, we performs adiabatic compression of a single gaseous phase, the drying air, formed by 3-species: N_2, O_2, H_2O. The total shaft work is a function of the air flow rate, the pressure head, and the fan efficiency $\eta_f = 0.8$

$$\dot{W}_1^{\leftarrow} = \frac{\dot{W}_{rev}^{\leftarrow}(\dot{m}_1, \Delta p_1)}{\eta_f}, \quad \text{(A17)}$$

where $\dot{W}_{rev}^{\leftarrow}$ refers to an ideal component when $\eta_f = 1$. The airflow rate is specific for each operative set, while Δp_1 is calculated in each scenario to balance the pressure losses of all components.

Appendix A.3. Combustion Chamber Equations

An adiabatic combustion chamber covers the heat demand of the drying cycles: after fuel injection, the airflow reaches the desired temperature by an instantaneous combustion reaction. The running set gives the outlet temperature T_H of open cycles; in closed cycles, the outlet temperature T'_H is calculated over the heat demand of both units HE0 and HE3.

The two phases processed by this unit are drying air (Phase 1) which reacts with a hydrocarbon $C_l H_n$ (Phase 2) as

$$C_l H_n + (l + \frac{n}{4}) O_2 = l CO_2 + \frac{n}{2} H_2O. \quad \text{(A18)}$$

The combustion products are CO_2 and H_2O, and we can neglect the formation of other components, such as CO and NO_x because of the high air excess ($\lambda > 4$) and the low operative temperatures (<1250 K) [49]. Pure methane ($l =$

1, n = 4, indexed by j = 1) is used as fuel, and assuming its complete combustion, the amount (per moles of fuel) of j-species exchanged between two phases is given by the stoichiometric coefficients ν_{ij}, derived by Equation (A18):

$$\Delta \dot{m}_{12,j} = \epsilon_{21} \dot{m}_2 \nu_{ij} \frac{M_{ij}}{M_{21}} \tag{A19}$$

M_{ij} is molar mass. The energy and the entropy exchanged by combustion are respectively calculated by the standard enthalpy and the entropy of formation [49]

$$\Delta \dot{Q}_{12,j} = \Delta \dot{m}_{12,j} \Delta H^0_{f,j} \tag{A20}$$

$$\Delta \dot{S}_{12,j} = \Delta \dot{m}_{12,j} \Delta S^0_{f,j}. \tag{A21}$$

Appendix A.4. Drying Chamber Equations

In the drying chamber, the hot air (Phase 1) crosses the wet product (Phase 2) and evaporates its liquid fraction. We neglect the heat losses across the chamber walls (adiabatic chamber), and the Ergun equation [65] gives the pressure losses of the airflow across the bed. The evaporation process is described by the Page equation:

$$MR = exp(-kt^n), \tag{A22}$$

where MR is the moisture ratio, which is defined as

$$MR = \frac{u - u_0}{u_0 - u_{eq}}. \tag{A23}$$

The constants k, n depend on the nature of the product and drying conditions. In this work, we calculate the k, n values for rice paddy drying by the empirical correlations of Reference [50] given as function of the drying airflow rate ($\dot{m}_a$), the drying temperature (T_H), and the hold-up of drying bed. The k, n values of MSS derive from the experimental database of Reference [55], considering the appropriate drying temperature and ultrasound turned off. The thermo-physical properties of rice were taken from References [16,47,50] and those of MSS from References [12,48,66].

The moisture content (on dry basis) of the dried product along the bed is calculated by the velocity of solid flow v_2:

$$u(x) = u_{eq} + (u_0 - u_{eq}) exp(\frac{-kx^n}{v_2^n}), \tag{A24}$$

where u_{eq} is the equilibrium moisture content of drying air, calculated using Laithong equation [67], and u_0 is the initial moisture content of the dried product. Only water changes phases, and assuming j=3 for water vapor and j=2 for liquid water, the mass exchanged by phase transition (i.e., moisture evaporation rate) can be written in these two equivalent forms:

$$\Delta \dot{m}_{12} = (\epsilon_{11} + \epsilon_{12}) \dot{m}_1 [u(x + dx) - u(x)] \tag{A25}$$

$$\Delta \dot{m}_{21} = \epsilon_{21} \dot{m}_2 [u(x + dx) - u(x)]. \tag{A26}$$

The evaporated moisture instantaneously reaches equilibrium with the air flow (perfect mixing assumption [68, 69]), and because $u_{eq} << u_0$, the evaporation immediately starts at the chamber inlet. Hence, the heat exchanged between two phases is exclusively in the latent form:

$$\Delta \dot{Q}_{12} = \Delta \dot{m}_{12} \Delta \dot{H}_{fg,H_2O} \tag{A27}$$

$$\Delta \dot{S}_{12} = \Delta \dot{m}_{12} \Delta \dot{S}_{fg,H_2O}. \tag{A28}$$

Appendix A.5. Heat Exchangers: Modeling Approach

The current practice distinguishes two approaches for designing heat exchangers: the rating and sizing problem [70]. The latter consists of calculating the heat exchange surface that meets the target thermodynamic states of exchanging fluids: the heat exchanger setup is unknown, and the fluid temperatures and heat transfer rate are given. Our off-design analysis studies the effects on the thermodynamic performance of a reference drying system by varying the operating conditions and adding new components. Therefore, we model the heat exchangers by the sizing approach: we calculate the heat exchange surface to transform the drying air at target states. Inlet conditions and target states depend on the parameters presented in Table 1 and the specific purpose of each unit reported in following paragraphs.

Appendix A.6. Heat Exchangers: Air-To-Air Units HE0, HE1, and HE3

Air-to-air heat exchangers transfer heat between the drying air (Phase 1) and another gaseous flow (Phase 2); the moisture content and chemical mixture of both phases is different in each unit. Air-to-air heat exchangers are adiabatic and present specific inlet conditions and design target:

- HE1 preheats the compressed air by the drying exhaust. Input temperatures T_E and T_F derive from the drying chamber and fan, and the target temperature is $T_P = T_F + 10$ K (i.e., HE1 preheats the compressed air by $10K$). The value $\Delta T = 10$ K is coherent with other works focused on heat recovery in drying system [71–73]. However, this parameter could be easily changed according to targets and limitations set by the designer: for larger ΔT, the HE1 size (costs) and benefits on global performance increase; in the opposite case, the heat recovery is cheaper, but the benefits on system performance decrease.
- HE0 heats the air to the drying temperature of the cycle by combustion exhausts (see T_H in Table 1); the inlet temperature T_P derives from HE1 and the temperature T_H' is fixed to 500 K. The temperature $T_H' = 500$ K ensures the combustion chamber to cover the heat demand of both HE0 and HE3 in all *sets*. Results show that the heating air presents a residual heat fraction when it leaves HE3, and therefore the T_H' could be reduced in future applications, promising an improvement of the closed cycle performance.
- The unit HE3 restores the initial temperature of the air. Hence, the target temperature is T_0 (i.e., the dead-state temperature), and no assumptions have been made for inlet temperatures that derive from the cooling units and HE0.

All air-to-air heat exchangers are banks of aluminum tubes with an inner diameter $d = 7.5 \times 10^{-3}$ m and a shell $t = 2 \times 10^{-3}$ m. The tube lenght is 3 m (along z-direction), and the spacing between two near tubes is set equal in both direction $s_x = s_y = 1.25$ d. Each column consists of fixed-line numbers L and is dx wide, and the total heat exchanged within the single column is given as:

$$\Delta \dot{Q}_{12} = U_{t,he} \Delta A_{he} (T_2 - T_1), \tag{A29}$$

where ΔA_{he} is the exchange surface of the column and $U_{t,he}$ is a constant heat transfer coefficient (HTC), which is calculated by the following resistance scheme:

$$U_{t,he} = \frac{1}{\frac{1}{k_1} + \frac{t}{\lambda_{Al}} + \frac{1}{k_2}}, \tag{A30}$$

where t is the tube wall thickness, λ_{Al} is the tube shell thermal conductivity and k_1, k_2 derive from empirical correlations representing the forced convection of an air flow on a tube bank (Phase 1 side) and the forced convection of an airflow within a horizontal tube (Phase 2 side) [74]. The pressure drop across the air-to-air heat exchangers are calculated by the model of Beale for tube banks [75].

Appendix A.7. Heat Exchangers: Air-To-Water Units HE2 and Cooling Tower

The HE2 and CT are indirect and direct cooling systems, respectively. Both units are adiabatic, and they restore the dead-state moisture content of the drying air (Phase 1) by pure water flow (Phase 2); thus, the target state of these units is the moisture content $\omega_D = \omega_0$. The inlet temperature T_C derives from heat recovery, and the inlet temperature of the cooling water is fixed to 280.15 K according to values prescribed by the European standard EN 14511:2018 [76] to rate the performance of process chillers. However, the water leaving the cooling units presents a residual cooling capacity; therefore, the inlet water temperature could be reduced in future applications, promising performance optimization for closed cycles. Air cooling and moisture condensation involve heat and mass exchange, split into two subsequent steps: (i) saturation of the airflow and (ii) condensation of its water fraction; the exchanges defined below are referred to in each step by superscripts *i* and *ii*.

HE2 is a bank of serpentine tubes with similar features to the air-to-air units (equals d, t, s_x, s_y). In this unit, the saturation occurs with no mass exchange and the total exchanged heat becomes

$$\Delta \dot{Q}^{i}_{12} = U_{t,he2} \Delta A_{he2} (T_2 - T_1), \tag{A31}$$

where ΔA_{he2} is the exchange surface of a single column and $U_{t,he2}$ is a constant HTC calculated as:

$$U_{t,he2} = \frac{1}{\frac{1}{k_{1,he2}} + \frac{t}{\lambda_{Al}} + \frac{1}{k_{2,he2}}}, \tag{A32}$$

where $k_{1,he2}$ is the condensing side transfer coefficient (Phase 1 side), t is the tube wall thickness, λ_{Al} is the tube shell thermal conductivity, and $k_{2,he2}$ models the forced convection of water within a serpentine tube (Phase 2 side) [74]. By cooling the drying air, we can decrease its saturation moisture content

$$\omega_{sat} = 0.622 \frac{p_{sat}}{p_0 - p_{sat}}, \tag{A33}$$

where p_{sat} is given by Tetens equation [77] as a function of T_1. When ω_{sat} is lower than the effective moisture content of drying air (Equation (A34)), the water vapor condenses and the total exchanged heat is in the sensible and latent form (Equations (A35) and (A36)):

$$\omega_1 = \frac{\epsilon_{1,3}}{(1 - \epsilon_{1,3})} \tag{A34}$$

$$\Delta \dot{m}^{ii}_{12} = U_{m,he2} \Delta A_{he2} (\omega_{sat} - \omega_1) \tag{A35}$$

$$\Delta \dot{Q}^{ii}_{12} = U_{t,he2} \Delta A_{he2} (T_2 - T_1) + \Delta \dot{m}_{12} \Delta \dot{H}_{fg,H_2O}, \tag{A36}$$

where $U_{m,he2}$ is the mass transfer coefficient (MTC), derived from $U_{t,he2}$ by the Lewis approximations for air-water systems [78]. The HE2 is configured as a tube bank; therefore, we calculate the pressure losses of this unit by the same models used for air-to-air units [75].

The cooling tower presents a circular section with a 1m diameter. The airflow is directly mixed to the cooling water in a porous packed bed with a specific surface ratio of 300 m^2/m^3, and therefore an inter-phase mass and heat exchange occurs since the saturation step:

$$\Delta \dot{m}^{i}_{12} = U_{m,ct} \Delta A_{ct} (\omega_{sat} - \omega_1) > 0 \tag{A37}$$

$$\Delta \dot{Q}^{i}_{12} = U_{t,ct} \Delta A_{ct} (T_2 - T_1) + \Delta \dot{m}^{i}_{12} \Delta \dot{H}_{fg,H_2O}. \tag{A38}$$

The drying air is gradually saturated and cooled, and when its moisture content becomes higher than ω_{sat}, water vapor condensation occurs:

$$\Delta \dot{m}^{ii}_{12} = U_{m,ct} \Delta A_{ct} (\omega_{sat} - \omega_1) < 0 \tag{A39}$$

$$\Delta \dot{Q}^{ii}_{12} = U_{t,ct} \Delta A_{ct} (T_2 - T_1) + \Delta \dot{m}^{ii}_{12} \Delta \dot{H}_{fg,H_2O}. \tag{A40}$$

The term $U_{t,ct}$ derives from $U_{m,ct}$ by the Lewis approximations. The exchange surface ΔA_{ct} in a tower element of height dx depends upon the specific surface of porous packed bed. The MTC and pressure losses of the packed bed are calculated by the empirical correlations presented by Reference [61].

Appendix A.8. Heat Exchangers: Geometrical Setup

The tables below resume the geometrical features of the tube banks and cooling tower distinguished for single *sets*.

Table A1. The lines (L) and columns (C) of the air-to-air heat exchangers.

	SET	Scenario 1		Scenario 2		Scenario 3	
		L	C	L	C	L	C
HE1	1	100	39	100	41	100	40
	2		29		30		30
	3		36		38		37
	4		34		36		35
HE0	1	-	-	110	58	110	59
	2	-	-		142		144
	3	-	-		72		73
	4	-	-		59		59
HE3	1	-	-	100	13	100	13
	2	-	-		19		19
	3	-	-		4		4
	4	-	-		12		13

Table A2. The lines (L) and columns (C) of the HE2 units and the heights (H) of the cooling tower.

	SET	Scenario 2			SET	Scenario 3
		L	C			H(m)
HE2	1	100	41	CT	1	1.5
	2	100	30		2	2
	3	100	38		3	2.2
	4	100	36		4	1.5

Appendix A.9. Validation Cases

Table A3. The validation cases of the reference drying system.

Study	Case	Dead-State Conditions			Drying Conditions			Results	
		u_0	T_0 [K]	ω_0	$\dot{m}_a$ [$\frac{kg}{s}$]	$\dot{m}_s$ [$\frac{kg}{s}$]	T_H [K]	u_f	T_E [K]
[50]	1	0.362	306.9	0.0114	10.82	2.36	363.1	0.330	332.8
	2	0.365	307.2	0.0116			363.1	0.309	334
	3	0.331	304.1	0.0009			363.1	0.282	335.8
	4	0.348	303.5	0.0009			363.1	0.306	334.9
	5	0.360	303.1	0.0009			363.1	0.311	333.7
	6	0.308	304.8	0.0101			372.1	0.253	333.7
	7	0.331	300.5	0.0079			372.1	0.285	332.9
	8	0.323	300.5	0.0079			372.1	0.262	340.3
	9	0.309	300.2	0.0079			372.1	0.245	347.1
[51]	1	0.280	305.1	0.0193	10.82	2.36	391	0.236	358
	2	0.308	304.6	0.0187			396	0.308	363
	3	0.304	304.1	0.0182			398	0.258	363
	4	0.325	304.1	0.0182			390	0.255	351
	5	0.332	305.1	0.0193			393	0.274	361
	6	0.312	304.6	0.0187			389	0.241	351
	7	0.310	304.1	0.0182			393	0.243	354
	8	0.275	302.1	0.0162			385	0.232	359
	9	0.282	302.1	0.0162			380	0.230	351
	10	0.281	303.1	0.0172			378	0.237	353
	11	0.289	303.1	0.0172			383	0.234	351
	12	0.280	303.1	0.0172			376	0.233	347
	13	0.292	303.1	0.0172			386	0.240	355
	14	0.280	304.15	0.0182			378	0.234	352

Table A4. The validation cases on off-design conditions; (*) we reconstruct the experimental drying curve measured at $u_0 = 4$.

Study	Case	Dead-State Conditions			Drying Conditions			Results	
		u_0	T_0 [K]	ω_0	$\dot{m}_a$ [$\frac{kg}{s}$]	$\dot{m}_s$ [$\frac{kg}{s}$]	T_H [K]	u_f	T_E [K]
[54]	1	0.218	300	0.01	6.74	4.78	388.1	0.194	n.a.
	2	0.225						0.196	
	3	0.220						0.198	
	4	0.218						0.197	
[53]	1	0.283	300	0.01	12.37	4.48	418	0.236	n.a.
	2	0.280			11.39			0.235	
	3	0.284			11.39			0.238	
	4	0.302			12.99			0.219	
	5	0.314			13.11			0.230	
[55] *	1	4	300	0.01	10.98	0.036	363	3.78	n.a.
	2							3.46	
	3							3.16	
	4							2.85	
	5							2.55	
	6							2.26	
	7							1.98	
	8							1.70	
	9							1.43	
	10							1.17	
	11							0.95	
	12							0.73	
	13							0.53	
	14							0.19	

References

1. Lewis, W.K. The rate of drying of solid materials. *Ind. Eng. Chem.* **1921**, *13*, 427–432. [CrossRef]
2. Castro, A.; Mayorga, E.; Moreno, F. Mathematical modelling of convective drying of fruits: A review. *J. Food Eng.* **2018**, *223*, 152–167. [CrossRef]
3. Mikhailov, M.; Shishedjiev, B. Temperature and moisture distributions during contact drying of a moist porous sheet. *Int. J. Heat Mass Transf.* **1975**, *18*, 15–24. [CrossRef]
4. Sandu, C. Infrared radiative drying in food engineering: A process analysis. *Biotechnol. Prog.* **1986**, *2*, 109–119. [CrossRef] [PubMed]
5. Behjat, Y.; Shahhosseini, S.; Hashemabadi, S.H. CFD modeling of hydrodynamic and heat transfer in fluidized bed reactors. *Int. Commun. Heat Mass Transf.* **2008**, *35*, 357–368. [CrossRef]
6. Alagha, M.S.; Szentannai, P. Analytical review of fluid-dynamic and thermal modeling aspects of fluidized beds for energy conversion devices. *Int. J. Heat Mass Transf.* **2020**, *147*, 118907. [CrossRef]
7. Daud, W.R.W. Fluidized bed dryers—Recent advances. *Adv. Powder Technol.* **2008**, *19*, 403–418. [CrossRef]
8. Turchiuli, C.; Smail, R.; Dumoulin, E. Fluidized bed agglomeration of skim milk powder: Analysis of sampling for the follow-up of agglomerate growth. *Powder Technol.* **2013**, *238*, 161–168. [CrossRef]
9. Kemp, I.C. Fundamentals of energy analysis of dryers. *Mod. Dry. Technol.* **2012**, *4*, 1–46.
10. Erbay, Z.; Hepbasli, A. Exergoeconomic evaluation of a ground-source heat pump food dryer at varying dead state temperatures. *J. Clean. Prod.* **2017**, *142*, 1425–1435. [CrossRef]
11. Milota, M.; Mosher, P. Emissions of hazardous air pollutants from lumber drying. *Facilities* **2007**, *58*, 50–55.
12. Deng, W.Y.; Yan, J.H.; Li, X.D.; Wang, F.; Zhu, X.W.; Lu, S.Y.; Cen, K.F. Emission characteristics of volatile compounds during sludges drying process. *J. Hazard. Mater.* **2009**, *162*, 186–192. [CrossRef] [PubMed]
13. Liu, W.; Xu, J.; Liu, J.; Cao, H.; Huang, X.F.; Li, G. Characteristics of ammonia emission during thermal drying of lime sludge for co-combustion in cement kilns. *Environ. Technol.* **2015**, *36*, 226–236. [CrossRef] [PubMed]

14. Tsatsaronis, G. Recent developments in exergy analysis and exergoeconomics. *Int. J. Exergy* **2008**, *5*, 489–499. [CrossRef]
15. Syahrul, S.; Hamdullahpur, F.; Dincer, I. Thermal analysis in fluidized bed drying of moist particles. *Appl. Therm. Eng.* **2002**, *22*, 1763–1775. [CrossRef]
16. Tohidi, M.; Sadeghi, M.; Torki-Harchegani, M. Energy and quality aspects for fixed deep bed drying of paddy. *Renew. Sustain. Energy Rev.* **2017**, *70*, 519–528. [CrossRef]
17. Dincer, I.; Sahin, A. A new model for thermodynamic analysis of a drying process. *Int. J. Heat Mass Transf.* **2004**, *47*, 645–652. [CrossRef]
18. Akpinar, E.K. Thermodynamic analysis of strawberry drying process in a cyclone type dryer. *J. Sci. Ind. Res.* **2007**, *66*, 152–161.
19. Yogendrasasidhar, D.; Setty, Y.P. Drying kinetics, exergy and energy analyses of Kodo millet grains and Fenugreek seeds using wall heated fluidized bed dryer. *Energy* **2018**, *151*, 799–811. [CrossRef]
20. Aviara, N.A.; Onuoha, L.N.; Falola, O.E.; Igbeka, J.C. Energy and exergy analyses of native cassava starch drying in a tray dryer. *Energy* **2014**, *73*, 809–817. [CrossRef]
21. Icier, F.; Colak, N.; Erbay, Z.; Kuzgunkaya, E.H.; Hepbasli, A. A comparative study on exergetic performance assessment for drying of broccoli florets in three different drying systems. *Dry. Technol.* **2010**, *28*, 193–204. [CrossRef]
22. Xiang, F.; Wang, L.; Yue, X.F. Exergy analysis and experimental study of a vehicle-mounted heat pump–assisted fluidization drying system driven by a diesel generator. *Dry. Technol.* **2011**, *29*, 1313–1324. [CrossRef]
23. Cay, A.; Tarakçıoğlu, I.; Hepbasli, A. Exergetic analysis of textile convective drying with stenters by subsystem models: Part 1—Exergetic modeling and evaluation. *Dry. Technol.* **2010**, *28*, 1359–1367. [CrossRef]
24. Cay, A.; Tarakçıoğlu, I.; Hepbasli, A. Exergetic analysis of textile convective drying with stenters by subsystem models: Part 2—Parametric study on exergy analysis. *Dry. Technol.* **2010**, *28*, 1368–1376. [CrossRef]
25. Gungor, A.; Erbay, Z.; Hepbasli, A. Exergoeconomic analyses of a gas engine driven heat pump drier and food drying process. *Appl. Energy* **2011**, *88*, 2677–2684. [CrossRef]
26. Meyer, L.; Tsatsaronis, G.; Buchgeister, J.; Schebek, L. Exergoenvironmental analysis for evaluation of the environmental impact of energy conversion systems. *Energy* **2009**, *34*, 75–89. [CrossRef]
27. Morosuk, T.; Tsatsaronis, G. Advanced exergy-based methods used to understand and improve energy-conversion systems. *Energy* **2019**, *169*, 238–246. [CrossRef]
28. Tsatsaronis, G. Thermoeconomic analysis and optimization of energy systems. *Prog. Energy Combust. Sci.* **1993**, *19*, 227–257. [CrossRef]
29. Gidaspow, D. *Multiphase Flow and Fluidization: Continuum and Kinetic Theory Descriptions*; Academic Press: Cambridge, MA, USA, 1994.
30. Defraeye, T. Advanced computational modelling for drying processes–A review. *Appl. Energy* **2014**, *131*, 323–344. [CrossRef]
31. Ramachandran, R.P.; Akbarzadeh, M.; Paliwal, J.; Cenkowski, S. Computational fluid dynamics in drying process modelling—A technical review. *Food Bioprocess Technol.* **2018**, *11*, 271–292. [CrossRef]
32. Assari, M.; Tabrizi, H.B.; Saffar-Avval, M. Numerical simulation of fluid bed drying based on two-fluid model and experimental validation. *Appl. Therm. Eng.* **2007**, *27*, 422–429. [CrossRef]
33. Assari, M.; Tabrizi, H.B.; Najafpour, E. Energy and exergy analysis of fluidized bed dryer based on two-fluid modeling. *Int. J. Therm. Sci.* **2013**, *64*, 213–219. [CrossRef]
34. Li, M.; Duncan, S. Dynamic model analysis of batch fluidized bed dryers. *Part. Part. Syst. Charact.* **2008**, *25*, 328–344. [CrossRef]
35. Ranjbaran, M.; Emadi, B.; Zare, D. CFD simulation of deep-bed paddy drying process and performance. *Dry. Technol.* **2014**, *32*, 919–934. [CrossRef]
36. Rosli, M.I.; Nasir, A.; Mu'im, A.; Takriff, M.S.; Chern, L.P. Simulation of a Fluidized Bed Dryer for the Drying of Sago Waste. *Energies* **2018**, *11*, 2383. [CrossRef]
37. Jang, J.; Arastoopour, H. CFD simulation of a pharmaceutical bubbling bed drying process at three different scales. *Powder Technol.* **2014**, *263*, 14–25. [CrossRef]

38. Padoin, N.; Dal'Toé, A.T.; Rangel, L.P.; Ropelato, K.; Soares, C. Heat and mass transfer modeling for multicomponent multiphase flow with CFD. *Int. J. Heat Mass Transf.* **2014**, *73*, 239–249. [CrossRef]
39. Yiming, S.; Jinrong, S.; Ming, G.; Bin, C.; Yanhong, Y.; Xiaoxun, M. Modeling of Mass Transfer in Nonideal Multicomponent Mixture with Maxwell-Stefan Approach. *Chin. J. Chem. Eng.* **2010**, *18*, 362–371.
40. Cui, X.; Li, X.; Sui, H.; Li, H. Computational fluid dynamics simulations of direct contact heat and mass transfer of a multicomponent two-phase film flow in an inclined channel at sub-atmospheric pressure. *Int. J. Heat Mass Transf.* **2012**, *55*, 5808–5818. [CrossRef]
41. Defraeye, T.; Blocken, B.; Carmeliet, J. Analysis of convective heat and mass transfer coefficients for convective drying of a porous flat plate by conjugate modelling. *Int. J. Heat Mass Transf.* **2012**, *55*, 112–124. [CrossRef]
42. Erbay, Z.; Icier, F. A review of thin layer drying of foods: Theory, modeling, and experimental results. *Crit. Rev. Food Sci. Nutr.* **2010**, *50*, 441–464. [CrossRef] [PubMed]
43. Han, X.; Liu, M.; Zhai, M.; Chong, D.; Yan, J.; Xiao, F. Investigation on the off-design performances of flue gas pre-dried lignite-fired power system integrated with waste heat recovery at variable external working conditions. *Energy* **2015**, *90*, 1743–1758. [CrossRef]
44. Han, X.; Liu, M.; Wu, K.; Chen, W.; Xiao, F.; Yan, J. Exergy analysis of the flue gas pre-dried lignite-fired power system based on the boiler with open pulverizing system. *Energy* **2016**, *106*, 285–300. [CrossRef]
45. Manente, G.; Toffolo, A.; Lazzaretto, A.; Paci, M. An Organic Rankine Cycle off-design model for the search of the optimal control strategy. *Energy* **2013**, *58*, 97–106. [CrossRef]
46. Ypma, T.J. Historical development of the Newton–Raphson method. *SIAM Rev.* **1995**, *37*, 531–551. [CrossRef]
47. Mohapatra, D.; Bal, S. Determination of Specific Heat and Gelatinization Temperature of Rice using Differential Scanning Calorimetry. In Proceedings of the 2003 ASAE Annual Meeting, Las Vegas, NV, USA, 27–30 July 2003; American Society of Agricultural and Biological Engineers: St. Joseph, MI, USA, 2003; p. 1.
48. Faitli, J.; Magyar, T.; Erdélyi, A.; Murányi, A. Characterization of thermal properties of municipal solid waste landfills. *Waste Manag.* **2015**, *36*, 213–221. [CrossRef] [PubMed]
49. Gyftopoulos, E.P.; Beretta, G.P. *Thermodynamics: Foundations and Applications*; Courier Corporation: Chelmsford, MA, USA, 2005.
50. Sarker, M.; Ibrahim, M.; Aziz, N.A.; Punan, M. Application of simulation in determining suitable operating parameters for industrial scale fluidized bed dryer during drying of high impurity moist paddy. *J. Stored Prod. Res.* **2015**, *61*, 76–84. [CrossRef]
51. Sarker, M.S.H.; Ibrahim, M.N.; Aziz, N.A.; Punan, M.S. Energy and exergy analysis of industrial fluidized bed drying of paddy. *Energy* **2015**, *84*, 131–138. [CrossRef]
52. da Silva, F.R.G.B.; de Souza, M.; da Costa, A.M.D.S.; de Matos Jorge, L.M.; Paraíso, P.R. Experimental and numerical analysis of soybean meal drying in fluidized bed. *Powder Technol.* **2012**, *229*, 61–70. [CrossRef]
53. Prachayawarakorn, S.; Tia, W.; Poopaiboon, K.; Soponronnarit, S. Comparison of performances of pulsed and conventional fluidised-bed dryers. *J. Stored Prod. Res.* **2005**, *41*, 479–497. [CrossRef]
54. Wetchacama; Soponronnarit; Wangi. Mathematical Model for Industrial Fluidized Bed Dryer. In Proceedings of the 18th ASEAN Seminar on Grain Post Harvest Technology; Manila, Philippines, 11–13 March 1997.
55. Sun, G.; Chen, M.; Huang, Y. Evaluation on the air-borne ultrasound-assisted hot air convection thin-layer drying performance of municipal sewage sludge. *Ultrason. Sonochem.* **2017**, *34*, 588–599. [CrossRef] [PubMed]
56. Rosen, M.A.; Dincer, I. Exergy analysis of waste emissions. *Int. J. Energy Res.* **1999**, *23*, 1153–1163. [CrossRef]
57. McDonald, A.; Dare, P.; Gifford, J.; Steward, D.; Riley, S. Assessment of air emissions from industrial kiln drying of Pinus radiata wood. *Holz Als Roh-Und Werkst.* **2002**, *60*, 181–190. [CrossRef]
58. de Gouw, J.A.; Howard, C.J.; Custer, T.G.; Fall, R. Emissions of volatile organic compounds from cut grass and clover are enhanced during the drying process. *Geophys. Res. Lett.* **1999**, *26*, 811–814. [CrossRef]
59. Mistry, K.H.; Zubair, S.M. Effect of entropy generation on the performance of humidification-dehumidification desalination cycles. *Int. J. Therm. Sci.* **2010**, *49*, 1837–1847. [CrossRef]
60. Tondeur, D.; Kvaalen, E. Equipartition of entropy production. An optimality criterion for transfer and separation processes. *Ind. Eng. Chem. Res.* **1987**, *26*, 50–56. [CrossRef]

61. Goshayshi, H.; Missenden, J. The investigation of cooling tower packing in various arrangements. *Appl. Therm. Eng.* **2000**, *20*, 69–80. [CrossRef]
62. Xu, Z.; Wang, R.; Yang, C. Perspectives for low-temperature waste heat recovery. *Energy* **2019**, *176*, 1037–1043. [CrossRef]
63. Ertekin, C.; Firat, M.Z. A comprehensive review of thin-layer drying models used in agricultural products. *Crit. Rev. Food Sci. Nutr.* **2017**, *57*, 701–717. [CrossRef]
64. Thiel, G.P. Entropy generation in condensation in the presence of high concentrations of noncondensable gases. *Int. J. Heat Mass Transf.* **2012**, *55*, 5133–5147. [CrossRef]
65. Macdonald, I.; El-Sayed, M.; Mow, K.; Dullien, F. Flow through porous media-the Ergun equation revisited. *Ind. Eng. Chem. Fundam.* **1979**, *18*, 199–208. [CrossRef]
66. Houghton, J.I.; Burgess, J.E.; Stephenson, T. Off-line particle size analysis of digested sludge. *Water Res.* **2002**, *36*, 4643–4647. [CrossRef]
67. Soponronnarit, S. Fluidized bed paddy drying. *Sci. Asia* **1999**, *25*, 51–56. [CrossRef]
68. Chandran, A.; Rao, S.S.; Varma, Y. Fluidized bed drying of solids. *AIChE J.* **1990**, *36*, 29–38. [CrossRef]
69. Bizmark, N.; Mostoufi, N.; Sotudeh-Gharebagh, R.; Ehsani, H. Sequential modeling of fluidized bed paddy dryer. *J. Food Eng.* **2010**, *101*, 303–308. [CrossRef]
70. Kakac, S.; Liu, H.; Pramuanjaroenkij, A. *Heat Exchangers: Selection, Rating, and Thermal Design*; CRC Press: Boca Raton, FL, USA, 2020.
71. Oğulata, R.T. Utilization of waste-heat recovery in textile drying. *Appl. Energy* **2004**, *79*, 41–49. [CrossRef]
72. Aktaş, M.; Şevik, S.; Amini, A.; Khanlari, A. Analysis of drying of melon in a solar-heat recovery assisted infrared dryer. *Sol. Energy* **2016**, *137*, 500–515. [CrossRef]
73. El Fil, B.; Garimella, S. Waste Heat Recovery in Commercial Gas-Fired Tumble Dryers. *Energy* **2020**, *218*, 119407. [CrossRef]
74. Felli, M. *Lezioni di Fisica Tecnica II (Energetica-Meccanica)-Trasmissione del Calore, Acustica, Tecnica Dell'illuminazione*; Morlacchi Editore: Perugia, Italy, 2004.
75. Beale, S.B. Fluid Flow and Heat Transfer in Tube Banks. Ph.D. Thesis, Imperial College London (University of London), London, UK, 1992.
76. *UNI EN 14551:2018-1: Air Conditioners, Liquid Chilling Packages and Heat Pumps for Space Heating and Cooling and Process Chillers, with Electrically Driven Compressors-Part 1: Terms and Definitions*; Ente Nazionale Italiano di Unificazione (UNI): Milano, Italy, 2018.
77. Buck, A.L. New equations for computing vapor pressure and enhancement factor. *J. Appl. Meteorol.* **1981**, *20*, 1527–1532. [CrossRef]
78. Osterle, F. On the analysis of counter-flow cooling towers. *Int. J. Heat Mass Transf.* **1991**, *34*, 1313–1316. [CrossRef]

Article

Experimental Study on a Thermoelectric Generator for Industrial Waste Heat Recovery Based on a Hexagonal Heat Exchanger

Rui Quan [1,2,*], Tao Li [1], Yousheng Yue [1], Yufang Chang [2] and Baohua Tan [3]

1 Hubei Key Laboratory for High-efficiency Utilization of Solar Energy and Operation Control of Energy Storage System, Hubei University of Technology, Wuhan 430068, China; z2283902662@163.com (T.L.); David7689@163.com (Y.Y.)

2 Hubei Collaborative Innovation Center for High-efficiency Utilization of Solar Energy, Hubei University of Technology, Wuhan 430068, China; changyf@hbut.edu.cn

3 School of Science, Hubei University of Technology, Wuhan 430068, China; tan_bh@126.com

* Correspondence: quan_rui@126.com; Tel.: +86-27-5975-0858

Received: 22 March 2020; Accepted: 12 June 2020; Published: 17 June 2020

Abstract: To study on the thermoelectric power generation for industrial waste heat recovery applied in a hot-air blower, an experimental thermoelectric generator (TEG) bench with the hexagonal heat exchanger and commercially available Bi_2Te_3 thermoelectric modules (TEMs) was established, and its performance was analyzed. The influences of several important influencing factors such as heat exchanger material, inlet gas temperature, backpressure, coolant temperature, clamping pressure and external load current on the output power and voltage of the TEG were comparatively tested. Experimental results show that the heat exchanger material, inlet gas temperature, clamping pressure and hot gas backpressure significantly affect the temperature distribution of the hexagonal heat exchanger, the brass hexagonal heat exchanger with lower backpressure and coolant temperature using ice water mixture enhance the temperature difference of TEMs and the overall output performance of TEG. Furthermore, compared with the flat-plate heat exchanger, the designed hexagonal heat exchanger has obvious advantages in temperature uniformity and low backpressure. When the maximum inlet gas temperature is 360 °C, the maximum hot side temperature of TEMs is 269.2 °C, the maximum clamping pressure of TEMs is 360 kg/m^2, the generated maximum output power of TEG is approximately 11.5 W and the corresponding system efficiency is close to 1.0%. The meaningful results provide a good guide for the system optimization of low backpressure and temperature-uniform TEG, and especially demonstrate the promising potential of using brass hexagonal heat exchanger in the automotive exhaust heat recovery without degrading the original performance of internal combustion engine.

Keywords: industrial waste heat recovery; thermoelectric generator; hexagonal heat exchanger; temperature distribution; output performance

1. Introduction

The continuously updated development of green energy techniques is a good alternative to resolve the global energy crisis and increase environmental protection. Owing to several advantages, such as little vibration, high reliability and durability and no moving parts, thermoelectric modules (TEMs) have been widely developed in photovoltaic, automotive, military, aerospace, wearable devices, wireless sensor networks and microelectronic applications over the past years [1–9]. Jaziri et al. [1] described and concluded the exploitation of thermoelectric generators (TEGs) in various fields starting from low-power applications (medical and wearable devices, IoT: internet of things, and WSN: wireless sensor network) to high-power applications (industrial electronics, automotive engines and aerospace). Liu et al. [2] developed a TEG based on concentric filament architecture for low power aerospace microelectronic devices, which was able to produce an electrical voltage of 83.5 mVand an electrical power of 32.1 μW with a planar heat source and temperature of 398.15 K. Willars-Rodríguez et al. [3] created and studied a solar hybrid system including photovoltaic (PV) module, concentrating Fresnel lens, thermo-electric generator (TEG) and running water heat extracting unit. Demir et al. [4] proposed a novel system for recovering waste heat of the automobile by a system based thermoelectric generator, and presented the variations of material properties, efficiency and generated power with respect to temperature and position. Meng et al. [5] proposed a technical solution recycling exhaust gas sensible heat based on thermoelectric power generation, which can produce about 1.47 kW electrical energy per square meter and achieve a conversion efficiency of 4.5% for exhaust gas at 350 degrees. Proto et al. [6] analyzed the results of measurements on thermal energy harvesting through a wearable thermoelectric generator (TEG) placed on the arms and legs whose generated power values were in the range from 5 to 50 μW. Kim et al. [7] demonstrated a self-powered wireless sensor node (WSN) driven by a flexible thermoelectric generator (f-TEG) with a significantly improved degree of practicality, and developed a large-area f-TEG that can be wrapped around heat pipes with various diameters, improving their usability and scalability. Leonov et al. [8] studied a thermoelectric energy harvesting on people that generated power in a range of 5–0.5 mW at ambient temperatures of 15 °C–27 °C, respectively. Holgate et al. [9] conceptualized and modeled a newly designed enhanced multi-mission radioisotope thermoelectric generator (MMRTG) that utilized the more efficient skutterudite-based thermoelectric materials, and presented a discussion of the motivations, modeling results and key design factors. With the rapid development of the economy and society, there is considerable unused waste heat dissipated in industrial heat-generating processes such as power industrial boilers, steelmaking, central air-conditioning, heating equipment and so on.

Thermoelectric generator (TEG) is a promising energy source that includes a hot source (heat exchanger) and a cold source (cooling unit); it can convert the heat into electricity based on the temperature difference of the installed TEMs that are sandwiched between the heat exchanger and cooling unit. In the case of TEG for waste heat recovery, many studies have focused on the application and optimization of the flat-plate heat exchanger to obtain higher power output or evaluate its performance [10–14]. For instance, regarding automotive applications, Zhang et al. [10] developed an automotive exhaust thermoelectric generator (AETEG) with 300 TEGs and water cooling for recovering diesel engine exhaust heat, and obtained 1002.6 W power and 2.1% efficiency when the average exhaust temperature was 823 K and the mass flow rate was 480 g/s. Szybist et al. [11] conducted an experiment to investigate 72 TEGs installed on the downstream of three-way catalytic converter and cooled by engine coolant based on a medium-sized gasoline passenger sedan, and 50 W generation power could be generated in the three typical cycles such as FTP (Federal Test Protocol), HWFET (Highway Fuel Economy Test) and US06 cycles. Kim et al. [12] put forward a novel AETEG with heat pipes and measured a maximum power of 350W with 443 K at the evaporator surface of heat pipes. Merkisz et al. [13] placed 24 TEGs at a distance of about 1.5 m from the end of three-way catalytic converter for a gasoline engine and generated the maximum power (189.3W) and maximum efficiency (1.3%) of AETEG, corresponding to the engine speed of 2600 rpm and 2200 rpm, respectively. Fernandez-Yanez et al. [14,15] developed a diesel engine and a gasoline engine AETEGs with engine coolant cooling, and concluded that the maximum produced power was approximately 270 W for both engines if a bypass was not included. However, two non-negligible problems existing in the flat-plate heat exchanger are the heat uniformity and the unwanted backpressure. The former plays an important role in the average temperature difference and thermoelectric efficiency of TEGs, as it dominates the energy conversion efficiency of heat to electricity. The latter reduces the efficiency of the engine for it seriously affects the original fuel economy and emission performance, or even worse, the lost performance of the engine cannot be compensated by the small generated power of TEG. Therefore, an ideal heat exchanger used in a TEG especially in an AETEG, should provide sufficient surface area to install TEMs as much as possible, increase its surface temperature uniformity, make the backpressure as low as possible and ensure the waste heat gas flow at a high value.

In this paper, we expanded the previous study to develop a low-cost and symmetrical TEG based on a brass hexagonal heat exchanger and the commercially available Bi_2Te_3 TEMs to recover the waste heat from a hot-air blower, which can simulate the internal combustion engine used in the AETEG. This study aimed to validate the promising potential and advantages of using the hexagonal heat exchanger in waste heat recovery, and analyze the influences of several important influencing factors on the output performance of the TEG. This work can provide an application guideline for industrial heat-generating processes and automotive exhaust heat recovery.

2. Experimental Setup of a TEG System

A detailed schematic diagram of the designed TEG system is shown in Figure 1; it consists of a hot-air blower, a hexagonal heat exchanger, a pump, a radiator, a water tank, a pressure regulator, temperature and pressure sensors (P_1 denotes pressure sensor, T_1 denotes inlet temperature sensor, T_2 denotes outlet temperature sensor) and six groups of TEMs and cooling units. The hot-air blower provides waste heat to the hexagonal heat exchanger, which can also simulate different driving cycles of vehicles by adjusting the gas temperature, flow and pressure. When the hot-air blower works, exhaust gas flows into the hexagonal heat exchanger to provide the hot side temperature, and the cooling water (i.e., coolant) stored in the water tank is circulated with the pump among the cooling units to form the cold side. Therefore, electricity is generated due to the temperature difference between the hot side and cold side of each TEM. To obtain higher temperature difference and better performance, the rotate speed of radiator can be controlled to precool the inlet coolant in cooling units.

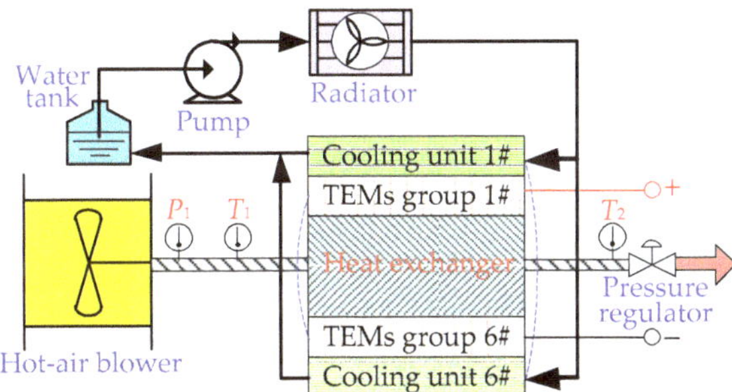

Figure 1. Schematic of an experimental setup of a TEG System. TEM: thermoelectric module.

The specific architecture of TEG and the dimensions of the hexagonal heat exchanger are shown in Figure 2. Each group of TEMs is sandwiched between the surface of the hexagonal heat exchanger and a cooling unit, and they are clamped with an adjustable force using five bolt and screw combinations. In total, there are 30 TEMs of Bi_2Te_3-based materials arranged on the six surfaces of the heat exchanger in five columns (i.e., five TEMs in each column are fixed with a common cooling box), and all the TEMs stalled above the surface of the hexagonal heat exchanger are electrically connected in series. Furthermore, to guarantee uniform cold side temperature of TEMs, all the cooling boxes are thermally connected in parallel (i.e., all the inlets of six cooling boxes are connected with the outlet of radiator, all the outlets of six cooling boxes are connected with the inlet of water tank). To evaluate the hot side temperatures distribution of the 30 TEMs, the corresponding K-type thermocouples below each TEM are pasted above each surface of the hexagonal heat exchanger. For the TEG, the Bi_2Te_3 TEMs (Model Name: TEHP1-1264-0.8) are manufactured by Thermonamic Electronics (Jiangxi, China) Corp. Ltd., whose high conductivity graphite paper is used as the thermally conductive interface material. The specific parameters of the principal components of the TEG system are provided in Table 1.

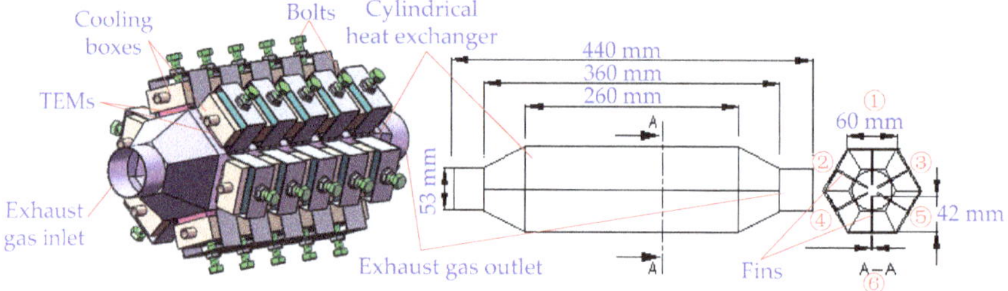

Figure 2. TEG architecture.

Table 1. Parameters of the principal components of a TEG system.

Parameters	Value
Dimension of TEM	40 mm × 40 mm × 40 mm
Maximum operation temperature of TEM	400 °C
Rated operation temperature of TEM	330 °C
Maximum conversion efficiency of TEM	6.5%
Dimension of cooling box	250 mm × 50 mm × 20 mm
Material of cooling box	Aluminum
Thickness of cooling box	1 mm
Thickness of heat exchanger	2 mm
Maximum power of pump	40 W
Maximum flow of pump	5000 L/H
Rated power of radiator	100 W (24V DC)
Rated power of hot-air blower	2000 W

The real experimental setup of a TEG system is shown in Figure 3; the output of TEG is connected with an adjustable electronic load, and the experiments described in the next section were carried out to evaluate several important influencing factors such as heat exchanger material, inlet gas temperature, backpressure, coolant temperature and external load current on its temperature distribution and output performance.

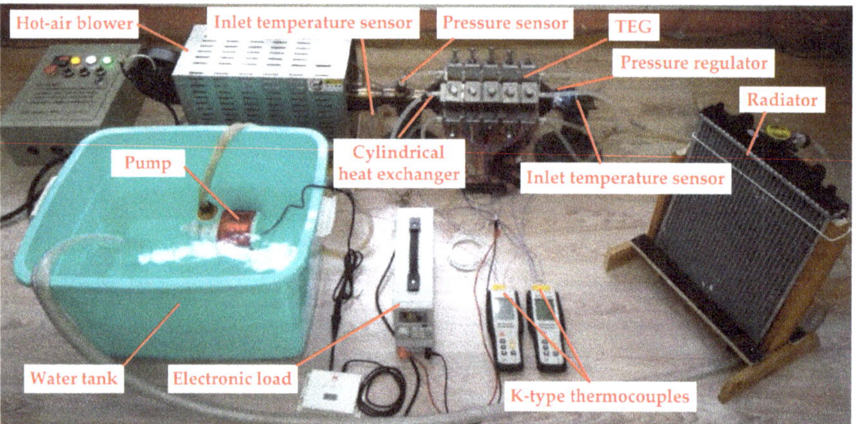

Figure 3. Real experimental setup of a TEG system.

3. Experimental Results and Discussions

3.1. Temperature Distribution

To analyze the temperature uniformity of the hexagonal heat exchanger, 30 independent K-type thermocouples pasted above its six surfaces are divided into five columns, and each K-type thermocouple is installed right below the central hot side of the corresponding single TEM. The specific surface temperature distribution of the 30 temperature detected locations corresponding to the 30 TEMs is shown in Figure 4. On this occasion, the hexagonal heat exchanger is made of stainless steel (Model Name: 304), its inlet temperature is 260 °C, the pressure regulator is fully open and the effective transfer size of its interior fins is 260 mm in length and 42 mm in width (1.5 mm in thickness). For the interior cavity of heat exchange is much larger than the inlet tube, the gas flows quickly across the inlet section. Furthermore, the gas flow decreases accordingly from column 1 to column 5 because of the pressure caused by the interior fins, which enhances the heat transfer between the hot gas and heat exchanger. Thus, it can be concluded that the section closer to the inlet has the lowest surface temperature, the temperatures of the detected locations marked with blue near the exhaust gas outlet is higher than those in the central area and closer to the exhaust gas inlet because of the gas eddy caused by the sudden narrow outlet tube and the temperatures of the six detected locations in the same column are almost the same (the maximum temperature is below 2 °C). Thus, compared with our previously designed chaos-shaped flat-plate heat exchanger [16,17], the temperature uniformity of the hexagonal heat exchanger is better, which is in good agreement with our expected results.

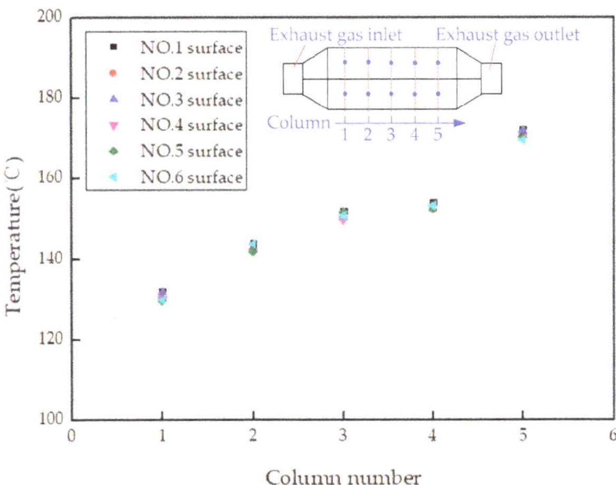

Figure 4. The surface temperature distribution of the hexagonal heat exchanger.

3.2. Influence of Heat Exchanger Material

For the above designed hexagonal heat exchanger shown in Figure 2, two kinds of different metal materials are adopted without changing its dimension to analyze their influences on the surface temperature distribution and the output performance of TEG system. The first one is made of the above stainless steel (Model Name: 304), and the other one is made of brass. The compared characteristic of TEG with the two different kinds of hexagonal heat exchangers is shown in Figure 5. In this case, the clamping pressure of TEMs above each surface increases is 120 kg/m^2, the inlet gas pressure is 70 Pa, the coolant is pumped without radiator, the coolant flow is 5000 L/h and its original temperature is equal to the ambient temperature (27 °C).

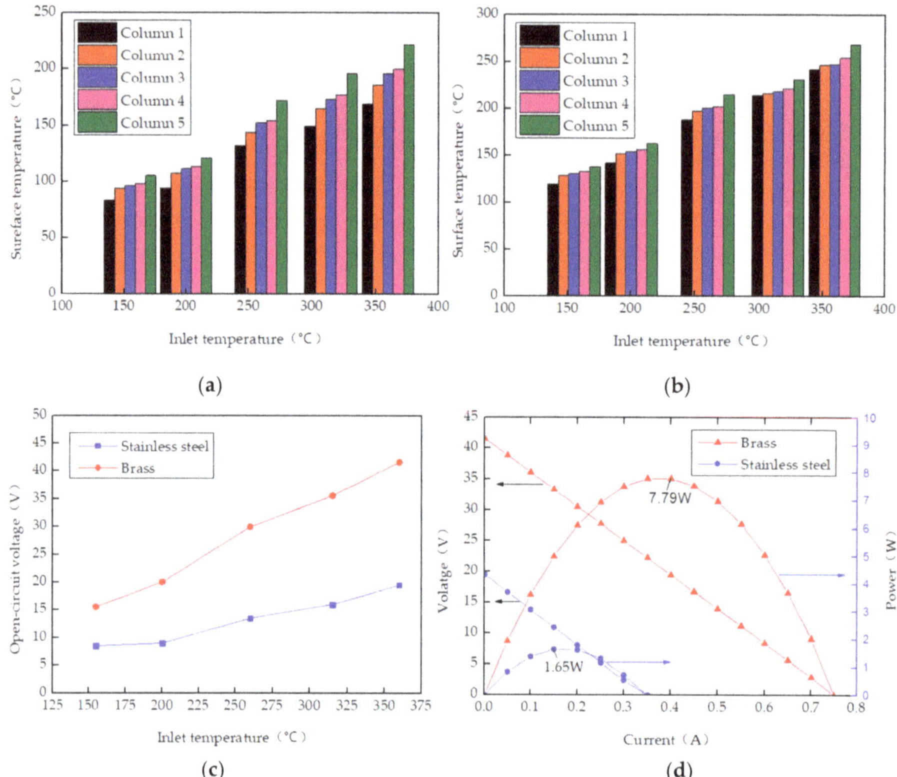

Figure 5. The compared characteristic performance of TEG with different heat exchanger materials. (**a**) Surface temperature distribution of the stainless steel heat exchanger. (**b**) Surface temperature distribution of the brass heat exchanger. (**c**) Open-circuit voltage of TEG with inlet gas temperatures. (**d**). Voltage and power versus current when the maximum inlet temperature is 360 °C.

Figure 5a,b demonstrate the surface temperature distribution of the two heat exchangers with different inlet temperatures of 155 °C, 200 °C, 260 °C, 315 °C and 360 °C, respectively. Considering the above uniform temperature distribution in each column shown in Figure 4, only the temperature detected locations in the NO.1 surface are selected as the average temperatures of the hexagonal heat exchangers (i.e., the hot side temperatures of TEMs) in each column. It is obvious that the average temperatures of both the stainless steel and brass hexagonal heat exchangers are in direct proportion to the inlet gas temperature. According to the Fourier equation, the absorbed heat (denoted Q) of heat exchanger can be calculated as follows:

$$Q = KA\Delta T/d \tag{1}$$

where K is the heat conductivity coefficient, A is the heat transfer area, ΔT is the temperature variation and d is the heat transfer distance. For the average heat conductivity of stainless steel 304 is about 16.2 W/m.K, while the one of brass is about 120 W/m.K, which is almost seven times of the former. The detected locations temperatures in the same column validated that the brass hexagonal heat exchanger is much better than the stainless steel hexagonal heat exchanger in heat transfer with the same inlet gas temperature. Properly speaking, when the maximum inlet gas temperature is 360 °C, the maximum temperature of the stainless steel hexagonal heat exchanger (in column 5) is 222.1 °C, while the one of the brass hexagonal heat exchanger (in column 5) is 269.2 °C, which is an increase of

21.2%. According to Equation (1), it can be calculated that the absorbed heat of brass heat exchanger is increased by about 930% because of its higher heat conductivity coefficient.

Figure 5c shows the corresponding open-circuit voltage of TEG with the above two kinds of hexagonal heat exchangers, the open-circuit voltage of TEG based on the two different hexagonal heat exchangers increases with the augment of its inlet gas temperature. Figure 5d provides the measured characteristics curves of voltage and power versus current of TEG with two different hexagonal heat exchangers when the maximum inlet gas temperature is 360 °C. It can be seen that the open-circuit voltage of TEG with the stainless steel hexagonal heat exchanger is only 19.5 V, while the one with the brass hexagonal heat exchanger is 41.6 V, which is an increase of 113.3%. In addition, the maximum power of TEG with the stainless steel hexagonal heat exchanger is only 1.65 W when the current is 0.15 A, while the one with the brass hexagonal heat exchanger is 7.79 W, which is an increase of nearly four times.

For the above coolant of ambient temperature is adopted, the cold side temperatures of TEMs in the same column of both the two kinds of different hexagonal heat exchangers can be seen similar. Furthermore, as shown in Figure 5b, the TEMs installed in the same column above the brass hexagonal heat exchanger have much higher hot side temperatures with the same inlet gas temperatures, it can be concluded that the temperature difference of TEMs with the brass hexagonal heat exchanger is much larger than that with the stainless steel hexagonal heat exchanger for more hot gas heat is absorbed on the same occasion because of the higher heat transfer coefficient of brass. Thus, the brass hexagonal heat exchanger has overwhelming heat-conducting property advantage over the stainless steel hexagonal heat exchanger despite its high cost, and the TEG system with the brass hexagonal heat exchanger has a better output performance.

3.3. *Influence of Backpressure*

Considering the advantage of the above brass hexagonal heat exchanger used in TEG, the influence of inlet gas backpressure on TEG based on the brass hexagonal heat exchanger is shown in Figure 6. On this occasion, the clamping pressure of TEMs above each surface increases is 120 kg/m^2, the coolant is pumped without radiator, the coolant flow is 5000 L/h and its original temperature is equal to the ambient temperature (27 °C), the inlet gas backpressure is adjusted by the regulator opening. For the detected temperature locations in the 5th column of the brass hexagonal heat exchanger have the highest temperature values, Figure 6a,b show the maximum surface temperatures of the brass hexagonal heat exchanger and the open-circuit voltage of TEG corresponding to different inlet gas temperatures with different inlet gas backpressures of 70 Pa, 180 Pa and 220 Pa, respectively. Figure 6c,d show the voltage and output power versus current characteristics with the above different inlet gas backpressures when the maximum inlet gas temperature is 360 °C.

Larger inlet gas backpressure means lower maximum surface temperature and open-circuit voltage with the same inlet gas temperature; the reason of this is that large inlet gas backpressure will decrease the rotation speed of hot-air blower's fan and the corresponding flow of hot gas. In this case, the heat transfer between hot gas and heat exchanger is reduced, which leads to a lower temperature difference of TEMs based on the same cooling condition. When the maximum inlet gas temperature is 360 °C, the maximum surface temperature of the brass hexagonal heat exchanger is 268.2 °C (70 Pa), 250.1 °C (180 Pa) and 241.3 °C (220 Pa); the open-circuit of TEG is 41.6 V (70 Pa), 35.1 V (180 Pa) and 29.9 V (220 Pa). Furthermore, the open-circuit voltage and output power of TEG are in inverse proportion to the inlet gas backpressures, decreased both output voltage and power with the same output current are accompanied with the augment of inlet gas backpressures. Since the cold side temperature of TEMs in each column with different inlet gas backpressures can be regarded as the same because of the ambient temperature coolant, it can be concluded that the lower inlet gas backpressure is, the larger inlet gas flow will be. Lower inlet gas backpressure contributes to higher temperature differences and better output performance of TEMs, since it ensures higher rotating speed of hot-air blower, and efficient heat conduction between the gas and brass hexagonal heat exchanger on the same occasion.

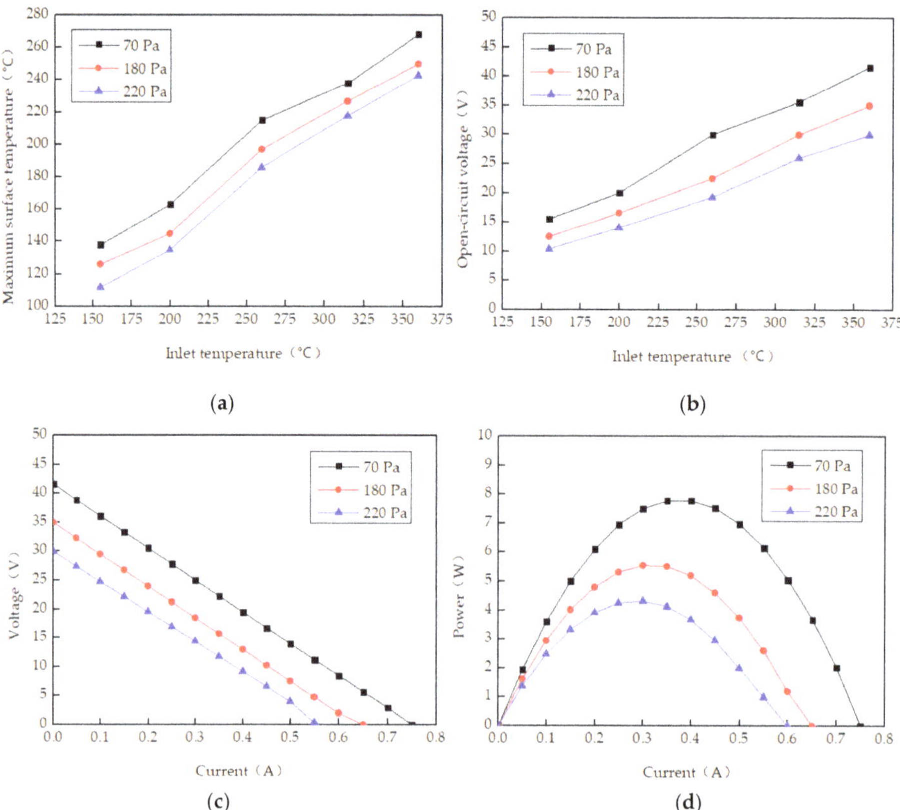

Figure 6. The compared characteristic performance of TEG based on the brass hexagonal heat exchanger with different inlet gas backpressures. (**a**) Maximum surface temperatures of the brass heat exchanger with inlet gas temperatures. (**b**) Open-circuit voltage of TEG with inlet gas temperatures. (**c**) Voltage versus current when the maximum inlet temperature is 360 °C. (**d**) Power versus current when the maximum inlet temperature is 360 °C.

3.4. Influence of Coolant Temperatures

To ensure lower cold side temperatures and higher temperature differences of TEMs with the same hot side temperatures, three different kinds of coolant were used in the test as follows: case 1: pumped coolant of ambient temperature without radiator; case 2: pumped coolant of ambient temperature with radiator of 100% speed; case 3: pumped coolant of ice water mixture (0 °C) without radiator. Figure 7 shows the characteristic performance of TEG based on the brass hexagonal heat exchanger in the above three cases when the clamping pressure of TEMs above each surface increases is 120 kg/m^2, the coolant flow is 5000 L/h, the ambient temperature is 27 °C and the inlet gas backpressure is maintained at 70 Pa.

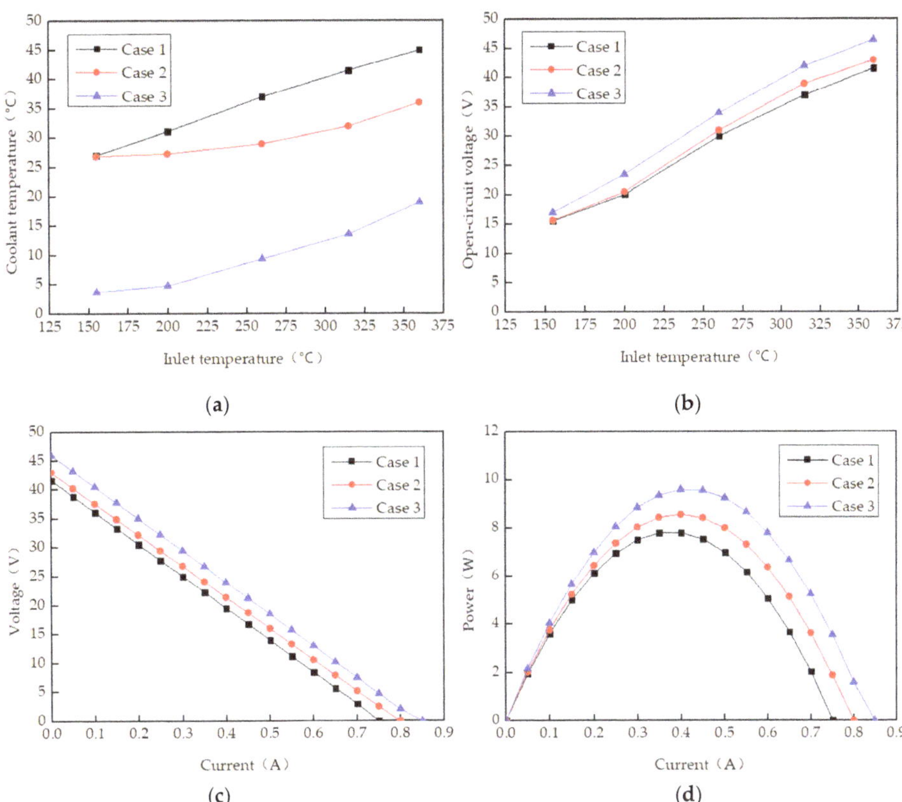

Figure 7. The compared characteristic performance of TEG based on the brass hexagonal heat exchanger in different coolant cases. (**a**) Coolant temperatures of TEG with inlet gas temperatures. (**b**) Open-circuit voltage of TEG with inlet gas temperatures. (**c**) Voltage versus current when the maximum inlet temperature is 360 °C. (**d**) Power versus current when the maximum inlet temperature is 360 °C.

For Figure 7a, the inlet gas temperatures are changed every five minutes to ensure the steady heat conduction among TEMs, heat exchanger and cooling boxes. The coolant temperatures rise as the inlet gas temperatures increase, the coolant temperature in case 1 is the highest, the coolant temperature in case 2 takes second place, while the coolant temperature in case 3 is the lowest. When the maximum inlet gas temperature is 360 °C, the maximum coolant temperature in case 1 is 45.9 °C, the one in case 2 is 36.6 °C and the one in case 3 is 20.2 °C. Figure 7b shows the open-circuit voltage of TEG with different inlet gas temperatures based on the above three different kinds of coolant. It is obvious that the open-circuit voltage of TEG in case 3 is the largest, the one in in case 2 takes second place, while the one in case 1 is the lowest with the same inlet gas temperature. Figure 7c,d show the performance characteristics of TEG in the above three cases with different load currents when the maximum inlet gas temperature is 360 °C. It can be seen that both the output voltage and power of TEG corresponding to the same load current increase evidently with lower coolant temperatures, the maximum power of TEG in case 1, 2 and 3 is 7.79 W, 8.56 W and 9.65 W, respectively. Additionally, the coolant temperature will increase evidently without a radiator because of the large thermal conductivity of TEMs, and the coolant temperature with radiator can be maintained in a relatively stable range. To ensure larger output power and higher output voltage, a coolant of ice water mixture is recommended in the cooling unit of TEG, as it can evidently enlarge the temperature difference of TEM without consuming extra radiator power.

3.5. Influence of Clamping Pressure

Considering the thermal contact resistance caused by the clamping pressure may affect the hot side and cold side temperatures of TEG, the effect of different clamping pressures on the overall performance of TEG is shown in Figure 8. On this occasion, the coolant of ice water mixture (0 °C) is pumped without radiator, the coolant flow is 5000 L/h, the ambient temperature is 27 °C, the inlet gas backpressure is maintained at 70 Pa and the maximum inlet temperature of heat exchanger is 360 °C. The maximum output power of TEG is 9.65 W, 10.32 W and 11.49 W with different clamping pressures of 120 kg/m^2, 240 kg/m^2 and 360 kg/m^2, respectively. The corresponding open-circuit voltage of TEG is 46.2 V, 49.1 V and 53.3 V, respectively.

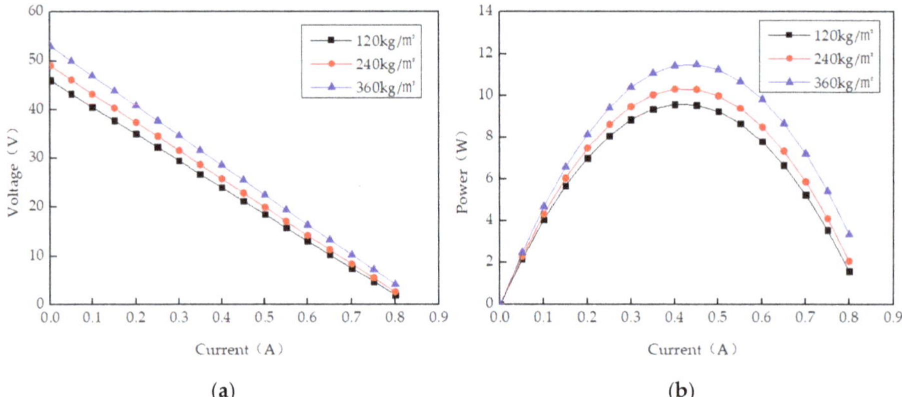

Figure 8. TEG performance with clamping pressure of 120 kg/m^2, 240 kg/m^2 and 360 kg/m^2. (**a**) Voltage versus current curves, (**b**) power versus current curves.

From the separate characteristic curves shown in Figure 8a, the slope of voltage versus current decreases with the increased clamping pressure, which means that the inner resistance of TEG increased accordingly. Combined with the above presented results, it demonstrates that the thermal contact resistance of TEG can be reduced for the empty air gap is decreased with large clamping pressure. Thus, much more inlet gas heat from the brass heat exchanger can be absorbed by the hot sides of TEMs, and increasing heat from the cold sides of TEMs can be brought off with large clamping pressure. The larger clamping pressure is, the lower thermal insulator is, which contributes to the enhanced output performance becaused of the lower thermal contact resistance. Therefore, to ensure larger output power and high efficiency, AETEG should be clamped as tight as possible within the allowable pressure of each TEM.

3.6. System Efficiency

For the 30 TEMs of Bi_2Te_3 based materials used in TEG, the output of TEG can be expressed as follows [1,18,19]:

$$V_{TEG} = \sum_{i=1}^{240} n\alpha_{PNi}\Delta T_i = n\left(\alpha_{pi} - \alpha_{ni}\right)\left(T_{Hi} - T_{Li}\right) \tag{2}$$

$$R_{TEG} = \sum_{i=1}^{240} R_i = 240\left(nl_p/(\sigma_p A_p) + nl_n/(\sigma_n A_n)\right) \tag{3}$$

$$\alpha_{PNi} = \left(22224 + 930.6 \times 0.5 \times \Delta T_i - 0.9905 \times (0.5 \times \Delta T_i)^2\right) \times 10^{-9} \tag{4}$$

where n is the p-type and the n-type semiconductor galvanic arms number in each TEM and V_{TEG} and R_{TEG} are the open circuit voltage and internal resistance of TEG, respectively. α_{PNi} is the relative Seebeck coefficient (V/K), α_{Pi} and α_{ni} are the Seebeck coefficients of the p-type and the n-type semiconductor galvanic arms, respectively. T_{Hi} and T_{Li} are the hot side and cold side temperature (K), respectively. l_p, σ_P and A_p are the leg length (m), electricity resistivity (Ωm) and cross-sectional area (m^2) of a p-type semiconductor galvanic arm, respectively, while l_n, σ_n and A_n are the leg length, electricity resistivity and cross-sectional area of an n-type semiconductor galvanic arm, respectively.

As shown in Figures 5d–8b the output power of the TEG reaches its maximum value (denoted P_{TEG}) when the external load resistance (denoted R_m) is equal to its internal resistance, and is expressed as:

$$P_{TEG} = U_{TEG}^2/(4R_m) \tag{5}$$

The heat input to the hexagonal heat exchanger is obtained as [20]:

$$Q_h = G_h \rho_h C_h \Delta T_h = G_h \rho_h C_h (T_{hi} - T_{ho}) \tag{6}$$

where G_h is the volume flow rate of inlet gas, ρ_h is the density of inlet gas, C_h is the heat capacity of inlet gas at constant pressure, while T_{hi} and T_{ho} are the inlet gas temperature and outlet gas temperature of hot gas, respectively. In this case, the maximum TEG system efficiency η can be calculated as follows:

$$\eta = P_{TEG}/Q_h \tag{7}$$

Figure 9 shows the temperature drop and outlet temperature of hot gas along the brass hexagonal heat exchanger when the inlet gas temperature changes from 155 °C to 360 °C. On this occasion, the clamping pressure of TEMs above each surface increases to 360 kg/m^2, the pumped coolant of ice water mixture (0 °C) without radiator is adopted, the ambient temperature is 27 °C and the inlet gas backpressure is 70 Pa (the corresponding flow rate is 0.18 m^3/h). It can be seen that both the temperature drops and outlet temperature increases with increasing inlet gas temperatures; the maximum temperature drop between the inlet and outlet gas is within 50 °C.

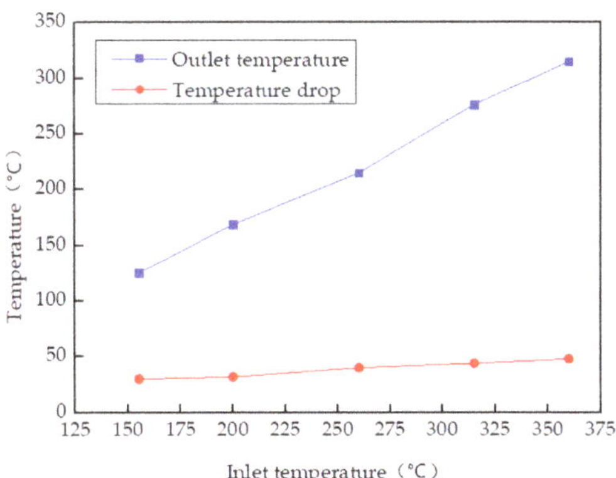

Figure 9. The decrease in temperature and the outlet gas temperature for different inlet gas temperatures.

Experimental results of both the maximum output power and corresponding system efficiency of TEG are shown in Figure 10 with respect to the inlet gas temperatures. As shown in Figure 10, for fixed inlet gas flow rates and backpressures, the maximum output power and system efficiency of TEG based on the coolant of ice water mixture increase with increasing inlet gas temperature. When the maximum inlet gas temperature is 360 °C, the outlet gas temperature is 315.1 °C, the generated maximum output power of TEG is 11.49 W, and the corresponding system efficiency is 0.96%. For the maximum temperature limitation (360 °C) of hot-air blower outlet, the maximum surface temperature of the brass hexagonal heat exchanger shown in Figure 6 is only 269.2 °C, which is much lower than the operated operation temperature of TEM (330 °C). Thus, if the maximum surface temperature of the brass hexagonal heat exchanger can be raised to 330 °C, it can be deduced that the maximum output power of TEG will approach 20 W, and the corresponding system efficiency will be above 1.5%.

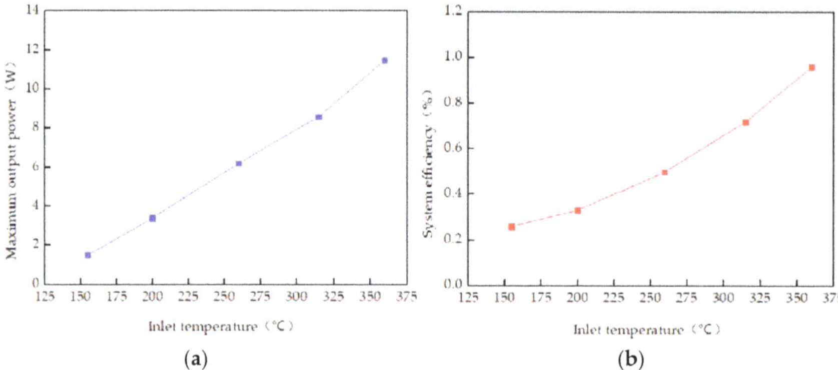

Figure 10. The overall output performance of TEG with respect to the inlet gas temperature. (**a**) Maximum output power; (**b**) system efficiency.

Furthermore, to validate the above numerical model of TEG, Figure 11 shows the comparison between the predicted and measured maximum power and corresponding system efficiency with different matched load resistance, the clamping pressure of TEMs above each surface increases is 360 kg/m^2, the pumped coolant of ice water mixture (0 °C) without radiator is adopted, the coolant flow is 5000 L/h, the ambient temperature is 27 °C and the inlet gas backpressure is 70 Pa (the corresponding flow rate is 0.18 m^3/h). It can be seen that the above numerical model can predict the performances of TEG when the inlet gas temperature changes from 125 °C to 360 °C. At low temperatures, there is a good agreement between the experimental performances and the predicted results. However, at high temperatures, the discrepancy between the experimental performances and predicted values increases evidently and is also acceptable. The main reason is that the heat losses from the hot sides to cold sides of TEMs are not considered in the numerical model, which plays a more important role in the heat transfer at high temperatures. Overall, to estimate the performance and establish the precise model of TEG, the properties of the thermoelectric materials used in TEMs should be assumed to be variables, the heat losses from the uncovered surface, TEMs gap and hot sides to cold sides should be taken into account.

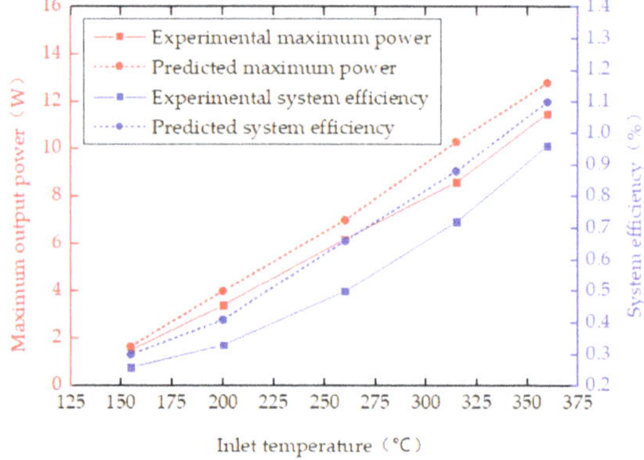

Figure 11. Comparison between the predicted and measured maximum power and corresponding system efficiency of TEG.

4. Conclusions

Waste heat recovery based on TEMs presents a promising research focus worldwide, but enhancing the output performance and system efficiency of TEG remains a significant challenge. In this study, a TEG system with the low-temperature and common commercially available Bi_2Te_3 TEMs and a hexagonal heat exchanger is designed for the waste heat recovery of an industrial hot-air blower. The influences of different operating conditions such as material, backpressure, inlet gas temperature, clamping pressure and coolant temperature on the temperature distribution, open-circuit voltage, the maximum output power and the system efficiency of TEG are reported from the experimental measurements. The experimental results demonstrate that the surface temperature distribution of hexagonal heat exchanger is uniform in each column, and is greatly affected by the adopted material of heat exchanger and the backpressure of inlet gas.

The comparisons indicate that the brass hexagonal heat exchanger has better heat conduction, lower backpressure can enhance the gas flow rate and the average surface temperature of heat exchanger (i.e., hot side temperature of TEMs) and the coolant of ice water mixture contributes to lower cold side temperature of TEMs. Furthermore, the maximum output power and system efficiency of TEG is proportional to the practical temperature differences of TEMs caused by inlet gas temperatures, backpressures, clamping pressures and coolant temperature. The designed brass hexagonal heat exchanger has low pressure drop, and it is very suitable for the automotive exhaust thermoelectric generator. Further experiments are planned to optimize the TEG and apply it in the automotive exhaust heat recovery.

Author Contributions: T.L. and Y.Y. performed the experiments and data analysis. R.Q. designed the experiments and wrote the manuscript. Y.C. and B.T. offered valuable discussions in analyses and revised the manuscript. All authors have read and agreed to the published version of the manuscript.

Funding: This research received no external funding.

Acknowledgments: "This work was supported by the National Natural Science Foundation of China (No. 51977061, 61903129, 51407063)"and "This work was supported by the Doctor Scientific Research Foundation of Hubei University of Technology (No. BSQD13064)".

Conflicts of Interest: The authors declare no conflict of interest.

List of Notations

P_1	inlet gas temperature sensor
T_1	inlet gas temperature sensor
T_2	outlet gas temperature sensor
n	p-type and the n-type semiconductor galvanic arms number
V_{TEG}	open circuit voltage of TEG
R_{TEG}	internal resistance of TEG
α_{PNi}	relative Seebeck coefficient
α_{pi}	Seebeck coefficients of the p-type semiconductor galvanic arms
α_{ni}	Seebeck coefficients of the n-type semiconductor galvanic arms
T_{Hi}	hot side temperature of TEM
T_{Li}	cold side temperature of TEM
l_{p}	leg length of a p-type semiconductor galvanic arm
σ_{p}	electricity resistivity of a p-type semiconductor galvanic arm
A_{p}	cross-sectional area of a p-type semiconductor galvanic arm
l_{n}	leg length of a n-type semiconductor galvanic arm
σ_{n}	electricity resistivity of a n-type semiconductor galvanic arm
A_{n}	cross-sectional area of a p-type semiconductor galvanic arm
P_{TEG}	maximum output power of TEG
R_{m}	external load resistance
Q_{h}	heat input to the cylindrical heat exchanger
G_{h}	volume flow rate of inlet gas
ρ_{h}	density of inlet gas
C_{h}	heat capacity of inlet gas at constant pressure
T_{hi}	inlet gas temperature
T_{ho}	outlet gas temperature
η	maximum TEG system efficiency

References

1. Jaziri, N.; Boughamoura, A.; Müller, J.; Mezghani, B.; Tounsi, F.; Ismail, M. A comprehensive review of thermoelectric generators: Technologies and common applications. *Energy Rep.* **2019**. [CrossRef]
2. Liu, K.; Tang, X.; Liu, Y.; Yuan, Z.; Li, J.; Xu, Z.; Zhang, Z.; Chen, W. High performance and integrated design of thermoelectric generator based on concentric filament architecture. *J. Power Sources* **2018**, *393*, 161–168. [CrossRef]
3. Willars-Rodríguez, F.J.; Chávez-Urbiola, E.A.; Vorobiev, P.; Vorobiev, Y.V. Investigation of solar hybrid system with concentrating Fresnel lens, photovoltaic and thermoelectric generators. *Int. J. Energy Res.* **2017**, *41*, 377–388. [CrossRef]
4. Demir, M.E.; Dincer, I. Performance assessment of a thermoelectric generator applied to exhaust waste heat recovery. *Appl. Therm. Eng.* **2017**, *120*, 694–707. [CrossRef]
5. Meng, F.K.; Chen, L.G.; Feng, Y.L.; Xiong, B. Thermoelectric generator for industrial gas phase waste heat recovery. *Energy* **2017**, *135*, 83–90. [CrossRef]
6. Proto, A.; Bibbo, D.; Cerny, M.; Vala, D.; Kasik, V.; Peter, L.; Conforto, S.; Schmid, M.; Penhaker, M. Thermal energy harvesting on the bodily surfaces of arms and legs through a wearable thermo-electric generator. *Sensors* **2018**, *18*, 1927. [CrossRef] [PubMed]
7. Kim, Y.; Gu, H.M.; Kim, C.; Choi, H.; Lee, G.; Kim, S.; Yi, K.; Lee, S.; Cho, B. High-performance self-powered wireless sensor node driven by a flexible thermoelectric generator. *Energy* **2018**, *162*, 526–533. [CrossRef]
8. Leonov, V. Thermoelectric energy harvesting of human body heat for wearable sensors. *IEEE Sens. J.* **2013**, *13*, 2284–2291. [CrossRef]
9. Holgate, T.-C.; Bennett, R.; Hammel, T.; Caillat, T.; Keyser, S.; Sievers, B. Increasing the efficiency of the multi-mission radioisotope thermoelectric generator. *J. Electron. Mater.* **2015**, *44*, 1814–1821. [CrossRef]
10. Zhang, Y.L.; Cleary, M.; Wang, X.W.; Kempf, N.; Schoensee, L.; Yang, J.; Joshi, G.; Meda, L. High-temperature and high-power-density nanostructured thermoelectric generator for automotive waste heat recovery. *Energy Convers. Manag.* **2015**, *105*, 946–950. [CrossRef]
11. Szybist, J.; Davis, S.; Thomas, J.F.; Kaul, B.C. Performance of a Half-Heusler thermoelectric generator for automotive application. *SAE Tech. Pap.* **2018**, 1. [CrossRef]
12. Kim, S.K.; Won, B.C.; Rhi, S.H.; Kim, S.H.; Yoo, J.H.; Jang, J.C. Thermoelectric power generation system for future hybrid vehicles using hot exhaust gas. *J. Electron. Mater.* **2011**, *40*, 778–783. [CrossRef]
13. Merkisz, J.; Fuc, P.; Lijewski, P.; Ziolkowski, A.; Galant, M.; Siedlecki, M. Analysis of an increase in the efficiency of a spark ignition engine through the application of an automotive thermoelectric generator. *J. Electron. Mater.* **2016**, *45*, 4028–4037. [CrossRef]
14. Fernandez-Yanez, P.; Armas, O.; Capetillo, A.; Martinez-Martinez, S. Thermal analysis of a thermoelectric generator for light-duty diesel engines. *Appl. Energy* **2018**, *226*, 690–702. [CrossRef]
15. Fernandez-Yanez, P.; Armas, O.; Kiwan, R.; Stefanopoulou, A.G.; Boehman, A.L. A thermoelectric generator in exhaust systems of spark-ignition and compression-ignition engines. A comparison with an electric turbo-generator. *Appl. Energy* **2018**, *229*, 80–87. [CrossRef]
16. Quan, R.; Zhou, W.; Yang, G.Y.; Quan, S.H. A hybrid maximum power point tracking method for automobile exhaust thermoelectric generator. *J. Electron. Mater.* **2017**, *46*, 2676–2683. [CrossRef]
17. Quan, R.; Liu, G.Y.; Wang, C.J.; Zhou, W.; Huang, L.; Deng, Y.D. Performance investigation of an exhaust thermoelectric generator for military SUV application. *Coatings* **2018**, *8*, 45. [CrossRef]
18. Fraisse, G.; Ramousse, J.; Sgorlon, D.; Goupil, C. Comparison of different modeling approaches for thermoelectric elements. *Energy Convers. Manag.* **2013**, *65*, 351–366. [CrossRef]
19. Quan, R.; Wang, C.J.; Wu, F.; Chang, Y.F.; Deng, Y.D. Parameter matching and optimization of an isg mild hybrid powertrain based on an automobile exhaust thermoelectric generator. *J. Electron. Mater.* **2019**. [CrossRef]
20. Niu, X.; Yu, J.L.; Wang, S.Z. Experimental study on low-temperature waste heat thermoelectric generator. *J. Power Sources* **2009**, *188*, 621–626. [CrossRef]

Article

Simulation of the GOx/GCH4 Multi-Element Combustor Including the Effects of Radiation and Algebraic Variable Turbulent Prandtl Approaches

Evgenij Strokach [1], Igor Borovik [1,*] and Oscar Haidn [2]

[1] Institute No. 2 "Aviation, rocket engines and power plants", Moscow Aviation Institute, Volokolamskoe shosse 4, Moscow 125993, Russia; strokachea@mai.ru

[2] Department of Aerospace and Geodesy, Technical University of Munich, Boltzmannstr. 15, 85748 Garching b. Munich, Germany; oskar.haidn@ltf.mw.tum.de

* Correspondence: borovik.igor@mai.ru

Received: 18 August 2020; Accepted: 18 September 2020; Published: 23 September 2020

Abstract: Multi-element thrusters operating with gaseous oxygen (GOX) and methane (GCH4) have been numerically studied and the results were compared to test data from the Technical University of Munich (TUM). A 3D Reynolds Averaged Navier–Stokes Equations (RANS) approach using a 60° sector as a simulation domain was used for the studies. The primary goals were to examine the effect of the turbulent Prandtl number approximations including local algebraic approaches and to study the influence of radiative heat transfer (RHT). Additionally, the dependence of the results on turbulence modeling was studied. Finally, an adiabatic flamelet approach was compared to an Eddy-Dissipation approach by applying an enhanced global reaction scheme. The normalized and absolute pressures, the integral and segment averaged heat flux were taken as an experimental reference. The results of the different modeling approaches were discussed, and the best performing models were chosen. It was found that compared to other discussed approaches, the BaseLine Explicit Algebraic Reynolds Stress Model (BSL EARSM) provided more physical behavior in terms of mixing, and the adiabatic flamelet was more relevant for combustion. The effect of thermal radiation on the wall heat flux (WHF) was high and was strongly affected by spectral models and wall thermal emissivity. The obtained results showed good agreement with the experimental data, having a small underestimation for pressures of around 2.9% and a good representation of the integral wall heat flux.

Keywords: combustor; turbulent Prandtl approaches; Navier–Stokes simulation

---

1. Introduction

The design and optimization of rocket propulsion devices is a complex procedure that nowadays also includes the numerical simulation of flow and tough physical phenomena as a very important part of the process. This is mainly due to the significant reduction of the cost and time needed for overall testing, production, and development cycle. The main criteria that show if the code can be applied for rocket thrusters are wall heat transfer, combustion efficiency, specific impulse, and pressure representation. The tool should both accurately predict the parameter distribution inside the combustor and satisfy typical needs of design engineers such as robust and stable operation and as small as possible calculation times. Numerical codes, therefore, should be validated for the criteria above-mentioned and the most suitable models of physical phenomena should be chosen.

Conversely, the methane–oxygen fuel pair is now considered as prospective for rocket propulsion. In recent years, the Space Propulsion Division of TUM has been working experimentally and numerically on various aspects of methane/oxygen combustion. Among these, a GCH4/GOx multi-element injector test case was chosen for the RANS-based numerical assessment in our study. The reason to choose this experiment was not only the well-described setting that is well suited for numerical verification, but also the fact that numerous research groups have also used this data, which is extremely useful to efficiently compare the different methodologies and their results, and finally choose appropriate approaches for our problems. A paper by Silvestri et al. [1,2] gives detailed data concerning the test bench and experimental setting and therefore only a short description is given here in the "test case" section.

Recently, some numerical studies have already been made to examine the effect of modeling approaches using both a single-injector test case [3–5] and a multi-element setting [6–9]. Furthermore, some extensive analyses have been performed concerning the PennState and Mascotte test cases [10–12]. All authors have given great attention to the turbulence modeling and state its high influence on the mixing intensity and therefore on the predicted heat fluxes and pressures. Another obvious and important statement is the high impact of the approach to model the interaction of turbulence and chemistry and of the chemical kinetic scheme itself.

Additionally, some research has been made aiming at the influence of the coefficients that account for turbulent diffusion such as turbulent Schmidt and Prandtl numbers [13–19]. An essential impact of these parameters was observed on the wall heat fluxes and pressures, coupled with the variation of the thermodynamic and reactive parameters inside the combustor. The papers either provide a sensitivity study to choose the correct values or propose methods for their local variation.

Additionally, computational fluid dynamics (CFD) is now extremely widespread in all fluid dynamics applications, especially in industrial purposes of rocket combustion modeling. The industry requires such predictions to be fast, but still sufficiently precise to allow their implementation into a routine design optimization process. Such approaches are needed during the early phases of the design process. Therefore, in our paper, the cheapest and most robust techniques were tested to establish a modeling methodology of such complex phenomena applicable for industrial applications and purposes. Such industry demands also drive the usage of most universal approaches for any processes described by the computational model, as changes in the geometry and mass flows during the optimization might change the parameter field significantly. This makes the application of the approaches for locally variable diffusion coefficients even more urgent.

The number of papers accounting for radiative heat transfer in such studies is limited. Zhukov [10] used a P1 Gray radiation heat transfer (RHF) model and noted some effect on the heat flux, whereas due to some uncertainty in the existing wall roughness, no reliable conclusions could be drawn. Thellmann [20] studied the hydrogen–oxygen and methane–oxygen combustion at high pressures, which resulted in radiative to total heat flux ratios up to 9%. Leccesse et al. [21] also studied these fuel pairs and found that the contribution of the RHF was up to 15% for the described cases. Some other studies have referred to general RHF evaluation in turbulent combustion tests [22–25] and showed significant influences that should be accounted for.

Most of the research activities that focused on single and multi-injector test cases had a grid resolution near the wall, which met the $y^+ \approx 1$ condition [3,4,6–8,25]. Some groups have focused on the development of special wall-functions that they implemented in their large Eddy simulations (LES) [26–32]. However, the usage of wall functions for RANS simulations can both accelerate and stabilize the optimization workflow. As wall function-based simulations combined with a wide variety of modeling approaches have not yet been performed on these cases, it was considered as important in this study.

Nearly all described studies used two-equation linear turbulence models to account for the turbulence phenomena. The significant influence of turbulence modeling on the mean pressure was noticed, which was due to improved or worsened mixing efficiency. Generally, a $k - \varepsilon$ model is considered to be giving more physical values of turbulence viscosity, and therefore, the mixing intensity. In this paper, the turbulence field was estimated by three models—the popular Menter's $k - \omega$ Shear stress transport (SST), the $k - \varepsilon$, and the BSL EARSM [33]—to include the anisotropy effects. The use of the BSL EARSM approach is driven by its satisfactory convergence behavior and reduced CPU demands compared to straight Reynolds stress models.

Therefore, the motivation and goals of this study follow the industrial simulation requirements for robustness, low calculation time, and sufficiently reliable simulation results, especially in terms of pressure and wall heat flux inside the combustor, and can be summarized as follows: (1) implementation of locally variable turbulent Prandtl approaches and comparison with its constant values; (2) further estimation of turbulence modeling methods based on the comparison of commonly used two-equation models and a BSL EARSM method; (3) estimation of the radiative heat transfer impact on the wall heat flux; and (4) all these being coupled with a wall-function based near-wall modeling approach for both the velocity and thermal boundary layer to test the capability of such methods in the high pressure and high thermal gradient environment.

2. The Test Case

The chosen test case was developed as a part of the Sonderforschungsbereich Transregio 40 (SFB-TRR 40) program. It includes seven coaxial gaseous methane and gaseous oxygen injectors, which allowed us to study effects such as the injector–injector and injector–wall interaction, which is highly important for turbulence–chemistry interaction and the wall heat transfer. The chamber diameter is 30 mm, whereas each injector consists of a central oxygen injector, a post wall, and a concentric methane injector. The oxygen injector has a diameter of 4 mm while the methane injector has a 5 mm inner and 6 mm outer diameter, respectively. The peripheral injectors had their axis 9 mm off the main axis. The throat diameter was equal to 19 mm, determining the chosen contraction ratio of 2.5. The view of the combustor is shown in Figure 1. One of the main features of the design is that it contains four cylindrical water-cooled segments and a nozzle segment, allowing the determination of the integral wall heat flux. The total length of 381 mm makes it possible for the coaxially injected jets to mix properly and react.

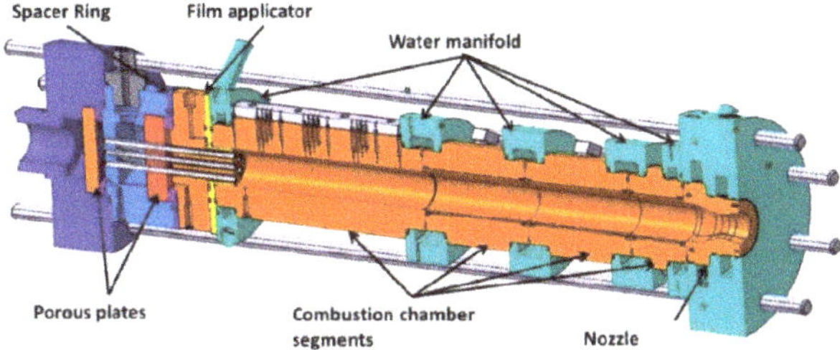

Figure 1. View of the studied thrust chamber [1,2].

For the present study, a load point corresponding to the mixture ratio of 2.65 and the mean pressure of 18.3 bar was chosen. Therefore, the total mass flows were 0.211 kg/s and 0.08 kg/s for oxygen and methane, respectively, while the injection temperatures were 259.4 K for oxygen and 237.6 K for methane. The experimentally determined data included heat flux for each segment, mean pressure, wall pressure distribution, wall temperatures, and methane/oxygen flow rates. References [1,2,6–8] give the full information about the experimental data available.

3. Numerical Setup

The numerical simulations were carried out via the Ansys CFX [33] three-dimensional coupled algebraic multigrid solver. The Favre-averaged equations were solved in a steady-state setting.

3.1. Simulation Domain

The domain chosen for the computations resolved a 60° sector of the combustor, which included one peripheral injector and 1/6 of the central injector. To account for the velocity profile at the injection point, the injector grid domains were also included. As described in Figure 2, the symmetry boundaries were applied to the planes corresponding to ±30° from the injector's radial position. This symmetrical approach has already been widely used to study rocket combustors numerically, both in this particular case and in others [3–8]. The experimental mass flows and temperatures were set for the propellant inlets and an "opening" type for the nozzle orifice boundary condition was set. At the thruster wall, the non-slip condition with the experimental temperature profile shown in Figure 3 was reproduced. A no-slip adiabatic approach was applied to other walls. Due to the previously detected ambiguities in the nozzle heat flux estimations and the need to account for the conjugate heat transfer [6–8], no comparison between the experimental and CFD data was applied. The nozzle temperature was set equal to the last value of the temperature profile in the cylindrical part. The fuel and products were assumed as ideal gases, whereas the kinetic theory based transport properties were used for O_2 and CH_4.

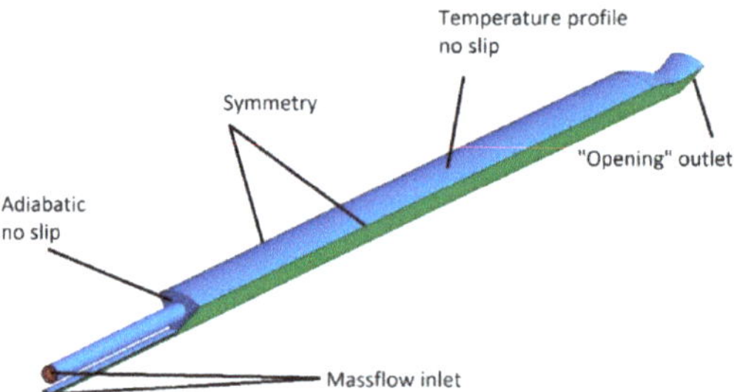

Figure 2. Domain and boundary conditions.

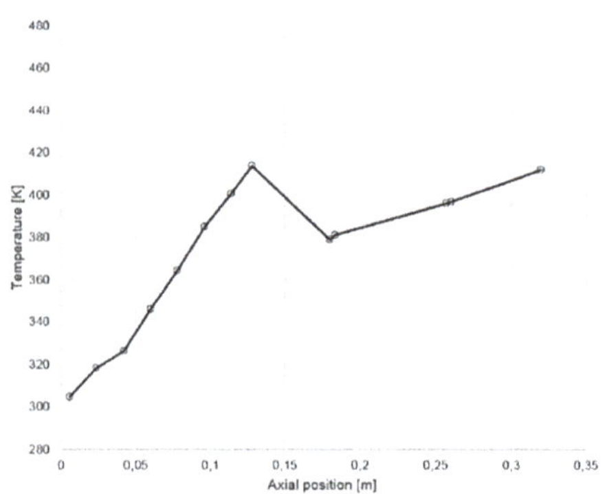

Figure 3. Axial wall temperature profile [1,2].

The approximate grid size of 3.3 mio. (4th mesh) excluding injectors, corresponding to a hexahedral mesh, was chosen after a convergence study outlined in Table 1, based on the maximal pressure and integral wall heat flux criteria. The relative error was calculated as in Equation (1).

$$\varepsilon = \frac{\varphi_i - \varphi_{i-1}}{\varphi_i} * 100\% \quad (1)$$

where i is the number of the grid studied and φ is the studied parameter. The adiabatic flamelet approach for combustion, shear-stress transport for turbulence, and default turbulent Prandtl combined with a P1 Weighted-Sum-of-Gray-Gases (WSGG) radiation approach were used for grid study calculations.

Table 1. Mesh independence study data.

Criterion	Coarse (1st)	2nd	3rd	4th	Fine (5th)
Num. of cells, mio.	0.8017	1.708	2.503	3.2803	6.92
Max. pressure, bar	16.38	17.32	17.59	17.65	17.72
Integral WHF, MW	0.151	0.157	0.1612	0.1621	0.1625
Relative error, max pressure %	—/—	5.43	1.53	0.34	0.40
Relative error, integral WHF %	—/—	3.82	2.61	0.56	0.25

One of the main issues of the current study was to check if the general wall function approach can be coupled with RANS for the wall heat flux estimation, which would make the optimization routine computations in the industry faster and easier. Thus, a mesh corresponding to the non-dimensional wall distance values of $y^+ \approx 45 \ldots 410$ was applied, making use of the default wall-function based near-wall treatment of the numerical models. In Figure 4, a front view of the mesh is shown.

Figure 4. Mesh front view.

3.2. Numerical Models

As for the turbulent Prandtl number, mostly variations of the constant values or use of the differential models are present in the combustion-related studies [13–19] among the observed papers. The second type of approach can be numerically expensive in some cases, and so this was not included in our study. The first, in contrast, does not account for turbulent diffusion effects variation along with the computational domain, which can be crucial for the variations of the geometry and operational point during optimization and for the development of a relatively universal approach for turbulent diffusion coefficients. Thus, two algebraic models were implemented and compared to three constant values of turbulent Prandtl: 0.3, 0.6, and 0.9. These two models, namely the Wassel–Catton (WC) (2) and the Kays–Crawford (KC) (3), had been previously studied by D. Yoder [34–36] and showed satisfactory performance (compared to the differential models studied) for near-wall, jet, and pipe flows. Additionally, both models can be considered as promising, as they are faster and more robust than the differential ones, while still accounting for the locally variable turbulent Prandtl.

$$Pr_t = \frac{C_3}{C_1 Pr}[1 - \exp\left(\frac{-C_4}{\mu_t/\mu_l}\right)]\left[1 - \exp\left(\frac{-C_2}{\left(\frac{\mu_t}{\mu_l}\right)Pr}\right)\right]^{-1} \tag{2}$$

$$Pr_t = \left\{\frac{1}{2Pr_{T\infty}} + \frac{CPe_t}{\sqrt{Pr_{T\infty}}} - (CPe_t)^2\left[1 - \exp\left(\frac{-1}{CPe_t\sqrt{Pr_{T\infty}}}\right)\right]\right\}^{-1} \tag{3}$$

where C_1 = 0.21; C_2= 5.25; C_3 = 0.2; C_4 = 5; Pr is the molecular Prandtl number; μ_t is the eddy viscosity; μ_l is the dynamic viscosity; C = 0.3; $Pr_{T\infty}$ = 0.85; and $Pe_t = \frac{\mu_t}{\mu_l}Pr$ is the turbulent Peclet number.

The high and non-homogenous temperature values inside the rocket combustors compel researchers to account for radiative heat transfer. Here, to study its influence and the impact of a spectral modeling approach, several cases were numerically compared: the non-radiative and two radiative ones. The radiative simulations were done with the Gray spectral model and the WSGG method. The P1 differential model was used due to its applicability to a wide range of optical thicknesses and its robust behavior during simulations. No scattering was considered in this study as no particles were introduced; studies of radiation heat transfer without scattering have previously been done for gaseous fuels and have given relevant results [20]. The default Gray spectral model was used, whereas the implemented WSGG refers to the paper by Centeno et al. [37]. In this paper, the authors introduced a four-gray-gases WSGG model with weighting factors for each gray gas derived as a polynomial function of temperature. The absorption coefficients depend upon the partial pressures of most radiating species, H_2O and CO_2. For the absorbing and emitting, non-scattering case, the radiation transfer equation would have the form [33]:

$$\frac{dI_\upsilon(\boldsymbol{r},\boldsymbol{s})}{ds} = -K_a * I_\upsilon(\boldsymbol{r},\boldsymbol{s}) + K_{a\upsilon} * I_b(\upsilon, T) \tag{4}$$

where υ is the frequency; $\boldsymbol{r}$ is the position vector; $\boldsymbol{s}$ is the direction vector; s is the path length; K_a is the absorption coefficient; $K_{a\upsilon}$ is the blackbody absorption coefficient; I_υ is the spectral radiation intensity; and I_b is the blackbody emission intensity. The weighting factors were determined by the relation below:

$$a_j(T) = \sum_{i=1}^{5} b_{j,i}T^{i-1} \tag{5}$$

where $b_{j,i}$ are the polynomial coefficients taken from reference [37]. The absorption coefficients were introduced by the equation:

$$k_j = \left(p_{H_2O} + p_{CO_2}\right) * k_{p,j}, \tag{6}$$

where p_{CO_2} and p_{H_2O} were CO_2 and H_2O partial pressures, respectively, and $k_{p,j}$ was taken from [37].

It is important to note that nevertheless, the WSGG coefficients were obtained for constant p_{H_2O} to p_{CO_2} ratio ($\frac{p_{H_2O}}{p_{CO_2}} = 2$) and the dependence of the absorption coefficient upon the partial pressures of species where the concentrations are derived locally, make this approach applicable for non-homogenous combustion studies. The comparison of results with the non-homogenous ratio based constants simulation gave good agreement and the constant ratio approach was recommended by the developers to save computational resources. In general, this model can be enhanced or revised to accomplish best fit to the test measurements. The Gray spectral model instead assumes that the radiative properties are equal for each gas. Another important point is the setting of the wall emissivity, as this value identifies the amount of radiative heat absorbed and reflected by the wall. Though it is known that the inner part of the chamber consisted of a CuCrZr alloy [7], the exact absorbing-emitting properties could not be obtained in the experiments. Furthermore, the absorbing-emitting properties can vary along the thruster during testing because of the temperature differences and the formation of oxides on the surface. Therefore, three values of wall emissivity were implied to get a sensitivity picture for the final estimation of radiative heat flux impact. Additionally, in some simulations, a unity emissivity value was set for both spectral models to get a higher limit of radiative wall heat flux. During all studies, a unity wall radiation diffuse fraction was set, which means that the reflected energy is spread diffusely.

In this paper, two general approaches were used to model chemical reactions. One is the adiabatic flamelet approach, whose underlying theory follows the diffusion laminar flame basis coupled with the turbulent field by a presumed probability density function and is now a widely applied method for combustion simulation [33,38,39]. Its main advantage is high computational performance while still considering full kinetics due to only two scalars to transport: the mixture fraction and mixture fraction variance. The turbulence–chemistry non-equilibrium interaction was accounted for by the scalar dissipation rate. The C1 CH_4/O_2 kinetics was used to create a flamelet database in CFX-RIF software.

Another combustion model used was an enhanced eddy-dissipation based approach [33,40]. The global reaction includes several species—OH, CO, H, H_2, O, O_2, CO_2, CH_4, H_2O. The procedure to derive the global reaction equation is as follows: (a) a thermodynamic 1-dimensional calculation is done in a gas-thermodynamic code (here a rocket propulsion analysis software [41]); (b) the species with the largest molar fractions are taken and molar fractions are used as the initial stoichiometric coefficients for the global reaction, (c) the left-side (CH_4 and O_2) and the right-side coefficients (for H or others) are changed to satisfy the molar balance. Due to only one global reaction, it remains fast and relatively robust while still accounting for high-temperature dissociation and the production of most participating species. The final global reaction is presented below (Equation (7)).

$$\begin{gathered}0.2789274\,[CH_4] + 0.5576753\,[O_2] = 0.147891\,[CO] + 0.131036\,[CO_2] + 0.028828\,[H] + \\ 0.0597107\,[H_2] + 0.433845\,[H_2O] + 0.0260146\,[O] + 0.072879\,[O_2] \\ +0.0997694\,[OH]\end{gathered} \tag{7}$$

4. Results and Discussion

4.1. Combustion Model Effect

The impact of two combustion models was studied using the SST turbulence approach, default turbulent Pr_t 0.9 while no radiation was accounted for. Change of the approach between the adiabatic flamelet and the enhanced eddy-dissipation did not have a high influence on the normalized pressure distribution, while it significantly impacted the absolute values (Figures 5 and 6). The flamelet approach gave the highest absolute pressures, which were around 0.7 bar (or <3.5%) lower than the experimental ones. The enhanced eddy dissipation model (EDM) approach with the global kinetics gave 1.5 bar (or 7.9%) less than the experimental pressures on average. Such low pressure cannot be addressed to wall heat loss: the EDM approach provides only slightly higher integral heat flux than flamelet, and the overall inconsistency between flamelet and EDM is more quantitative than qualitative (Figure 7). It is likely that the underestimation of heat flux (HF) is either due to unconsidered radiative heat transfer in this section, which would contribute into the total amount of heat flux, or poor performance of the turbulence model, which is discussed later. Another reason for wall heat flux inaccuracy might be the mechanism used for flamelet library formation, which was, however, not the focus of the study in the present paper. Additionally, the poor performance of the EDM is supposed to be due to the simplicity of the model formulation. One alternative for better accuracy could be the eddy-dissipation concept approach, which is now used for more and more reactive flow modeling studies. However, it is still not always as fast and robust as the mixture fraction type of models, which means that it can be hard to apply in some routine engineering simulations and therefore was not observed in the paper. As the library used in the flamelet approach accounts for many reactions compared to the EDM approach and is also more robust and showed better results for absolute pressures in this case, it was proposed for further use among these two. However, the applied eddy-dissipation methodology can still be utilized for the early stages of the design process.

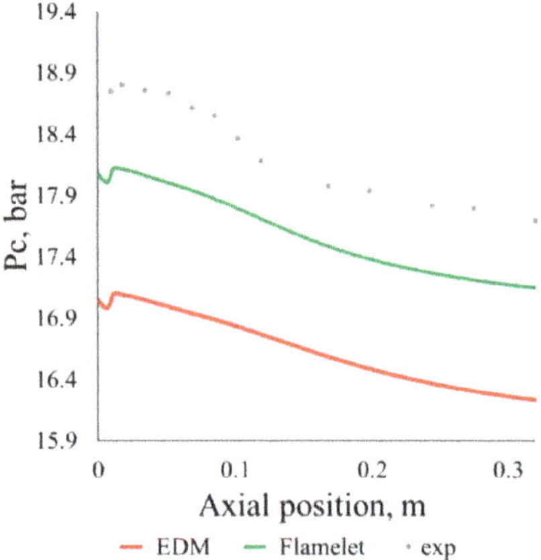

Figure 5. Pressure axial distribution.

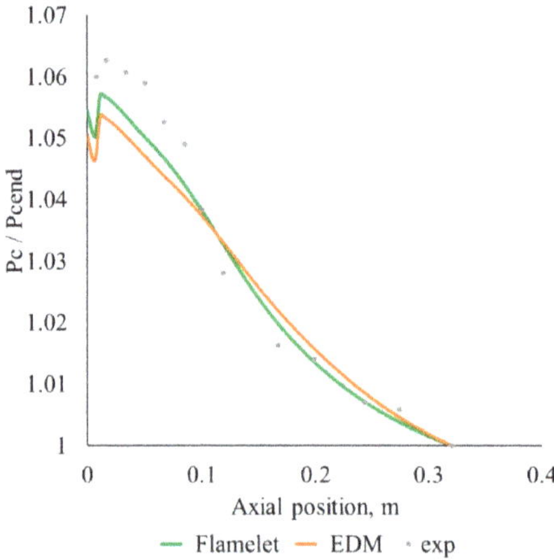

Figure 6. Normalized axial pressure distribution.

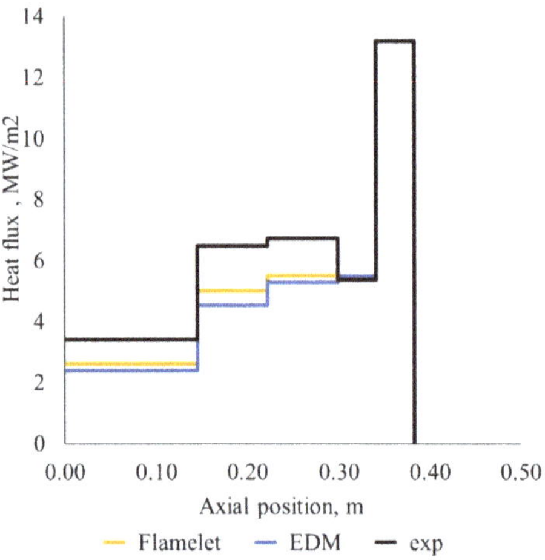

Figure 7. Segment averaged heat flux density distribution.

4.2. Radiative Heat Transfer Modeling

As already mentioned, the Gray and the WSGG spectral models were used to account for radiative heat flux (RHF). A flamelet combustion model, default turbulent Prandtl number (0.9), and the SST model were used in this study. The wall emissivity for the first estimations was set to 1, which means that the full amount of RHF emitted in the flow was absorbed by the wall. The normalized pressure plot showed no effect of RHF, nor with the Gray or WSGG approach, on the axial distributions. In contrast, inclusion of RHF changed the absolute pressure values by 0.25 bar and by 0.5 bar for the WSGG and the Gray spectral models, respectively. This was due to the shift of heat loss from the flame front and transport to the wall, which is shown on the heat flux plot (Figures 8–10). A notable thing is that the Gray model gave integral heat fluxes overestimated by 12.5%, whereas a no-radiation case led to underestimating it up to 15%. Some kind of compromise is found by the WSGG approach, which gives an integral HF value that is very close to the experimental, though the profile has some discrepancies in the first and last segments (Table 2). The overestimation of the heat flux by the Gray approach can also be seen from the radiative/total heat flux ratio. This value reached 32% for the Gray model, while only 20% was given by the WSGG. This value also corresponds to the amount of RHF that the authors [37] used in their verification study. Thellmann in his thesis [20] pointed out that the radiative amount of heat flux calculated with the WSGG type of models would be around 9–10% when estimating it for the space shuttle main engine (SSME) nozzle.

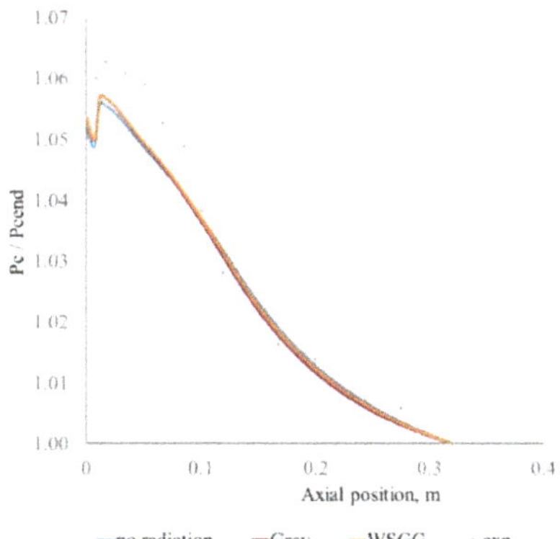

Figure 8. Normalized axial pressure distribution.

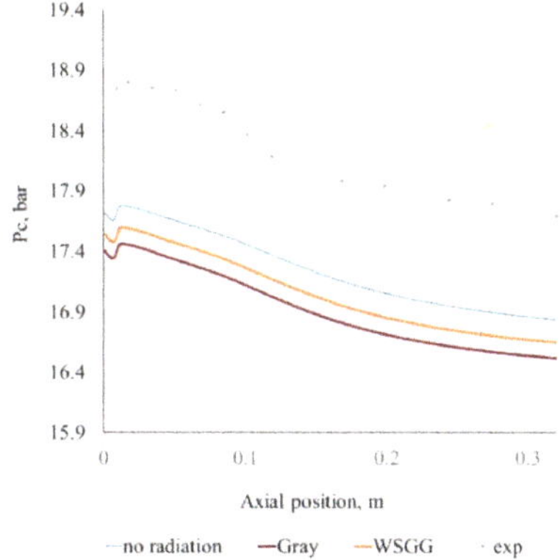

Figure 9. Absolute pressure axial distribution.

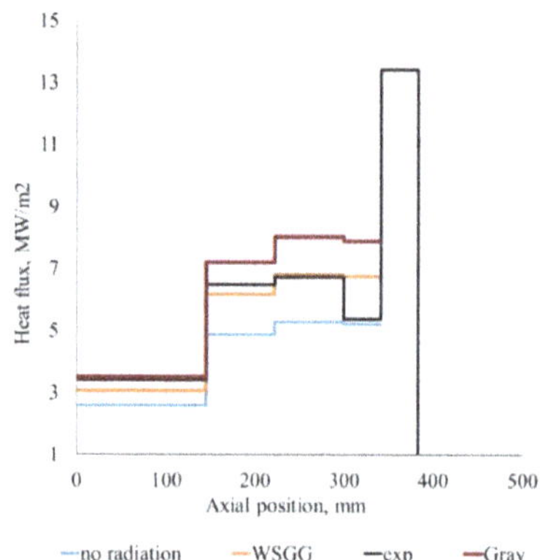

Figure 10. Segment average heat flux distribution.

Table 2. Influence of the radiative heat transfer spectral approach.

	Radiation Model			
	No Rad	**Gray**	**WSGG 4 Gases**	**Exp**
Integral, MW	0.1293	0.193	0.16542	0.16
Radiative/Total	—	0.3221	0.208	—

On the other hand, following the considerations presented in the paper by Leccese et al. [21], the radiative/total wall heat flux approximate ratio for such pressures and chamber diameter should fall in the range between 10 and 20%. It is also important to mention that the wall emissivity might be significant in determining the heat absorbed by the wall. In this sense, as no specific value of emissivity is known, it is worth providing a sensitivity study for the emissivity. Such study for three values of wall emissivity is discussed further in Section 4.5.

Despite the fact that the relative amount of RHF is still questionable, the WSGG model gives a better fit than the unphysical Gray spectral model and should be used for further simulations. It is necessary to note that future studies should be focused on the comparison of present results with: (a) usage of the same modeling approaches implying low-Reynolds grid wall resolution; and (b) computations using higher-fidelity combustion and turbulence-chemistry interaction models. Both would lead to better estimation of the low-temperature and near-wall effects, which may contribute to the combustion efficiency and thermal emission. These could give valuable information on the behavior and the modeling limits of the "engineering-oriented" methods, which are considered in the present paper.

The inclusion of the RHF decreases the flame temperature, regardless of the spectral model used and does not affect the flame shape. The WSGG model not only predicts much lower heat flux values, but also the most emitting area is sufficiently smaller. The radiation intensity fields also show that the radiative emission is highest in the second half of the cylindrical part for both Gray and WSGG cases while the WSGG showed some variation in the radial direction starting at the half-radius. This indicates that the radiation intensity is mainly dependent on the axial coordinate, is most intensive in the core, and is not that sensitive to the near-wall concentration fields. However, it needs to be studied in future if the effect is maintained for other geometries, mass flows, and pair of fuels. These effects derive from the CO_2 and H_2O mass fractions as the WSGG radiation intensity is dependent on these species' concentrations (Figures 11–17). It was also found that these effects are maintained for the enhanced EDM combustion model, different turbulence models, and turbulent Prandtl numbers. A relatively high stratification of the mass fraction fields is suggested to originate from the SST model performance, which gives lower mixing rates compared to other turbulence models, as was also found by other authors [4–6]. Through this section, the SST model was chosen for its best convergence stability.

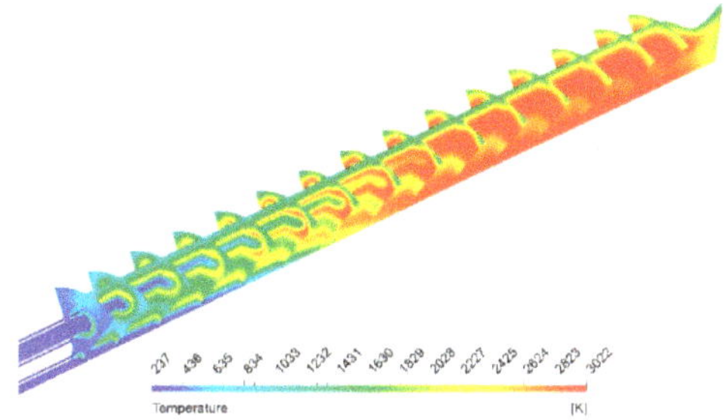

Figure 11. The no-radiating temperature field (50% axial scaling).

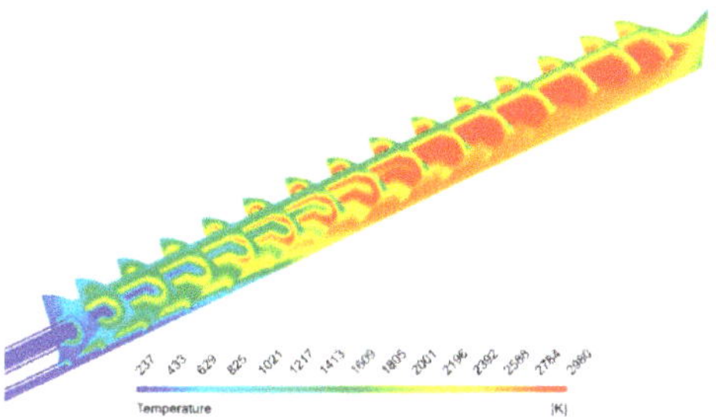

Figure 12. The P1 Gray model temperature field (50% axial scaling).

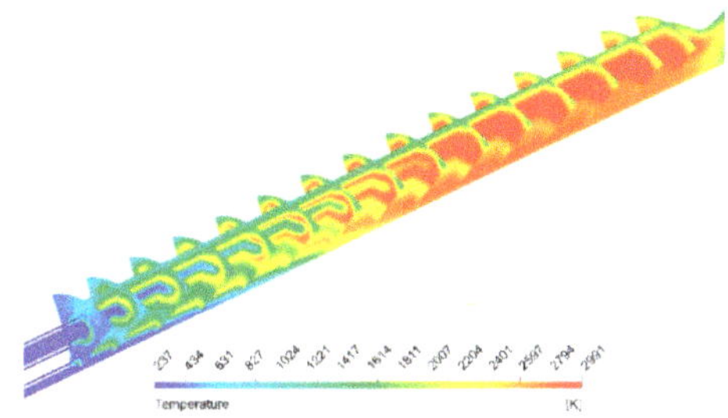

Figure 13. The P1 WSGG model temperature field (50% axial scaling).

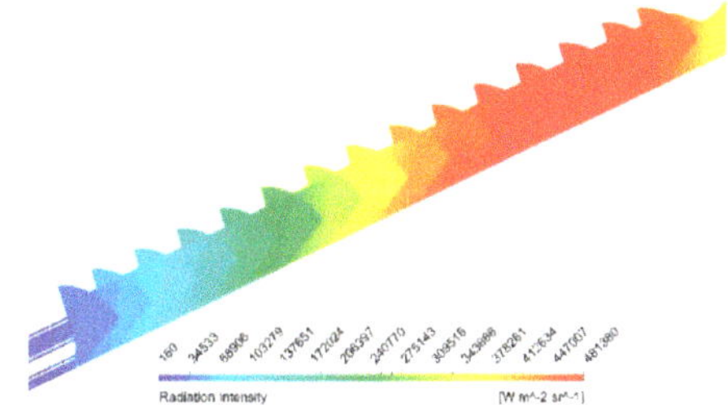

Figure 14. The Gray model radiation intensity (50% axial scaling).

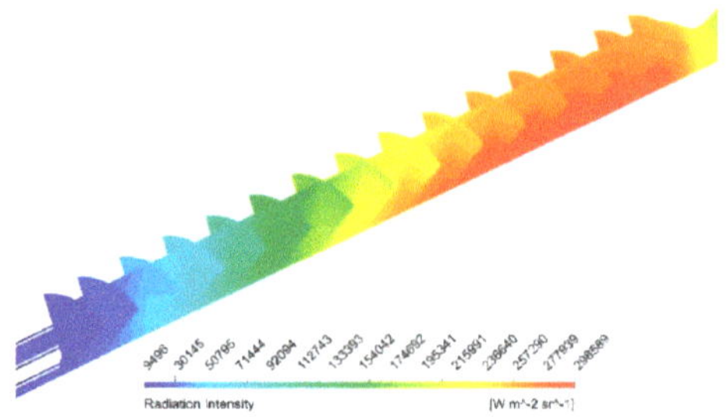

Figure 15. The WSGG model radiation intensity (50% axial scaling).

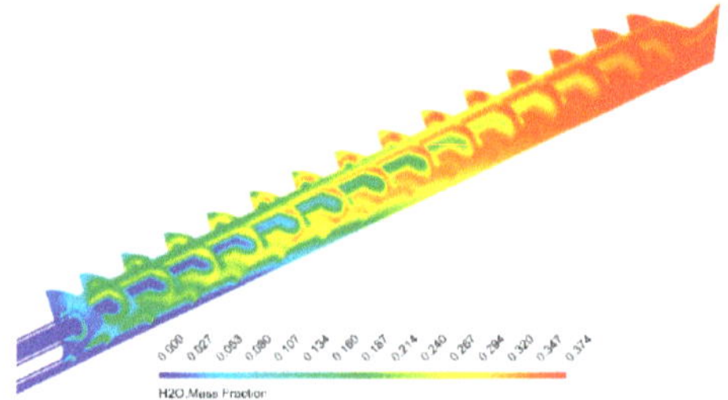

Figure 16. The H_2O mass fraction field (50% axial scaling).

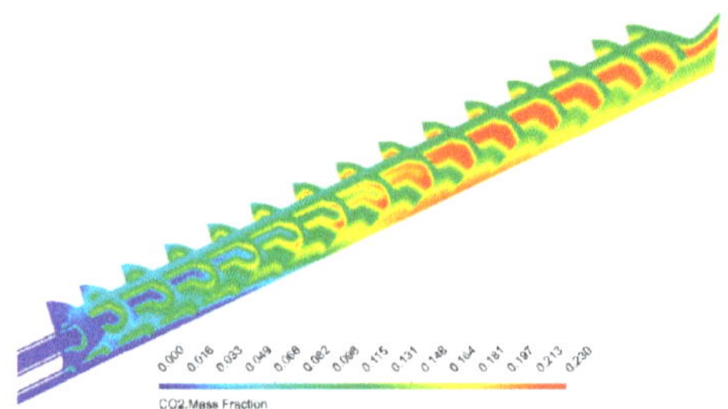

Figure 17. The CO_2 mass fraction field (50% axial scaling).

4.3. Turbulent Prandtl Number Effect

The computed average heat flux profile and pressure distribution for three constant turbulent Prandtl values and two algebraic approaches were compared to the measured values. Figures 18 and 19 shows the axial absolute and normalized pressure values for a band of models. During the study, a flamelet model for combustion, the SST turbulence model, and the P1 model with the WSGG spectral approach for radiative heat transfer were used. The unity wall emissivity was set for these simulations. The turbulent Prandtl value mostly affects the normalized lines and the lower the Pr_t, the more concave the profile along the thruster. $Pr_t = 0.9$, however, gave quite similar pressure values as the variable turbulent Prandtl models. The segment averaged heat flux profile comparison (Figure 20) also shows some interesting aspects. Heat flux increases with decreasing turbulent Prandtl and the results derived with the constant value of 0.9 were in the vicinity of the heat fluxes obtained with the algebraic variable models. Additional studies showed that this effect remains for cases without radiation modeling and for any turbulence models. The integral HF and radiative-to-total heat flux ratio (RHF/THF) outlined in Table 3 indicate this dependence; another interesting thing is that the RHF fraction increases with increasing Pr_t and at the level of 0.9, it is the same as for the algebraic Pr_t models. The increasing tendency is easily explained by the change of the convective heat flux fraction due to Prandtl increase, whereas the similarity of the Pr_t value of 0.9 with the algebraic locally variable approaches showed that this value (and values in the vicinity) is dominant in the studied thruster. These values also give the best fit for the test measurements.

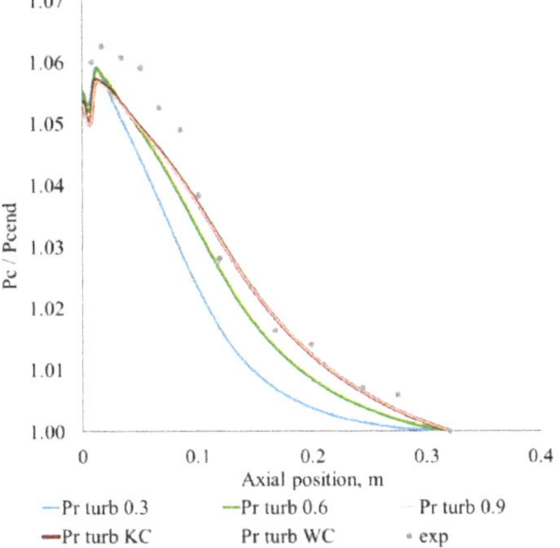

Figure 18. Axial normalized pressure.

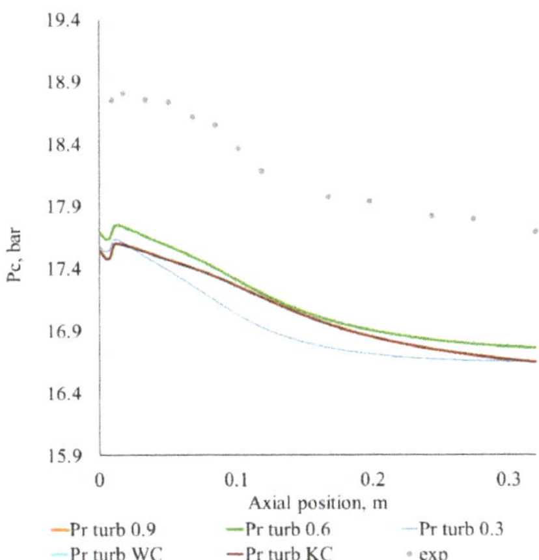

Figure 19. Absolute pressure axial distribution.

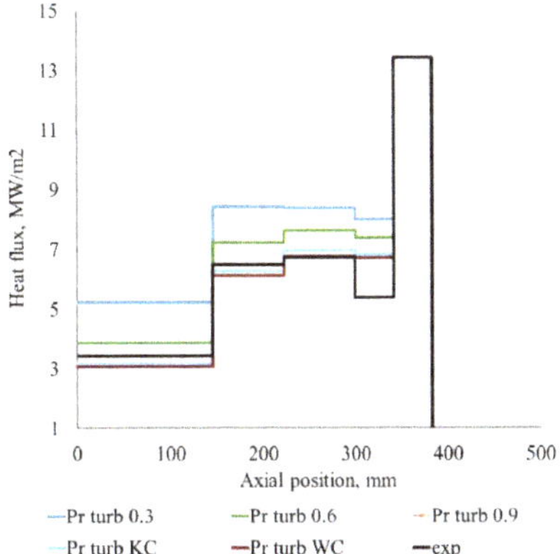

Figure 20. Segment averaged heat flux distribution.

Table 3. Influence of the turbulent Prandtl number approach.

Criterion	Turbulent Prandtl Number					
	0.3	0.6	0.9	KC	WC	Exp
Integral, MW	0.2247	0.189	0.16248	0.16542	0.1614	0.16
Radiative/Total	0.184	0.198	0.208	0.208	0.209	-

This was proved by turbulent Prandtl fields for the Kays and Crawford and Wassel and Catton formulations (Figures 21 and 22). The KC model gives the turbulent Prandtl range of 0.84–0.89 while the WC Prandtl number varied from 0.87 to 0.94. In general, these results correlate with the approximations taken by other authors [3–8] and show that the assumptions to take values in the range of 0.85–0.9 were valid. However, the variable models had large advantages over the constant values approaches—these are universal for a vast variety of injection schemes and are still robust—there had been no convergence difficulties or slowdown during simulations. Therefore, both KC and WC approaches can be recommended for engineering implementation and for future advanced research.

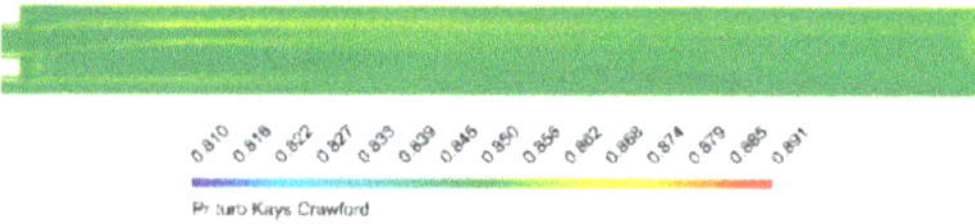

Figure 21. Pr_t field in the cylindrical part given by the Kays and Crawford model.

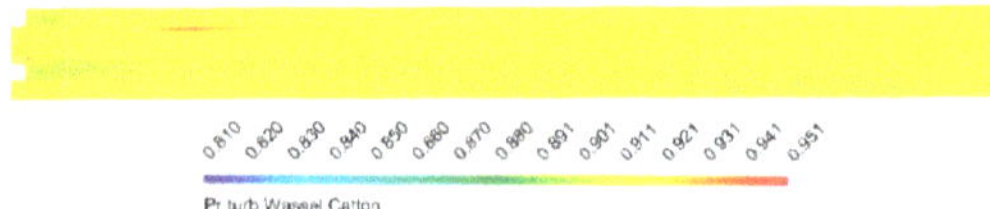

Figure 22. Pr_t field in the cylindrical part given by the Wassel and Catton model.

4.4. Turbulence Modeling Study

The study of turbulence model influence was done with the following modeling approaches fixed for all calculations: a flamelet combustion model, a default turbulent Prandtl number (0.9) to minimize the possible convergence issues, and radiation was neglected in these simulations. The three turbulence models studied were the k–ω SST, a k–ε, and the BSL EARSM model to account for turbulence anisotropy. The SST and BSL EARSM used a default automatic wall treatment approach that is based on a switch between the low-Reynolds and high-Reynolds modeling, while the k–ε model applied a default scalable wall function [33]. However, as it above-mentioned, in this case, the first grid node was intentionally located in the logarithmic sublayer, therefore all models used a default wall function approach implemented in CFX. Initially, the k–ω based BaseLine Reynolds stress model (BSL RSM) and the k–ε based Speziale–Sarkar–Gatski (SSG) Reynolds stress models were also planned to be studied, but the SSG model showed highly unstable convergence behavior and made it challenging to get the converged data in a reasonable amount of time. The BSL RSM simulations, in contrast, converged smoothly but resulted in one order of magnitude unphysically higher eddy viscosity for the central injector, which had not been noticed for other turbulence models (even for the partially converged SSG case). The cause of this is still under investigation, but the probable reason is the symmetry boundary conditions for the 1/6 sector of the central injector or ambiguity in the turbulence boundary conditions. In future, periodic boundary conditions should be applied and a study of boundary conditions should be planned to explore this behavior.

The normalized pressure profiles (Figure 23) showed a better prediction of the pressure peak in the first 100 mm after the injector by the SST model. The k–ε and BSL EARSM models similarly calculated the absolute pressure, whereas the SST model gave 0.2 bar less pressure on average. This behavior correlates with the results of other authors [3–8] as the k–ε types of models usually showed better mixing and higher pressures. Here, this effect was also addressed by the higher eddy viscosity given by the k–ε model and thus more intensive mixing. The BSL EARSM model showed even higher eddy viscosities in the near-injector region and in the back of the cylindrical part, which resulted in augmented pressure in these areas. The BSL EARSM gave higher heat fluxes in the first segment and slightly fewer values in the middle. The heat production by the SST model was the smallest among all described models (Figures 24–28). Giving both more convenient results for the pressure and quite a nice fit for the heat flux among other models, the BSL EARSM model was recommended for further use.

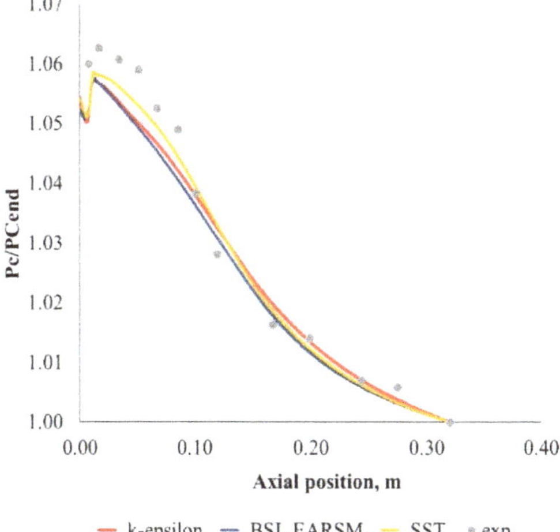

Figure 23. Normalized axial pressure distribution.

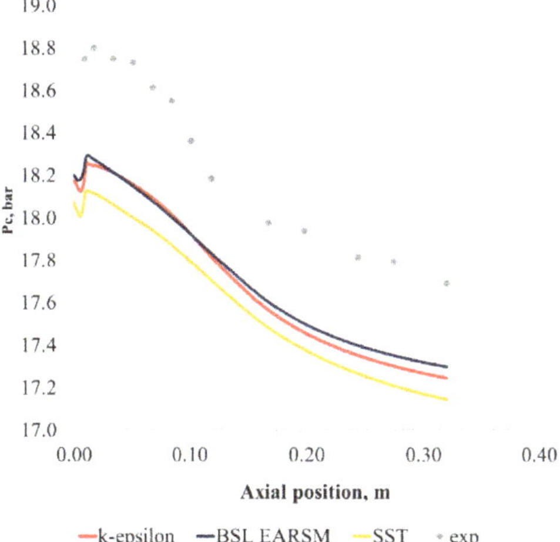

Figure 24. Absolute pressure distribution.

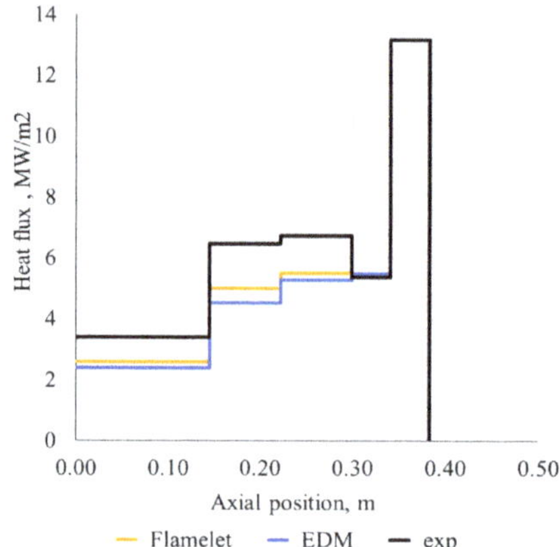

Figure 25. Segment averaged heat flux distribution.

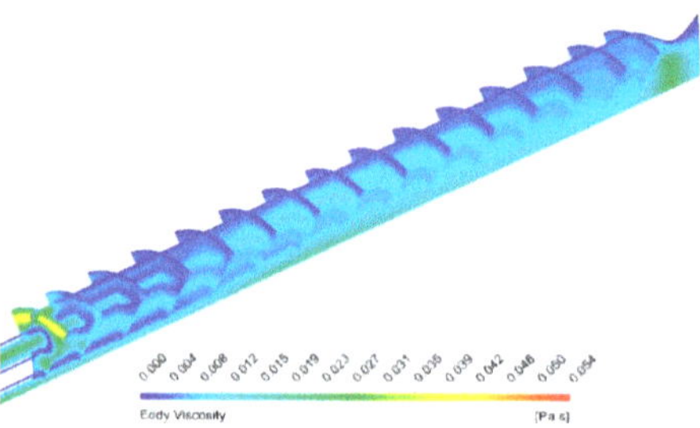

Figure 26. k–ε model eddy viscosity distribution (50% axial scaling).

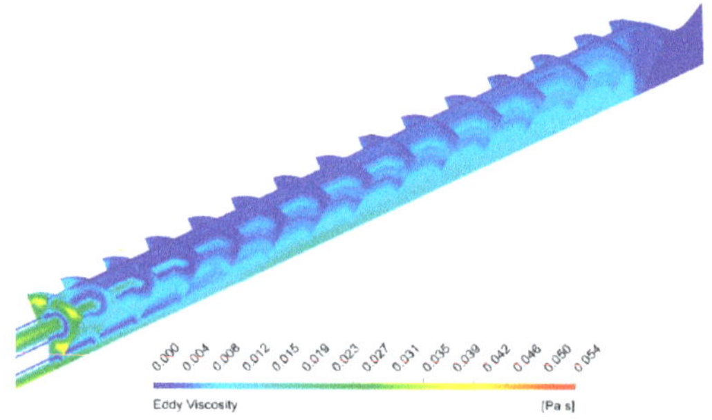

Figure 27. SST model eddy viscosity distribution (50% axial scaling).

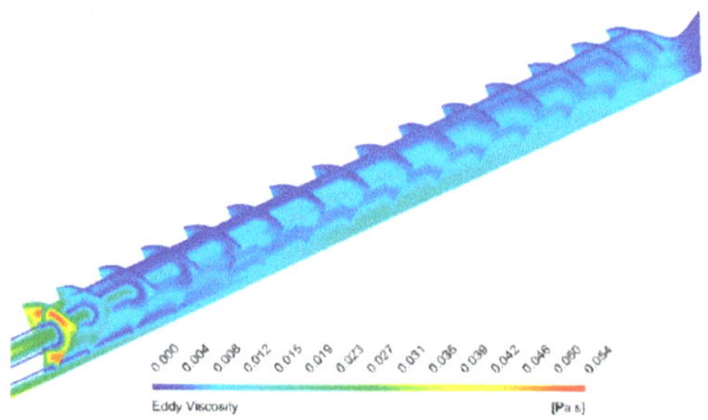

Figure 28. BSL EARSM model eddy viscosity distribution (50% axial scaling).

4.5. General Considerations

As was shown above, the BSL EARSM for turbulence and adiabatic flamelet for combustion performed best among the studied approaches. In this section, these models were combined with the P1 WSGG radiative transport and Wassel–Catton model for Pr_t. However, the amount of radiative heat flux impinged on walls can be very sensitive to the absorption and emission. This because during a firing test, the initial emissivity of the wall surface is not exactly known and may change along both the axis and in time due to varying temperatures and the formation of oxides, so is a source of ambiguity in the RHF estimation. Therefore, a sensitivity study for three values of wall emissivity was provided of ε = 0.1, 0.5, and 1. The results of the described cases coupled with the non-radiating simulations are presented below (Figures 29–31) and in Table 4.

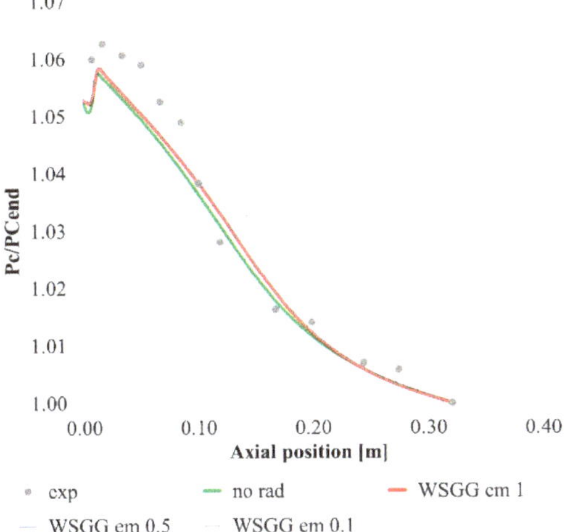

Figure 29. Normalized axial pressure distribution.

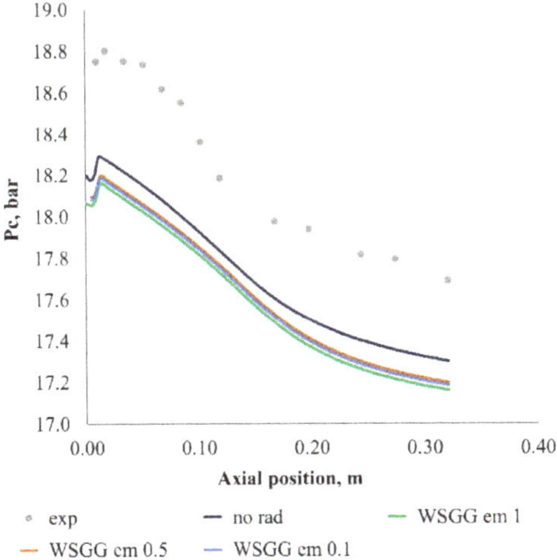

Figure 30. Pressure axial distribution.

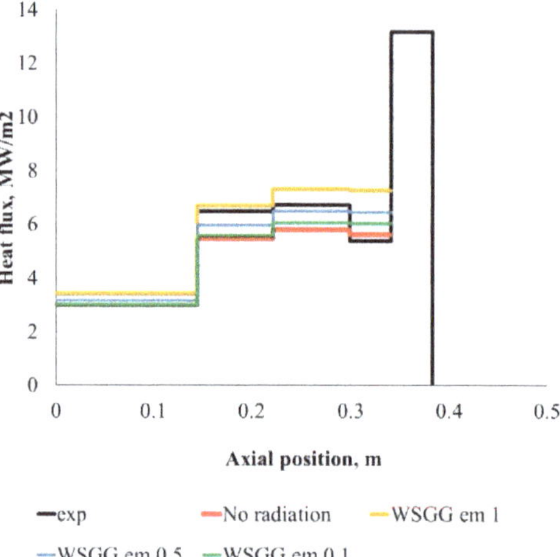

Figure 31. Segment average heat flux distribution.

Table 4. Influence of radiation and wall emissivity.

Criterion	Wall Emissivity				
	0.1	**0.5**	**1**	**No Radiation**	**Exp**
Integral, MW	0.1494	0.159060	0.177240	0.144780	0.16
Radiative/Total	0.0187	0.077	0.174	—	—

Again, radiation introduced no changes to normalized pressures. However, the RHF inclusion with $\varepsilon = 1$ lowered absolute pressures by around 0.3 bar, which also resulted in higher average and integral heat flux compared to the convective case. Due to a lower amount of energy absorbed by the wall when $\varepsilon = 0.1$ and $\varepsilon = 0.5$, those cases gave slightly higher pressures. The $\varepsilon = 1$ radiating case overestimated heat fluxes in the third and fourth segments, while the results also showed the high influence of the emissivity on wall RHF. The best fit with the experiment, both for the distribution and the integral values were given by $\varepsilon = 0.5$, which gave a RHF/THF ratio of about 7.7%. This value also correlated to estimations given by Thellmann [20] for the $CH_4 - O_2$ case. It should be noted again that the real value of emissivity is not known, and even for the known wall material, emissivity varied widely depending on the oxide formation, which is a function of temperature, composition, and therefore axial and tangential position. Thus, only sensitivity reproducing values could be taken and additional studies are needed to establish the true values. However, it can be drawn that the WSGG spectral approach still gives more physical values than the Gray model. All simulations overestimated the experimentally determined values in the fourth segment. Perakis et al. [6] showed that for the fourth and the nozzle segments, the experimental heat flux estimation had discrepancies between the coupled heat transfer calculations and the calorimetric method. This was attributed to a small deficiency of the experimental estimation. The overestimated heat flux values in the first, second, and third segments, however, were due to numerical errors in the present study. Overall, application of the BSL EARSM turbulence model and inclusion of the WSGG model with the average value of $\varepsilon = 0.5$ gave the best representation of the test results among the presented simulations.

During the numerical study, it was found that the applied eddy-viscosity based locally variable turbulent Prandtl models showed similar average values for the Kays and Crawford the turbulent Prandtl varied in the vicinity of 0.84–0.89, whereas for the Wassel and Catton, it was between 0.87–0.93. These values and the behavior of the results given were also close to those derived using the constant Pr_t 0.9. This corresponds to the approximations of other studies [3–8] and is considered to have a good fit with the experimental data. As already mentioned, the BSL EARSM showed the most physical behavior in the near-injector region and best agreement with the test data both for the pressures and the heat flux. This can be attributed to the nonlinear term present in the definition of the turbulent stresses, and therefore better representation of anisotropy in the near-wall and mixing areas. This is a question for further research and comparison with some other RSM models in the future. The k–ω shear-stress transport and k–ε models showed behavior similar to other studies [3–8], where the SST model provided less mixing intensity and therefore lower pressures.

The flamelet combustion model was both good at the representation of the pressure fields and the heat fluxes and gave robust and fast performance compared to the enhanced eddy-dissipation method used. The enhanced global reaction eddy-dissipation model gave an error of 8% in pressure, which is acceptable for first engineering estimations, but is still too coarse for the detailed design of a combustor.

The study also showed an overestimation of heat flux using the Gray spectral approximation. This can be improved by the application of the WSGG type of spectral approaches such as the one used in this study. The implementation of the four-gases WSGG approach with wall emissivity equal to unity gave a radiative/total heat flux ratio of about 17–20%, which is more reasonable than the 30% provided by the Gray approach. The wall emissivity average value showed high influence on the wall RHF, whereas the pressure distributions differed insignificantly. Although the used WSGG needs further verification studies and perhaps the inclusion of more gray gases to reproduce thin effects, it can be recommended for rocket propulsion coupled with at least approximate data for the wall emissivity coefficients.

Although some exact recommendations for the use of numerical models of the particular physical phenomena can be given as a result of the current study, additional research is needed to generalize the model choice for different environments (e.g., different mass flows or geometry). However, using the variable turbulent Prandtl approximations studied can be a source for generalization compared to the constant turbulent Prandtl as they include the effect of turbulent viscosity, which is a flow property, and originating from a turbulence model, is defined mostly by the geometry and injection parameters. The combined behavior of the turbulence models and the turbulent Prandtl approaches is anyway a point for further research.

It should also be especially mentioned that during the study, some results of previously observed papers were reaffirmed. For example, the difference between the numerical and experimental absolute pressures about 0.8 bar corresponded to the results presented by authors in other research papers [3–8] with differences varying from 0.4–1 bar approximately. Additionally, the comparative behavior of the k–epsilon and k–omega SST models showed similar performance concerning the pressure fields and species stratification. Furthermore, the influence of the turbulent Prandtl on the flowfield and resulting pressure and HF distributions was also as high as some other propulsion study papers have shown [13–19]. There were also differences compared to the mentioned papers, for example, the resulting absolute pressures for the k–epsilon and SST models were lower than those outlined by Perakis et al. [6,7]. This can be attributed to the wall function approach used in our study compared to the full boundary-layer resolving approach presented in this paper and to other minor differences between the setups. This outcome needs further extensive study.

In general, the results presented in the paper can contributed to the design and optimization of rocket propulsion devices in different ways. First, the exploration of the BSL EARSM (which keeps being a preferable choice compared to common RSM models with regard to calculation resources while still accounting for anisotropy) behavior and its possible performance in other propulsion environment can lower the time demands for the simulations while performing even better than some of the commonly used models studied here. Second, the application of the algebraic variable turbulent Prandtl approaches makes it possible to account for the geometry and injection parameters in the turbulent diffusion phenomena, which is very important for massive datasets of design data on the early design. The effect of radiation on the wall heat flux presented in the paper once again points out the importance of accounting for the radiative transfer in propulsion systems, which is not very resourceful compared to the many different physical phenomena encountered. It also notes that the WSGG type of models combined with the P1 model could be a prospective choice for such applications. All these outcomes can also be useful for everyday routine optimization simulations in the propulsion device design as, at least in the present study, they provide reasonable accuracy coupled with robustness and computational demands corresponding to such objectives.

5. Conclusions

Several 3D Favre-averaged Navier–Stokes simulations of a seven-element rocket thruster operating on GOX/GCH4 were carried out. During these simulations, a comparative study was produced for turbulence and combustion modeling approaches as well as for turbulent Prandtl approximations.

Both the Wassel–Catton and Kays–Crawford turbulent Prandtl approaches provided physical and reasonable results and did not result in any convergence difficulties or high resource requirements and can thus be recommended for future industrial simulations or numerical studies as they are suitable for any injection and mixing types.

The comparison of the combustion simulation methods showed that the adiabatic flamelet libraries generated using the C1 skeletal mechanism are more relevant for the present case than the EDM applied. The implied EDM approach, obviously, can be used for first approximations, but is not applicable for further comprehensive studies. The applied WSGG spectral model showed more reasonable behavior than the Gray model and is suggested for further usage and study. However, a more detailed survey of different WSGG models for methane–oxygen combustion is needed as well as a possible analysis of its performance based on other test cases with at least approximate values of wall emissivity provided from the experiment. Furthermore, usage of different radiation models (e.g., Monte-Carlo or discrete ordinates) with the WSGG type of spectral approach is suggested in future studies of propulsion devices. Another important outcome is the need to include the radiative heat flux in the simulations as the resulting wall heat flux field might be underestimated.

Finally, the combined application of the BSL EARSM, the adiabatic flamelet, the P1 radiation model with the WSGG approach with emissivity $\varepsilon = 0.5$, and the use of algebraic Pr_t models gave the best approximation of the experimental data. The resulting error in absolute pressures was around 2.9%, and the results for the wall heat flux were also sufficiently similar to the experimental ones, taking into account the discrepancies in the fourth segment. Another important finding of this research is that all these results were obtained on a relatively coarse grid having log-law area y^+ values, where default wall functions were applied. Though the reliable academic/scientific usage of wall functions combined with RANS algorithms for such applications is an object of further research, the applied and chosen approaches can be used for routine engineering optimization simulations, which would result in reasonable results for the pressures and heat flux parameters. Future work, as suggested, should concentrate on the extensive radiation modeling approaches by studying the turbulence modeling effects as well as more comprehensive analysis of the near-wall resolution influence on the resulting integral parameters.

Author Contributions: Conceptualization, methodology, writing—review and editing I.B., E.S., and O.H.; Writing—original draft preparation, investigation I.B. and E.S.; Project administration, funding acquisition, I.B. All authors have read and agreed to the published version of the manuscript.

Funding: This research was funded by the Russian Ministry of Science and Education (project FSFF-2020-0014).

Conflicts of Interest: The authors declare that they have no conflict of interest.

References

1. Silvestri, S.; Celano, M.P.; Schlieben, G.; Haidn, O.J. Characterization of a Multi-Injector Gox-Gch4 Combustion Chamber. In Proceedings of the 52nd AIAA/SAE/ASEE Joint Propulsion Conference, Salt Lake City, UT, USA, 25–27 July 2016.

2. Silvestri, S.; Kirchberger, C.; Schlieben, G.; Celano, M.P.; Haidn, O. Experimental and Numerical Investigation of a Multi-Injector GOX-GCH4 Combustion Chamber. *Trans. Jpn. Soc. Aeronaut. Space Sci. Aerosp. Technol. Jpn.* **2018**, *16*, 374–381. [CrossRef]
3. Roth, C.M.; Haidn, O.J.; Chemnitz, A.; Sattelmayer, T.; Frank, G.; Müller, H.; Zips, J.; Keller, R.; Gerlinger, P.M.; Maestro, D. Numerical investigation of flow and combustion in a single element GCH4/GOx rocket combustor. In Proceedings of the 52nd AIAA/SAE/ASEE Joint Propulsion Conference, Salt Lake City, UT, USA, 25–27 July 2016; p. 4995.
4. Chemnitz, A.; Sattelmayer, T.; Roth, C.; Haidn, O.; Daimon, Y.; Keller, R.; Gerlinger, P.; Zips, J.; Pfitzner, M. Numerical Investigation of Reacting Flow in a Methane Rocket Combustor: Turbulence Modeling. *J. Propuls. Power* **2018**, *34*, 864–877. [CrossRef]
5. Maestro, D.; Cuenot, B.; Chemnitz, A.; Sattelmayer, T.; Roth, C.; Haidn, O.J.; Daimon, Y.; Keller, R.; Gerlinger, P.M.; Frank, G.; et al. Numerical Investigation of Flow and Combustion in a Single-Element GCH4/GOX Rocket Combustor: Chemistry Modeling and Turbulence-Combustion Interaction. In Proceedings of the 52nd AIAA/SAE/ASEE Joint Propulsion Conference, Salt Lake City, UT, USA, 25–27 July 2016. [CrossRef]
6. Perakis, N.; Rahn, D.; Haidn, O.J.; Eiringhaus, D. Heat Transfer and Combustion Simulation of Seven-Element O2/CH4 Rocket Combustor. *J. Propuls. Power* **2019**, *35*, 1080–1097. [CrossRef]
7. Perakis, N.; Strauß, J.; Haidn, O.J. Heat flux evaluation in a multi-element CH_4/O_2 rocket combustor using an inverse heat transfer method. *Int. J. Heat Mass Transf.* **2019**, *142*, 118425. [CrossRef]
8. Perakis, N.; Haidn, O.J.; Eiringhaus, D.; Rahn, D.; Zhang, S.; Daimon, Y.; Karl, S.; Horchler, T. Qualitative and Quantitative Comparison of RANS Simulation Results for a 7-Element GOX/GCH4 Rocket Combustor. In Proceedings of the 54th AIAA/SAE/ASEE Joint Propulsion Conference, Cincinnati, OH, USA, 9–11 July 2018; p. 54.
9. Nasuti, F.; Concio, P.; Indelicato, G.; Lapenna, P.E.; Creta, F. Role of Combustion Modeling in the Prediction of Heat Transfer in LRE Thrust Chambers. In Proceedings of the 8th European Conference for Aeronautics and Aerospace Sciences (EUCASS), Madrid, Spain, 1–4 July 2019.
10. Zhukov, V.P. Computational Fluid Dynamics Simulations of a GO2/GH2 Single Element Combustor. *J. Propuls. Power* **2015**, *31*, 1–8. [CrossRef]
11. Zhukov, V.P.; Suslov, D.I. Measurements and modelling of wall heat fluxes in rocket combustion chamber with porous injector head. *Aerosp. Sci. Technol.* **2016**, *48*, 67–74. [CrossRef]
12. Poschner, M.; Pfitzner, M. Real Gas CFD Simulation of Supercritical H2-LOX in the MASCOTTE Single Injector Combustor Using a Commercial CFD Code. In Proceedings of the 46th AIAA Aerospace Sciences Meeting and Exhibit, Reno, Nevada, 7–10 January 2008; Volume 46. [CrossRef]
13. Molchanov, A.M.; Bykov, L.V.; Nikitin, P.V. Modeling High-Speed Reacting Flows with Variable Turbulent Prandtl and Schmidt Numbers. In Proceedings of the 9th International Conference on Heat Transfer, Fluid Mechanics and Thermodynamics, Valletta, Malta, 16–18 July 2012.
14. Yimer, I.; Campbell, I.; Jiang, L.-Y. Estimation of the Turbulent Schmidt Number from Experimental Profiles of Axial Velocity and Concentration for High-Reynolds-Number Jet Flows. *Can. Aeronaut. Space J.* **2002**, *48*, 195–200. [CrossRef]
15. He, G.; Guo, Y.; Hsu, A.T. The effect of Schmidt number on turbulent scalar mixing in a jet-in-crossflow. *Int. J. Heat Mass Transf.* **1999**, *42*, 3727–3738. [CrossRef]
16. Keistler, P.; Xiao, X.; Hassan, H.; Rodriguez, C. Simulation of Supersonic Combustion Using Variable Turbulent Prandtl/Schmidt Number Formulation. In Proceedings of the 36th AIAA Fluid Dynamics Conference and Exhibit, San Francisco, CA, USA, 5–8 June 2006.
17. Ivancic, B.; Riedmann, H.; Frey, M.; Knab, O.; Karl, S.; Hannemann, K. Investigation of Different Modeling Approaches for CFD Simulation of High Pressure Rocket Combustors. In Proceedings of the 5th European Conference for Aerospace Sciences (EUCASS), Munich, Germany, 1–5 July 2013.
18. Xiao, X.; Edwards, J.; Hassan, H.; Cutler, A. A Variable Turbulent Schmidt Number Formulation for Scramjet Application. In Proceedings of the 43rd AIAA Aerospace Sciences Meeting and Exhibit, Reno, Nevada, 10–13 January 2005. [CrossRef]
19. Goldberg, U.C.; Palaniswamy, S.; Batten, P.; Gupta, V. Variable Turbulent Schmidt and Prandtl Number Modeling. *Eng. Appl. Comput. Fluid Mech.* **2010**, *4*, 511–520. [CrossRef]

20. Thellmann, A. Impact of Gas Radiation on Viscous Flows, in Particular on Wall Heat Loads, in Hydrogen-Oxygen vs. Methane-Oxygen Systems, Based on the SSME Main Combustion Chamber. Ph.D. Thesis, Universitaet der Bundeswehr, Muenchen, Germany, 2010.
21. Leccese, G.; Bianchi, D.; Nasuti, F. Numerical Investigation on Radiative Heat Loads in Liquid Rocket Thrust Chambers. *J. Propuls. Power* **2019**, *35*, 930–943. [CrossRef]
22. Molchanov, A.M.; Bykov, L.V.; Yanyshev, D.S. Calculating thermal radiation of a vibrational nonequilibrium gas flow using the method of k-distribution. *Thermophys. Aeromech.* **2017**, *24*, 399–419. [CrossRef]
23. Mazzei, L.; Puggelli, S.; Bertini, D.; Pampaloni, D.; Andreini, A. Modelling soot production and thermal radiation for turbulent diffusion flames. *Energy Procedia* **2017**, *126*, 826–833. [CrossRef]
24. Poitou, D.; El Hafi, M.; Cuenot, B. Analysis of Radiation Modeling for Turbulent Combustion: Development of a Methodology to Couple Turbulent Combustion and Radiative Heat Transfer in LES. *J. Heat Transf.* **2011**, *133*, 062701. [CrossRef]
25. Daimon, Y.; Negishi, H.; Silvestri, S.; Haidn, O.J. Conjugated Combustion and Heat Transfer Simulation for a 7 element GOX/GCH4 Rocket Combustor. In Proceedings of the 54th AIAA/SAE/ASEE Joint Propulsion Conference, Cincinnati, OH, USA, 9–11 July 2018; pp. 2018–4553.
26. Muto, D.; Daimon, Y.; Shimizu, T.; Negishi, H. Wall modeling of reacting turbulent flow and heat transfer in liquid rocket engines. In Proceedings of the 54th AIAA/SAE/ASEE Joint Propulsion Conference, Cincinnati, OH, USA, 9–11 July 2018. [CrossRef]
27. Balaras, E.; Benocci, C.; Piomelli, U. Two-layer approximate boundary conditions for large-eddy simulations. *AIAA J.* **1996**, *34*, 1111–1119. [CrossRef]
28. Rani, S.L.; Smith, C.E.; Nix, A.C. Boundary-Layer Equation-Based Wall Model for Large-Eddy Simulation of Turbulent Flows with Wall Heat Transfer. *Numer. Heat Transfer Part B Fundam.* **2009**, *55*, 91–115. [CrossRef]
29. Kawai, S.; Larsson, J. Wall-modeling in large eddy simulation: Length scales, grid resolution, and accuracy. *Phys. Fluids* **2012**, *24*, 15105. [CrossRef]
30. Piomelli, U. Wall-layer models for large-eddy simulations. *Prog. Aerosp. Sci.* **2008**, *44*, 437–446. [CrossRef]
31. Larsson, J.; Kawai, S.; Bodart, J.; Bermejo-Moreno, I. Large eddy simulation with modeled wall-stress: Recent progress and future directions. *Mech. Eng. Rev.* **2016**, *3*, 1–23. [CrossRef]
32. Maheu, N.; Moureau, V.; Domingo, P.; Duchaine, F.; Balarac, G. Large-Eddy Simulations of flow and heat transfer around a low-Mach number turbine blade. Center for Turbulence Research. In Proceedings of the Summer Program, Stanford, CA, USA, 24 June–20 July 2012.
33. Ansys. *CFX Theory Guide*; ANSYS Inc.: Canonsburg, PA, USA, 2019; Volume 15317, p. R2.
34. Yoder, D.A. Comparison of Turbulent Thermal Diffusivity and Scalar Variance Models. In Proceedings of the 54th AIAA Aerospace Sciences Meeting, San Diego, CA, USA, 4–8 January 2016; p. 54. [CrossRef]
35. Kays, W.M.; Whitelaw, J.H. Convective Heat and Mass Transfer. *J. Appl. Mech.* **1967**, *34*, 254. [CrossRef]
36. Wassel, A.; Catton, I. Calculation of turbulent boundary layers over flat plates with different phenomenological theories of turbulence and variable turbulent Prandtl number. *Int. J. Heat Mass Transf.* **1973**, *16*, 1547–1563. [CrossRef]
37. Centeno, F.R.; Da Silva, C.V.; Brittes, R.; França, F.H.R. Numerical simulations of the radiative transfer in a 2D axisymmetric turbulent non-premixed methane–air flame using up-to-date WSGG and gray-gas models. *J. Braz. Soc. Mech. Sci. Eng.* **2015**, *37*, 1839–1850. [CrossRef]
38. Ansys. *Fluent Theory Guide*; ANSYS Inc.: Canonsburg, PA, USA, 2019; Volume 15317, p. R2.
39. Peters, N. Laminar diffusion flamelet models in non-premixed turbulent combustion. *Prog. Energy Combust. Sci.* **1984**, *10*, 319–339. [CrossRef]
40. Magnussen, B.; Hjertager, B. On Mathematical Modeling of Turbulent Combustion with Special Emphasis on Soot Formation and Combustion. *Symp. (Int.) Combust. Camb. Mass.* **1977**, *16*, 719–729. [CrossRef]
41. Rocket Propulsion Analysis, Version 2.3 User Manual. Cologne, Germany. 2017. Available online: http://www.propulsion-analysis.com/ (accessed on 5 April 2020).

Article

Thermal Analysis Strategy for Axial Permanent Magnet Coupling Combining FEM with Lumped-Parameter Thermal Network

Xikang Cheng, Wei Liu *, Ziliang Tan, Zhilong Zhou, Binchao Yu, Wenqi Wang, Yang Zhang and Sitong Liu

Key Laboratory for Precision and Non-Traditional Machining Technology of the Ministry of Education, Dalian University of Technology, Dalian 116024, China; xikangc@mail.dlut.edu.cn (X.C.); tanziliang2020@mail.dlut.edu.cn (Z.T.); zzl666@mail.dlut.edu.cn (Z.Z.); yubinchao@mail.dlut.edu.cn (B.Y.); wqwang19@mail.dlut.edu.cn (W.W.); zy2018@dlut.edu.cn (Y.Z.); liusitong@mail.dlut.edu.cn (S.L.)

* Correspondence: lw2007@dlut.edu.cn; Tel.: +86-131-3000-9890

Received: 2 September 2020; Accepted: 22 September 2020; Published: 24 September 2020

Abstract: Thermal analysis is exceptionally important for operation safety of axial permanent magnet couplings (APMCs). Combining a finite element method (FEM) with a lumped-parameter thermal network (LPTN) is an effective yet simple thermal analysis strategy for an APMC that is developed in this paper. Also, some assumptions and key considerations are firstly given before analysis. The loss, as well as the magnetic field distribution of the conductor sheet (CS) can be accurately calculated through FEM. Then, the loss treated as source node loss is introduced into the LPTN model to obtain the temperature results of APMCs, where adjusting conductivity of the CS is a necessary and significant link to complete an iterative calculation process. Compared with experiment results, this thermal analysis strategy has good consistency. In addition, a limiting and safe slip speed can be determined based on the demagnetization temperature permanent magnet (PM).

Keywords: thermal analysis; axial permanent magnet coupling (APMC); eddy current; finite element method (FEM); lumped-parameter thermal network (LPTN)

---

1. Introduction

Permanent magnet couplings (PMCs) as a new transmission topology can be employed in several industrial applications, such as conveyors, fans, pumps and braking devices [1–3]. Also, PMCs offer many advantages, such as no physical contact, soft starting, shock isolation, and misalignment tolerance, which all taken together provide better protection for mechanical systems [4,5]. Typically, there are two configurations of PMCs: axial [6–8] and radial [9–11] type, both with the above-mentioned advantages. In this work, we pay attention to axial permanent magnet coupling (APMC) as depicted in Figure 1. It is divided into two parts: one is the permanent magnet (PM) module including a PM holder and several PMs (generally Nd-Fe-B type), and the other is the conductor sheet (CS) module, generally manufactured with copper. Additionally, there are two other iron yokes to make the magnetic fluxes close corresponding the PM holder and the CS, respectively.

In [3,6,12,13], the operating principle of APMCs was introduced in detail. On account of having the slip ($s = n_{in} - n_{out}$) between the CS module and the PM module, the eddy currents, which are generated on the CS, interact with the original magnetic field of PMs and induce an effective torque. Meanwhile, the temperature rise, especially at the low-slip condition, is inevitable due to the generated-currents. Surpassing the thermal limit of PMs involves the risk of irreversible demagnetization, while the excessive temperature can also damage other components. Beside the precaution of each component destruction, intensive thermal stress can shorten equipment lifetime.

Therefore, thermal analysis is absolutely necessary both for the design stage and the monitoring stage of APMCs. Given the high-speed rotation characteristics of APMCs, analyzing the components' temperature imposes a greater challenge.

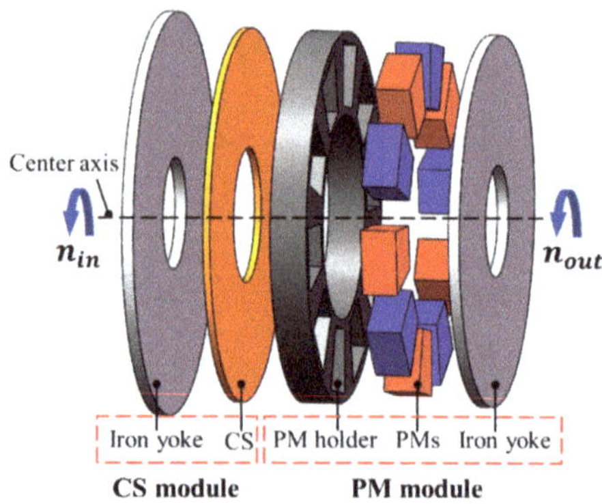

Figure 1. Configuration of an axial permanent magnet coupling.

Today, APMCs have been investigated for a long time, and a number of papers can be discovered in [14–22]. Research in [16] and [17], based on a two-dimensional (2-D) approximation of the magnetic field distribution, a practical and effective analytical calculation approach for the torque performance of APMCs, was proposed. Given the three-dimensional (3-D) edge effects and curvature effects, a 3-D analytical model was developed in [19] and [22] to compute the torque and the axial force. Obviously, all of studies mentioned above are focused on the torque analysis ignoring the thermal influence since APMCs belong to a kind of transmission device ultimately.

Unfortunately, thermal analysis to APMCs is not easy work due to the property variation of each component with temperature that result in mathematical difficulties. Hence, it is then acceptable to find very little literature about this. Here, good news is that there are some methods from PM motors to be borrowed [23–26]. In general, thermal analysis can be segmented into two primary means: numerical methods and thermal network. The former, such as finite-element method (FEM), not only can obtain the temperature distribution accurately, but also can take the actual 3D geometry and material properties into account. However, for the analysis objects, which have complex topologies or contain components equipped with highly diverse thermal characteristics, FEM is difficult to simplify [27]. Consequently, the mesh is refined and enormous, resulting in having a heavy expense in computer resources and being time-consuming, which then hampers the application of FEM in rapid optimization [28]. The latter is a lumped–parameter approach offering the advantages of economy, flexibility and simplicity. Nevertheless, it cannot acquire the temperature distribution in detail and heavily depends on the precise thermal parameters of the loss and heat coefficient (conduction, radiation and convection) [29,30]. For the aforementioned studies, however, the study methods are not fully applicable to the thermal analysis of APMCs.

From the perspective of engineering application and operation safety for APMCs, it is important to have a simplified and accurate thermal analysis strategy in order to quickly obtain the temperature distribution of APMCs. In this paper, FEM and lumped-parameter thermal network (LPTN) are combined to effectively calculate the temperature results of APMCs. The remainder of the paper is organized as follows: Section 2 presents the geometry of the studied APMC. In Section 3, the proposed strategy and assumptions are offered. The magnetic field model, including the loss results is established in Section 4. Then, Section 5 gives the LPTN model to obtain the temperature results. Finally, Section 6 concludes the work of this paper.

2. Geometry of the Studied APMC

Figure 2 provides the geometry of the studied APMC as well as its exploded view and geometrical parameters, where the pole-pairs number of the PMs is supposed to be 6. Here, l_g is adjustable and generally ranges from 3 mm to 8 mm. In the PM holder, the fan-shaped PMs are arranged in accordance with a N/S alternating sequence. The major parameters of the studied APMC are given in Table 1. These parameters are derived from some reliable engineering design experience and its detailed design thought can be referred to [3,4,6]. Given the adjustable air-gap and these parameters, the manufactured prototype of the studied APMC was built and put on the cast-iron platform, as shown in Figure 3.

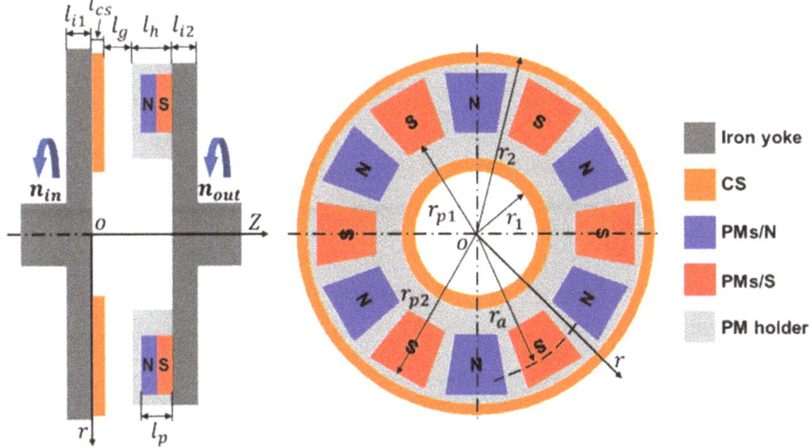

Figure 2. The geometry of the studied APMC ($p = 6$) with its exploded view and geometrical parameters.

Figure 3. The manufactured prototype of the studied APMC (l_g = 30 mm).

Table 1. Parameters of the studied APMC.

Symbol	Meaning	Value
l_{i1}	Thickness of the iron yoke (CS side)	10 mm
l_{i2}	Thickness of the iron yoke (PM side)	10 mm
l_{cs}	Thickness of the CS	6 mm
l_g	Thickness of the air-gap	3–8 mm
l_h	Thickness of the PM holder	26 mm
l_p	Thickness of the PM	25 mm
r_1	Inside radius of the CS	87.5 mm
r_2	Outside radius of the CS	187.5 mm
r_{p1}	Inside radius of the PM	115 mm
r_{p2}	Outside radius of the PM	165 mm
r_a	Average radius of the PM	140 mm
H_p	Coercive force of the PM	−900 KA/m
σ_{cs}	Conductivity of the CS	5.8×10^7 S/m (20 °C)

3. Proposed Strategy and Assumptions

3.1. Proposed Strategy

Figure 4 describes the procedure of the proposed thermal analysis strategy. Since the conductivity of the CS is closely related to temperature, this procedure is an iterative updating calculation. After initialization and referring to the parameters in Table 1, the magnetic field result is obtained as well as the loss employing FEM. Then, regarding the loss as the heat source, the LPTN model is employed to calculate the temperature distribution. Thanks to the temperature obtained each time being different, so in each iteration, the conductivity of the CS was updated according to its temperature characteristic. Based on the updated conductivity, the thermal analysis re-executed until the temperature convergence threshold is satisfied. Herein, ΔT is the percentage error of each component temperature relative to the previous calculation, and T_s is the set convergence threshold, such as 1%.

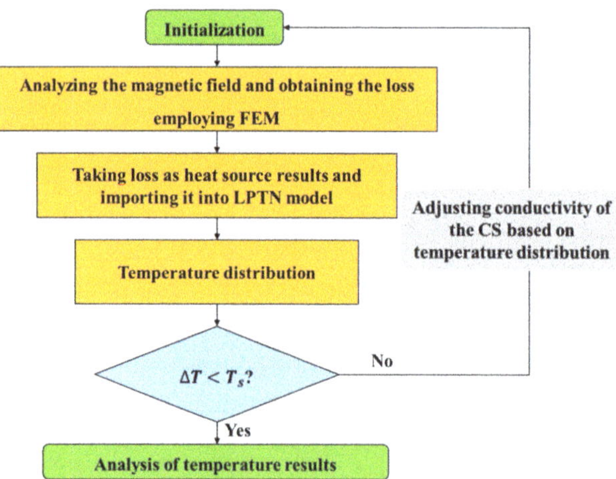

Figure 4. The procedure of the thermal analysis strategy.

3.2. Assumptions

1. From Figure 2, the geometry of the APMC is approximatively centrosymmetric. Also, the parameters of every component along the circumferential direction (r) were uniform. Therefore, the loss and temperature distributions were centrosymmetric as well.
2. The APMC operated in the steady-state with a certain slip speed (s), so it is reasonable to think the air in the air-gap as stable, whereby the temperature distribution in the air-gap was also the same.
3. Considering the skin effect, the loss is mainly concentrated upon the CS, and other losses, were ignored.

4. Magnetic Field Model

4.1. Key Considerations

In this section, we employ the FEM software Ansoft Maxwell 3D (v16, ANSYS Inc, Canonsburg, Pennsylvania, USA.) to model the magnetic field of the APMC. Assigning H_p and σ_{cs} in Table 1 to the PMs and the CS respectively is the first important consideration, where σ_{cs} is a parameter to be adjusted from Figure 4 and valued at room temperature (20 °C) initially. As important as the PM and the CS, the magnetic characteristic of the iron yokes (B-H curve) should also be considered in the FEM, as exposed in Figure 5. Additionally, Figure 6 shows the mesh size for all the components, following a certain proportion. Here, from the perspective of simulation precision, the CS and the PMs are meshed detailedly. Figure 7 shows the mesh in 3D of the FEM model, where this model includes 570,979 nodes and 119,316 elements.

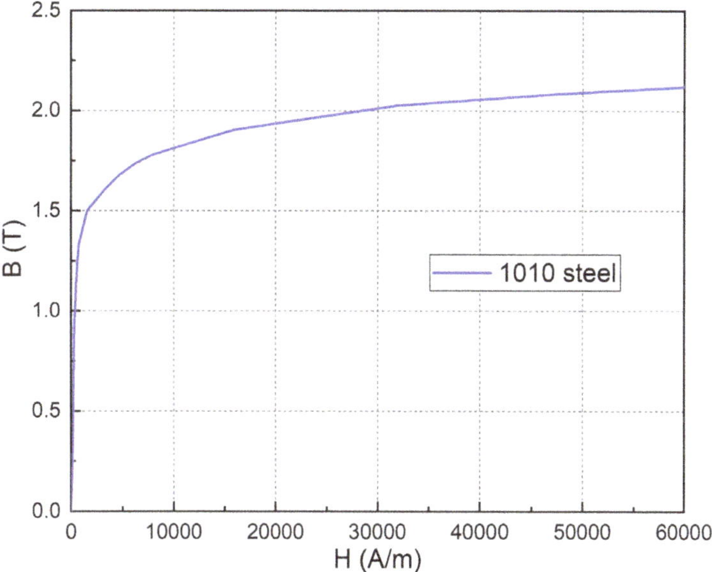

Figure 5. Nonlinear B–H curve (1010 steel) for the iron yokes.

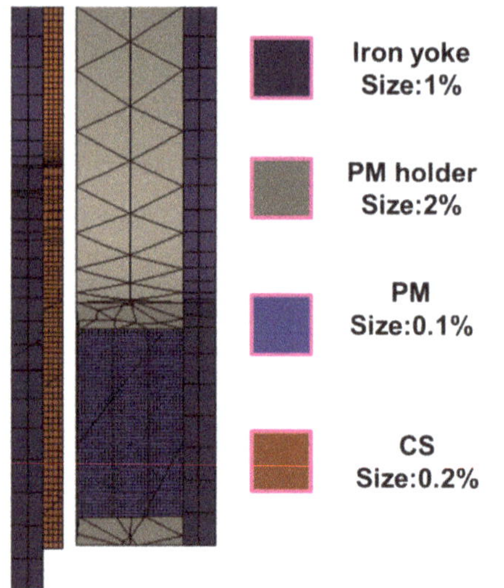

Figure 6. The mesh size for all the components used in the FEM.

Figure 7. Mesh in 3D of the FEM model.

Besides, considering the convergence of this FEM results, the simulation step can be given by

$$t_{step} = (1.2 \sim 1.5)\frac{60}{1000s} \tag{1}$$

wherein, the coefficient (1.2~1.5) indicates that the CS module has rotated 1.2 to 1.5 turns relative to the PM module.

4.2. Analysis of Magnetic Field on the CS

The magnetic flux density and the eddy current are significant for the loss analysis of the APMC. As presented in Figure 8a, the magnetic flux density distributions on the surface of the CS seen from z-axis correspond to the alternating distributions of the PMs (N/S), in which the slip speed is set to 50 r/min. Obviously, the peak value of the magnetic density on the surface of the CS is 0.62T, while the location of the minimum value is between N-pole and S-pole of the PMs. Also, we can obtain the eddy current distributions on the surface of the CS again this FEM model, as shown in Figure 8b. Different from the magnetic flux density, the center of the eddy currents is situated between N-pole and S-pole of the PMs. Corresponding to the number of the PMs, the number of the eddy currents is $2p$. In addition, the adjacent eddy currents are in opposite direction owing to the alternating magnetic poles.

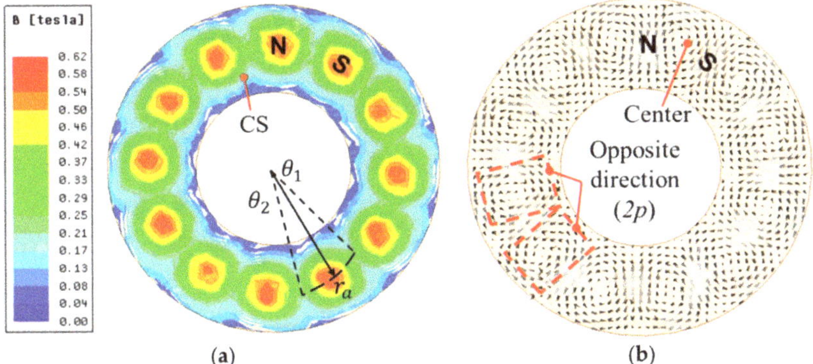

Figure 8. Magnetic field analysis (z-axis, p = 6 and s = 50 r/min): (**a**) the magnetic flux density distributions, (**b**) the eddy current distributions.

Clearly, corresponding to the distributions in Figures 8 and 9 depicts the relationship for the magnetic flux density (B_{cs}) and the eddy current density (J_{cs}) of the CS varying with different positions (θ) wherein θ_1 = 0° and θ_2 = 30° (360°/2p). Thanks to Faraday's law of electromagnetic induction, the region with the strongest magnetic flux density is the weakest eddy current density.

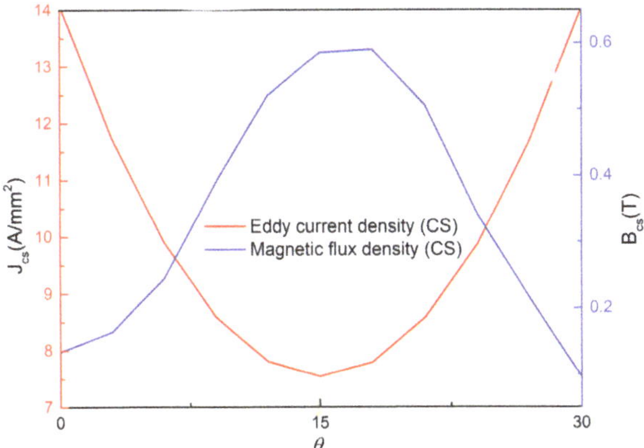

Figure 9. The curves for depicting the distributions on the CS.

4.3. Loss Calculation

Accurately calculating the loss is essential to study the temperature rise for ensuring reliable operation. On account of skin effect, the eddy currents mainly concentrated on the surface of the CS. Therefore, we only consider the eddy current loss generated on the CS and ignore other losses, like mechanical loss. Based on the magnetic field analysis and Equation (2) belonging to an embedded algorithm in the FEM, the loss results under different air-gap lengths and slips are displayed in Figure 10. Obviously, with the increase of the slip speed (s), the loss increases continuously, while the loss decreases gradually with the increase of the air-gap length.

$$P_{cs} = 1/\sigma_{cs} \int_V |\mathbf{J}| dV \tag{2}$$

where V is the volume of an integral region, and $\mathbf{J}$ is the vector of eddy current density.

$$\begin{cases} P_{loss} = k_c P_{cs} \\ k_c = 1 - \frac{\tanh(\pi w_{pm}/2\tau_p)}{(\pi w_{pm}/2\tau_p)[1+\tanh(\pi w_{pm}/2\tau_p)\tanh(\pi w_{cs}/\tau_p)]} \end{cases} \tag{3}$$

where $w_{pm} = r_{p2} - r_{p1}$ and $w_{cs} = (r_2 - r_1 - w_{pm})/2$.

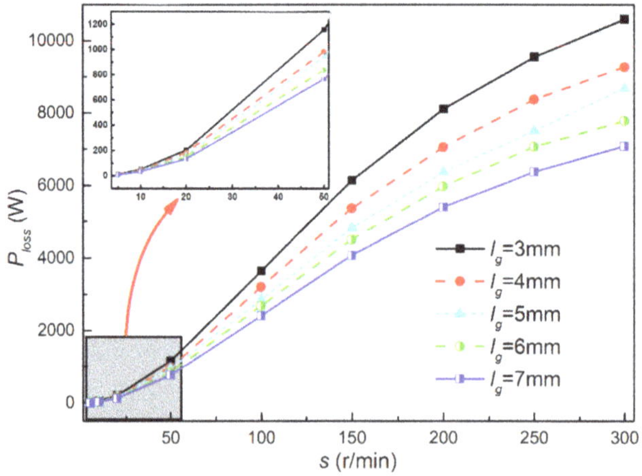

Figure 10. The loss results under different air-gap lengths and slips using FEM.

However, the loss results in Figure 9 do not take into account 3-D edge effect. In view of this, Figure 11 shows the real eddy current paths on the CS where only the central region plays a role. In order to solve this problem, the correction factor (k_c), also called Russell–Norsworthy correction factor [6,11], is adopted as Equation (3).

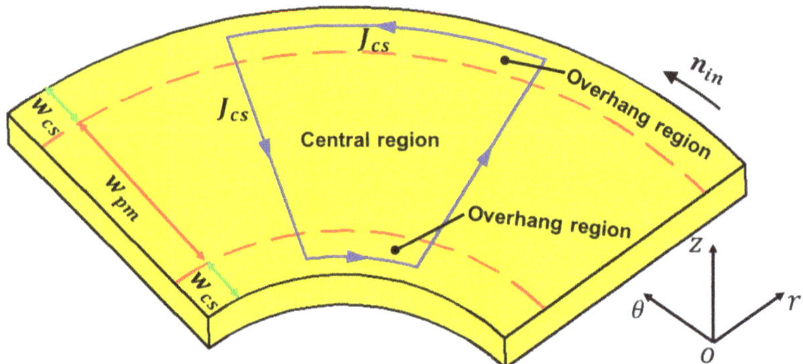

Figure 11. The real eddy current paths on the CS.

5. LPTN Model

5.1. LPTN Model

The proposed LPTN model, including thermal nodes distribution and equivalent LPTN are exposed in Figure 12. Because of the rotational axisymmetry, only three PMs are presented in Figure 12a, where the view is obtained from the average radius of the PM ($r = r_a$). In this model, the thermal modes are located at different components involving ambient, iron yoke, CS, air-gap, PM holder, PM an iron yoke. Totally, there are 22 nodes (0~22) in this thermal network. In addition, the thermal nodes in Figure 12b are divided into red and black types, in which the power loss computed in the previous section are injected into the red nodes. It is worth noting that only conduction heat transfer and convection heat transfer are considered, not existing radiation heat transfer nearly.

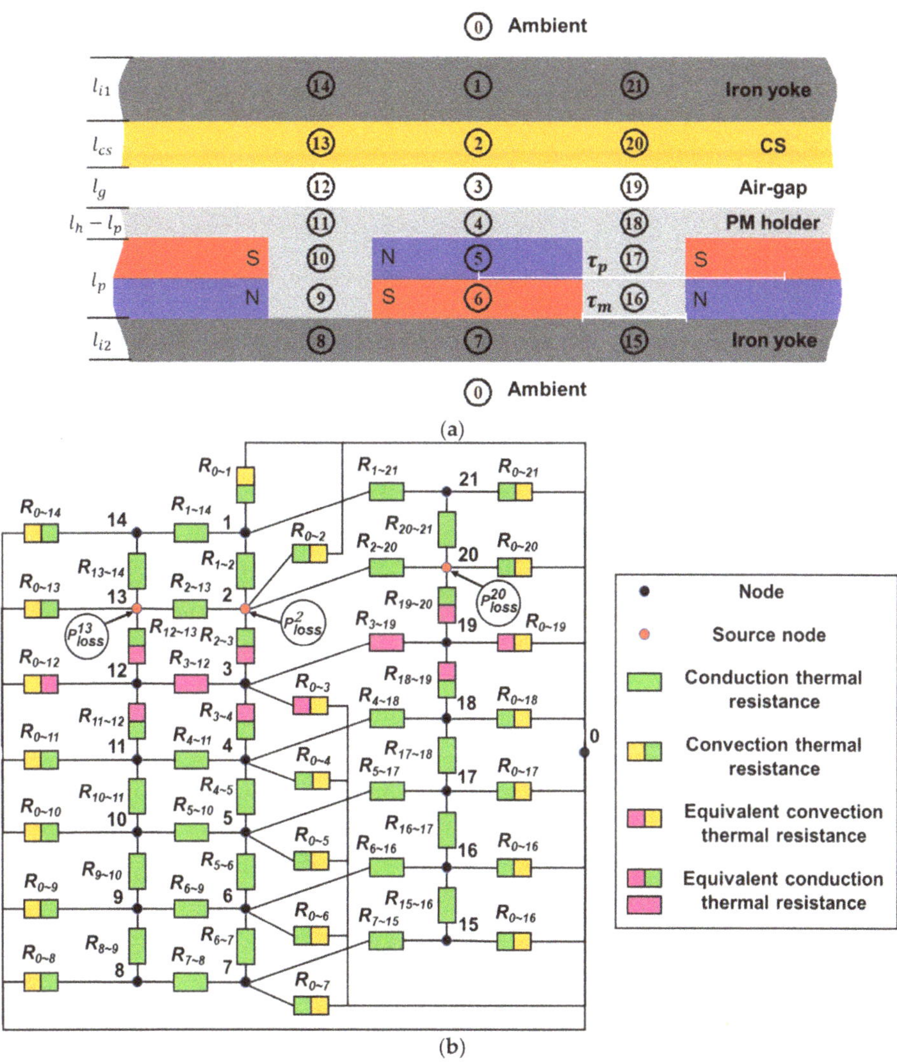

Figure 12. Proposed LPTN model: (**a**) thermal nodes distribution, (**b**) equivalent LPTN.

In addition, providing that the APMC is divided into $2p$ parts, the nodes 2, 13, and 20 in Figure 12b are given different proportional losses respectively, as follows:

$$\begin{cases} P_{loss}^{13} = P_{loss}^{20} = \alpha P_{loss}/2p \\ P_{loss}^{2} = (1-\alpha)P_{loss}/2p \end{cases} \quad (4)$$

where $\alpha = \tau_m/\tau_p$, representing the non-PM proportion after being divided into $2p$ parts.

5.2. Calculation for Thermal Resistances

1. Conduction heat transfer: The region not in contact with air belongs to conduction heat transfer, as follows: inside the iron yoke ($R_{1\sim14}$, $R_{1\sim21}$, $R_{7\sim8}$ and $R_{7\sim15}$); inside the CS ($R_{2\sim13}$ and $R_{2\sim20}$); between the iron yoke and the CS ($R_{1\sim2}$, $R_{13\sim14}$ and $R_{20\sim21}$); inside the PM holder ($R_{4\sim11}$, $R_{4\sim18}$, $R_{9\sim10}$, $R_{10\sim11}$, $R_{16\sim17}$ and $R_{17\sim18}$); between the PM and the PM holder ($R_{4\sim5}$, $R_{5\sim10}$, $R_{5\sim17}$, $R_{6\sim9}$ and $R_{6\sim16}$); inside the PM ($R_{5\sim6}$); between the PM and the iron yoke ($R_{6\sim7}$); between the PM holder and the iron yoke ($R_{8\sim9}$ and $R_{15\sim16}$). Conduction thermal resistances can be obtained as [23,29]

$$R = \frac{L}{kA} \quad (5)$$

where L (m) is the transfer path length, k (W/(m·°C) is the thermal conductivity of the material and A (m^2) is the transfer path area. Here, Table 2 presents the thermal conductivity of the materials.

Table 2. Thermal Conductivity of the Materials.

Material	Symbol	Value (W/(m·°C))
Steel	k_{st}	36
Copper	k_{cop}	390
Aluminum	k_{al}	237
Nd-Fe-B	k_{nd}	9
Air	k_a	0.026

Using Equation (5), all the conduction thermal resistances can be found in Appendix A.

2. Equivalent conduction/convection heat transfer: Since it is difficult to determine the flow condition in the air-gap, the air in the air gap can be regarded as a solid for convenience of calculation, that is, the convection heat transfer phenomenon in the air-gap can be simulated with a given air gap equivalent thermal conductivity. For the air-gap, corresponding to node 3, the equivalent heat transfer coefficient (h_{air}) can be calculated empirically as

$$\begin{cases} v_{air} = \pi(n_{in} - n_{out})(r_2 + r_{p2})/60 \\ \mathrm{Re}_{air} = v_{air} l_g / \gamma \\ h_{air} = 0.0019(l_g/r_2)^{-2.9084} \mathrm{Re}_{air}{}^{0.4614\ln[3.33361(l_g/r_2)]} \end{cases} \quad (6)$$

where v_{air} is the airflow velocity in the air gap, Re_{air} is the Reynolds number for the air-gap and γ is the kinematic viscosity of air.

Then, employing Equations (5) and (6), $R_{2\sim3}$, $R_{3\sim4}$, $R_{3\sim12}$, $R_{3\sim19}$, $R_{11\sim12}$, $R_{12\sim13}$, $R_{18\sim19}$ and $R_{19\sim20}$ can be found in Appendix B.

3. Convection heat transfer: In Figure 12, convection heat transfer certainly occurs between ambient node 0 and other nodes. Generally, as fluid flows, natural convection or forced convection would occur. The former is caused by the nonuniformity of the temperature or concentration; the latter relies on an external force, such as a pump or fan. For the studied APMC, although there is

no pump or fan, the heat transfer belongs to forced convection because of the rotational motion of the CS and the PM holder. Convection thermal resistances can be obtained as [29]

$$R = \frac{1}{hA} \tag{7}$$

where h (W/(m·°C)) is the convection heat transfer coefficient, and A (m2) is the transfer path area.

In the ambient, corresponding to node 0, the convection heat transfer coefficient (h_{am}) can be calculated by

$$\begin{cases} h_{am} = 13.3(1 + v_{am}{}^{0.5}) \\ v_{am} = \pi(n_{in} + n_{out})(r_2 + r_{p2})/120 \end{cases} \tag{8}$$

wherein v_{am} is the airflow velocity in the ambient.

Based on Equations (6) to (8), the convection thermal resistances can be found in Appendix C.

5.3. Calculation for Temperature Rise

For the LPTN model, the temperature rise of each node can be solved by

$$[G][T] = [P] \tag{9}$$

where $[P]$ is the column matrix of nodal power loss, $[G]$ is the thermal conductance matrix and $[T]$ is the temperature rise vector. The thermal conductance matrix can be defined by

$$[G] = \begin{bmatrix} \sum\limits_{i=1}^{n} \frac{1}{R_{1\sim i}} & -\frac{1}{R_{1\sim 2}} & \cdots & -\frac{1}{R_{1\sim n}} \\ -\frac{1}{R_{2\sim 1}} & \sum\limits_{i=1}^{n} \frac{1}{R_{2\sim i}} & \cdots & -\frac{1}{R_{2\sim n}} \\ \vdots & \vdots & \ddots & \vdots \\ -\frac{1}{R_{n\sim 1}} & -\frac{1}{R_{n\sim 2}} & \cdots & \sum\limits_{i=1}^{n} \frac{1}{R_{n\sim i}} \end{bmatrix} \tag{10}$$

where n is the number of all the nodes for the LPTN model, and $R_{i\sim j}$ represents the thermal resistance between adjacent nodes i and node j. Besides, $1/R_{i\sim j}$ is normally ignored because there is no heat exchange between itself and $R_{i\sim j} = R_{j\sim i}$.

Accounting for the effect of ambient temperature, Equation (9) is not perfect to solve the temperature rise of the APMC. Thus, a modified equation for Equation (9) is offered as

$$[T] = \{[P] - T_0[G_0]\}[G]^{-1} \tag{11}$$

where T_0 is the local ambient temperature in accordance with the room temperature of the experiment, and $[G_0] = [-1/R_{1\sim 0}, -1/R_{2\sim 0}, \ldots, -1/R_{n\sim 0}]$.

Besides, the node temperature in the same region is averaged as the calculation result. For example, the average value of nodes 2, 13 and 20 is taken as the temperature result of the CS.

5.4. Adjusting the Conductivity of the CS

Additionally, the column matrix of nodal power loss is temperature-dependent and require temperature update in the thermal analysis strategy. As a result, an iterative coupled electromagnetic and thermal modeling, corresponding to Figure 4, must be employed in this section. Here, the conductivity of the CS is the most crucial factor for updating the nodal power loss and temperature. Figure 13 gives the relationship between the conductivity of the CS (σ_{cs}) and different temperature (T_{cs}), which provides an effective reference for adjusting the conductivity of the CS. This iteration process is generally over until a set convergence threshold (T_s) is met, such as 1%.

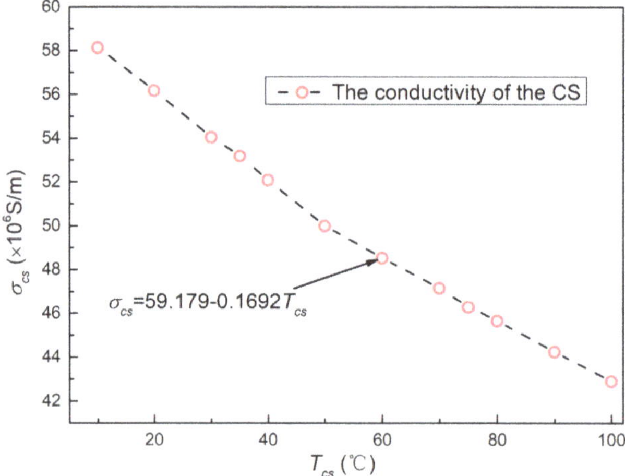

Figure 13. The relationship between the conductivity of the CS and different temperature.

6. Experiment Verification

In order to verify the thermal analysis strategy proposed in this paper, the experiment platform for the studied APMC is built in Figure 14. Since the APMC is in the state of high-speed rotation, the traditional static temperature measurement method cannot be carried out. Thus, a high-precision thermal camera, labeled as Telops FAST V100k, (Telops Inc, Montreal, Quebec, Canada.) is used to measure the temperature. The thermal camera was set up half a meter away from the APMC. After the APMC was in stable operation, the PM module and CS module were photographed. Since the interior of the components could not be photographed, surface temperature data was used as experimental results. However, the experimental results are generally obtained when the APMC operates for one hour to reach the thermal balance. Also, the cross-section of the LPTN model inside the components is about 10 mm away from the surface. Therefore, it is considered that the error between the internal temperature of this cross-section and surface temperature is not large.

Figure 14. The experiment platform for the studied APMC.

Besides, if the uniformity of measured temperature is well, and the fluctuation is less than 1%, the measured results can be used, as shown in Figure 15. Also, Figure 16 presents a measurement result by thermal camera in the case of 'l_g = 5 mm and s = 20 r/min'. The measured results can be obtained by taking the points. Specifically, in the post-processing of the thermal camera software, some points in the photographed area are selected as the temperature results of each component.

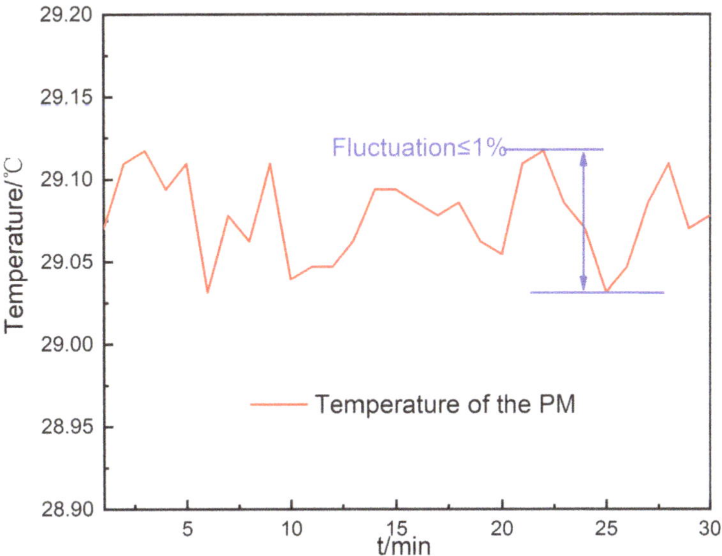

Figure 15. Temperature variation of the PM within 30 min (s = 25 r/min).

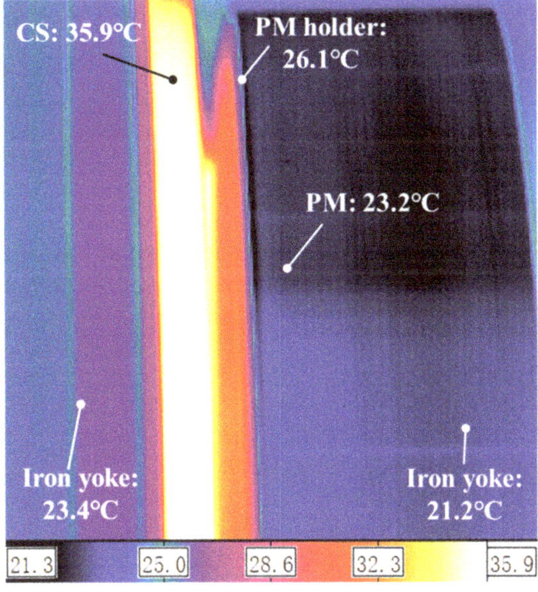

Figure 16. Temperature measurement results obtained by thermal camera (l_g = 5 mm and s = 20 r/min).

Here, the temperature of the CS and the PM is of greatest concern. As shown in Figure 17, compared with the experiment results at different slip speed, the proposed thermal analysis strategy for APMCs are in good agreement, and the relative error is within 6.7%. Moreover, the demagnetization temperature of the PM used in this paper is 180 °C. Therefore, the slip speed of 120r/min is the limit from Figure 17. If selected PM has a higher demagnetization temperature, a greater slip speed can be achieved.

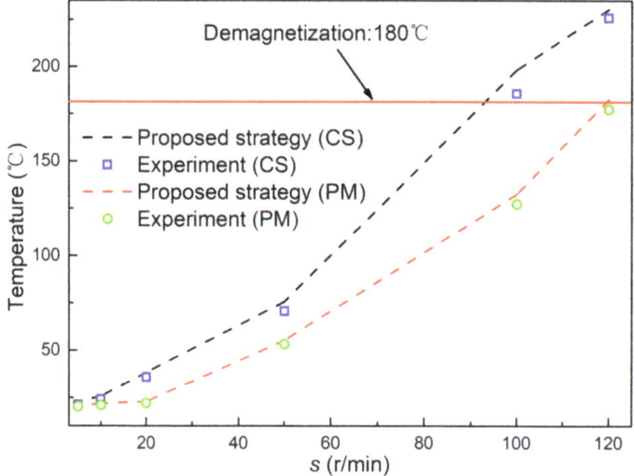

Figure 17. Comparison between the proposed strategy and experiment (l_g = 5 mm).

In order to further verify the accuracy of the strategy, Figure 18 presents the comparison between the proposed strategy and experiment under different air-gaps and slip speeds. With the increase of l_g, the temperature of PM decreases gradually. As can be seen from Figure 18, the strategy has a good coincidence with the experimental results, with the error less than 6.7%.

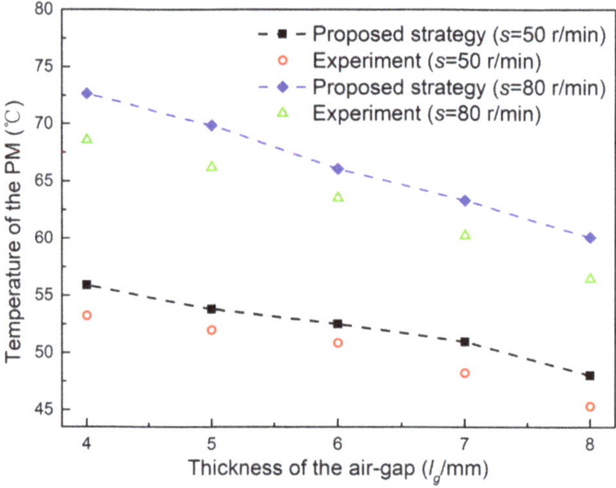

Figure 18. Comparison between proposed strategy and experiment under different air-gaps and slip speeds.

In the practical design and optimization of APMCs, the pole-pairs number of the PMs and the thickness of the CS are two important parameters. Besides, the temperature of the PM is key for the safe operation of APMCs due to demagnetization effect. Through the proposed strategy, Figure 19 presents the relationship between the temperature of the PM and the pole-pairs number of the PMs (p) at s = 120 r/min. From Figure 19, with the increase of p, the temperature of the PM increases gradually. Taking the demagnetization effect into account, selecting p as 6 is appropriate. Moreover, by the proposed strategy, Figure 20 presents the relationship between the temperature of the PM and the thickness of the CS (l_{cs}) at s = 120 r/min. From Figure 20, with the increase of l_{cs}, the temperature of the PM decreases gradually. Taking the demagnetization effect into account, selecting l_{cs} ≥6 is appropriate. However, in order to avoid excessive mass, selecting l_{cs} as 6 is optimal.

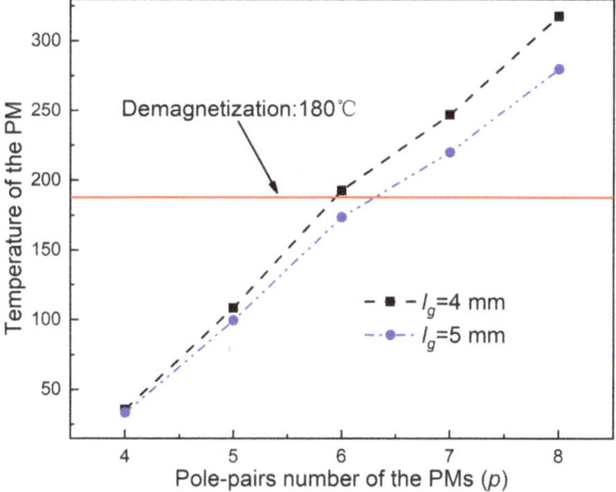

Figure 19. The relationship between the temperature of the PM and the pole-pairs number of the PMs. (s = 120 r/min).

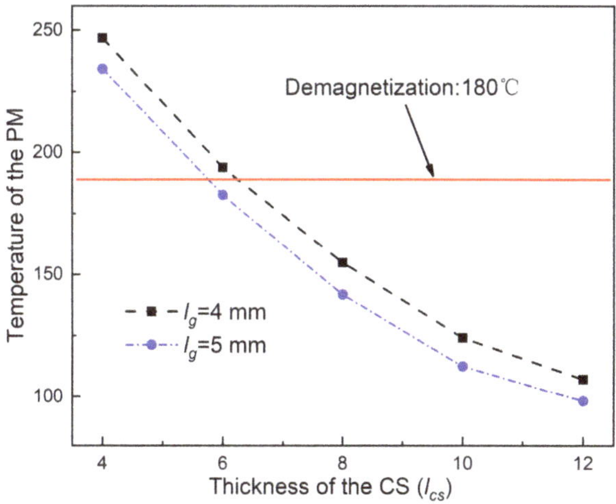

Figure 20. The relationship between the temperature of the PM and the thickness of the CS (s = 120 r/min).

7. Conclusions

A thermal analysis strategy is proposed to obtain the temperature results for APMCs, which combines FEM with LPTN. Firstly, the manufactured prototype of the studied APMC is built as well as giving its parameters. This proposed strategy is an iterative process considering some assumptions. Secondly, the magnetic field employing the FEM is offered to obtain the loss generated on the CS, where the magnetic field of the CS is analyzed. Then, the source nodes are assigned losses to calculate the matrix of the LPTN model and get the temperature results by adjusting the conductivity of the CS. Finally, the strategy is verified by experiment results where its relative error is less than 6.7%. In summary, the strategy developed in this paper can provide constructive references for operation safety of APMCs.

Author Contributions: Conceptualization, X.C. and W.L.; formal analysis, X.C. and Z.T.; funding acquisition, W.L.; investigation, X.C., Z.T., Z.Z., and B.Y.; methodology, X.C.; resources, B.Y., W.W., Y.Z., and S.L.; validation, X.C. and W.L.; writing—original draft, X.C. and W.L.; writing—review and editing, Z.Z., W.W., Y.Z., and S.L. All authors have read and agreed to the published version of the manuscript.

Funding: This research was jointly funded by the National Natural Science Foundation of China (No. U1808217), the LiaoNing Revitalization Talents Program (No. XLYC1807086, XLYC1801008) and the High-Level Personnel Innovation Support Program of Dalian (No. 2017RJ04).

Conflicts of Interest: The authors declare no conflict of interest.

Nomenclature

n_{in}	rotational speed of the CS module
n_{out}	rotational speed of the PM module
s	Slip speed ($n_{in} - n_{out}$)
p	pole-pairs number of the PMs
l_{i1}	thickness of the iron yoke (CS side)
l_{i2}	thickness of the iron yoke (PM side)
l_{cs}	thickness of the CS
l_g	thickness of the air-gap
l_h	thickness of the PM holder
l_p	thickness of the PM
r_1	inside radius of the CS
r_2	outside radius of the CS
r_{p1}	inside radius of the PM
r_{p2}	outside radius of the PM
r_a	average radius of the PM
H_p	coercive force of the PM
σ_{cs}	conductivity of the CS
R, $R_{i\sim j}$	thermal resistance
P_{cs}	loss generated on the CS
T, T_0	temperature, ambient temperature
τ_p	length between centers of adjacent PMs
τ_m	length between adjacent PMs

Appendix A

The conduction thermal resistances can be calculated in detail as:

$$\begin{cases} R_{1\sim 2} = \frac{l_{i1}/2}{k_{st}(\tau_p-\tau_m)(r_{p2}-r_{p1})} + \frac{l_{cs}/2}{k_{cop}(\tau_p-\tau_m)(r_{p2}-r_{p1})} \\ R_{13\sim 14} = R_{20\sim 21} = \frac{l_{i1}/2}{k_{st}\tau_m(r_{p2}-r_{p1})} + \frac{l_{cs}/2}{k_{cop}\tau_m(r_{p2}-r_{p1})} \\ R_{1\sim 14} = R_{1\sim 21} = \frac{(\tau_p-\tau_m)/2+\tau_m/2}{k_{st}l_{i1}(r_{p2}-r_{p1})} \\ R_{7\sim 8} = R_{7\sim 15} = \frac{(\tau_p-\tau_m)/2+\tau_m/2}{k_{st}l_{i2}(r_{p2}-r_{p1})} \\ R_{2\sim 13} = R_{2\sim 20} = \frac{(\tau_p-\tau_m)/2+\tau_m/2}{k_{cop}l_{cs}(r_{p2}-r_{p1})} \\ R_{4\sim 11} = R_{4\sim 18} = \frac{(\tau_p-\tau_m)/2+\tau_m/2}{k_{al}(l_h-l_p)(r_{p2}-r_{p1})} \\ R_{9\sim 10} = R_{16\sim 17} = \frac{l_p/2}{k_{al}\tau_m(r_{p2}-r_{p1})} \\ R_{10\sim 11} = R_{17\sim 18} = \frac{l_p/4+(l_h-l_p)/2}{k_{al}\tau_m(r_{p2}-r_{p1})} \\ R_{4\sim 5} = \frac{(l_h-l_p)/2}{k_{al}(\tau_p-\tau_m)(r_{p2}-r_{p1})} + \frac{l_p/4}{k_{nd}(\tau_p-\tau_m)(r_{p2}-r_{p1})} \\ R_{5\sim 10} = R_{5\sim 17} = R_{6\sim 9} = R_{6\sim 16} = \frac{(\tau_p-\tau_m)}{k_{nd}l_p(r_{p2}-r_{p1})} + \frac{\tau_m}{k_{al}l_p(r_{p2}-r_{p1})} \\ R_{5\sim 6} = \frac{l_p/2}{k_{nd}(\tau_p-\tau_m)(r_{p2}-r_{p1})} \\ R_{6\sim 7} = \frac{l_{i2}/2}{k_{st}(\tau_p-\tau_m)(r_{p2}-r_{p1})} + \frac{l_p/4}{k_{nd}(\tau_p-\tau_m)(r_{p2}-r_{p1})} \\ R_{8\sim 9} = R_{15\sim 16} = \frac{l_{i2}/2}{k_{st}\tau_m(r_{p2}-r_{p1})} + \frac{l_p/4}{k_{al}\tau_m(r_{p2}-r_{p1})} \end{cases} \tag{A1}$$

Appendix B

$R_{2\sim 3}$, $R_{3\sim 4}$, $R_{3\sim 12}$, $R_{3\sim 19}$, $R_{11\sim 12}$, $R_{12\sim 13}$, $R_{18\sim 19}$ and $R_{19\sim 20}$ can be calculated in detail as:

$$\begin{cases} R_{2\sim 3} = \frac{l_{cs}/2}{k_{cop}(\tau_p-\tau_m)(r_{p2}-r_{p1})} + \frac{l_g/2}{h_{air}(\tau_p-\tau_m)(r_{p2}-r_{p1})} \\ R_{3\sim 4} = \frac{l_g/2}{h_{air}(\tau_p-\tau_m)(r_{p2}-r_{p1})} + \frac{(l_h-l_p)/2}{k_{al}(\tau_p-\tau_m)(r_{p2}-r_{p1})} \\ R_{3\sim 12} = R_{3\sim 19} = \frac{(\tau_p-\tau_m)/2+\tau_m/2}{h_{air}l_g(r_{p2}-r_{p1})} \\ R_{11\sim 12} = R_{18\sim 19} = \frac{l_g/2}{h_{air}\tau_m(r_{p2}-r_{p1})} + \frac{(l_h-l_p)/2}{k_{al}\tau_m(r_{p2}-r_{p1})} \\ R_{12\sim 13} = R_{19\sim 20} = \frac{l_{cs}/2}{k_{cop}\tau_m(r_{p2}-r_{p1})} + \frac{l_g/2}{h_{air}\tau_m(r_{p2}-r_{p1})} \end{cases} \tag{A2}$$

Appendix C

The convection thermal resistances can be calculated in detail as:

$$\begin{cases} R_{0\sim 1} = \frac{l_{i1}/2}{k_{st}(\tau_p-\tau_m)(r_{p2}-r_{p1})} + \frac{1}{h_{am}(\tau_p-\tau_m)(r_{p2}-r_{p1})} \\ R_{0\sim 2} = \frac{1}{(1-\alpha)h_{am}\pi r_2 l_{cs}/p} \\ R_{0\sim 3} = \frac{1}{(1-\alpha)h_{am}\pi(r_2+r_{p2})l_g/2p} \\ R_{0\sim 4} = \frac{1}{(1-\alpha)h_{am}\pi r_{p2}(l_h-l_p)/p} \\ R_{0\sim 5} = R_{0\sim 6} = \frac{1}{(1-\alpha)h_{am}\pi r_{p2}l_p/2p} \\ R_{0\sim 7} = \frac{l_{i2}/2}{k_{st}(\tau_p-\tau_m)(r_{p2}-r_{p1})} + \frac{1}{h_{am}(\tau_p-\tau_m)(r_{p2}-r_{p1})} \\ R_{0\sim 8} = R_{0\sim 15} = \frac{l_{i2}/2}{k_{st}\tau_m(r_{p2}-r_{p1})} + \frac{1}{h_{am}\tau_m(r_{p2}-r_{p1})} \\ R_{0\sim 9} = R_{0\sim 10} = R_{0\sim 16} = R_{0\sim 17} = \frac{1}{\alpha h_{am}\pi r_{p2}l_p/2p} \\ R_{0\sim 11} = R_{0\sim 18} = \frac{1}{\alpha h_{am}\pi r_{p2}(l_h-l_p)/p} \\ R_{0\sim 12} = R_{0\sim 19} = \frac{1}{\alpha h_{am}\pi(r_2+r_{p2})l_g/2p} \\ R_{0\sim 13} = R_{0\sim 20} = \frac{1}{\alpha h_{am}\pi r_2 l_{cs}/p} \\ R_{0\sim 14} = R_{0\sim 21} = \frac{l_{i1}/2}{k_{st}\tau_m(r_{p2}-r_{p1})} + \frac{1}{h_{am}\tau_m(r_{p2}-r_{p1})} \end{cases} \tag{A3}$$

References

1. Ye, L.Z.; Li, D.S.; Ma, Y.J.; Jiao, B.F. Design and Performance of a Water-cooled Permanent Magnet Retarder for Heavy Vehicles. *IEEE Trans. Energy Convers.* **2011**, *26*, 953–958. [CrossRef]
2. Li, Y.B.; Lin, H.Y.; Huang, H.; Yang, H.; Tao, Q.C.; Fang, S.H. Analytical Analysis of a Novel Brushless Hybrid Excited Adjustable Speed Eddy Current Coupling. *Energies* **2019**, *12*, 308. [CrossRef]
3. Mohammadi, S.; Mirsalim, M.; Vaez-Zadeh, S.; Talebi, H.A. Analytical Modeling and Analysis of Axial-Flux Interior Permanent-Magnet Couplers. *IEEE Trans. Ind. Electron.* **2014**, *61*, 5940–5947. [CrossRef]
4. Lubin, T.; Mezani, S.; Rezzoug, A. Simple Analytical Expressions for the Force and Torque of Axial Magnetic Couplings. *IEEE Trans. Energy Convers.* **2012**, *27*, 536–546. [CrossRef]
5. Wang, S.; Guo, Y.C.; Cheng, G.; Li, D.Y. Performance Study of Hybrid Magnetic Coupler Based on Magneto Thermal Coupled Analysis. *Energies* **2017**, *10*, 1148. [CrossRef]
6. Wang, J.; Zhu, J.G. A Simple Method for Performance Prediction of Permanent Magnet Eddy Current Couplings Using a New Magnetic Equivalent Circuit Model. *IEEE Trans. Ind. Electron.* **2018**, *65*, 2487–2495. [CrossRef]
7. Dai, X.; Liang, Q.H.; Cao, J.Y.; Long, Y.J.; Mo, J.Q.; Wang, S.G. Analytical Modeling of Axial-Flux Permanent Magnet Eddy Current Couplings With a Slotted Conductor Topology. *IEEE Trans. Magn.* **2016**, *52*, 15. [CrossRef]
8. Lubin, T.; Mezani, S.; Rezzoug, A. Experimental and Theoretical Analyses of Axial Magnetic Coupling Under Steady-State and Transient Operations. *IEEE Trans. Ind. Electron.* **2014**, *61*, 4356–4365. [CrossRef]
9. Mohammadi, S.; Mirsalim, M. Double-sided permanent-magnet radial-flux eddy-current couplers: Three-dimensional analytical modelling, static and transient study, and sensitivity analysis. *IET Electr. Power Appl.* **2013**, *7*, 665–679. [CrossRef]
10. Ravaud, R.; Lemarquand, V.; Lemarquand, G. Analytical Design of Permanent Magnet Radial Couplings. *IEEE Trans. Magn.* **2010**, *46*, 3860–3865. [CrossRef]
11. Mohammadi, S.; Mirsalim, M.; Vaez-Zadeh, S. Nonlinear Modeling of Eddy-Current Couplers. *IEEE Trans. Energy Convers.* **2014**, *29*, 224–231. [CrossRef]
12. Fei, W.Z.; Luk, P.C.K. Torque Ripple Reduction of a Direct-Drive Permanent-Magnet Synchronous Machine by Material-Efficient Axial Pole Pairing. *IEEE Trans. Ind. Electron.* **2012**, *59*, 2601–2611. [CrossRef]
13. Min, K.C.; Choi, J.Y.; Kim, J.M.; Cho, H.W.; Jang, S.M. Eddy-Current Loss Analysis of Noncontact Magnetic Device with Permanent Magnets Based on Analytical Field Calculations. *IEEE Trans. Magn.* **2015**, *51*, 4. [CrossRef]
14. Choi, J.Y.; Jang, S.M. Analytical magnetic torque calculations and experimental testing of radial flux permanent magnet-type eddy current brakes. *J. Appl. Phys.* **2012**, *111*, 3. [CrossRef]
15. Dai, X.; Cao, J.Y.; Long, Y.J.; Liang, Q.H.; Mo, J.Q.; Wang, S.G. Analytical Modeling of an Eddy-current Adjustable-speed Coupling System with a Three-segment Halbach Magnet Array. *Electr. Power Compon. Syst.* **2015**, *43*, 1891–1901. [CrossRef]
16. Wang, J.; Lin, H.Y.; Fang, S.H.; Huang, Y.K. A General Analytical Model of Permanent Magnet Eddy Current Couplings. *IEEE Trans. Magn.* **2014**, *50*, 9. [CrossRef]
17. Lubin, T.; Rezzoug, A. Steady-State and Transient Performance of Axial-Field Eddy-Current Coupling. *IEEE Trans. Ind. Electron.* **2015**, *62*, 2287–2296. [CrossRef]
18. Dolisy, B.; Mezani, S.; Lubin, T.; Leveque, J. A New Analytical Torque Formula for Axial Field Permanent Magnets Coupling. *IEEE Trans. Energy Convers.* **2015**, *30*, 892–899. [CrossRef]
19. Lubin, T.; Rezzoug, A. 3-D Analytical Model for Axial-Flux Eddy-Current Couplings and Brakes Under Steady-State Conditions. *IEEE Trans. Magn.* **2015**, *51*, 12. [CrossRef]
20. Wang, J.; Lin, H.Y.; Fang, S.H. Analytical Prediction of Torque Characteristics of Eddy Current Couplings Having a Quasi-Halbach Magnet Structure. *IEEE Trans. Magn.* **2016**, *52*, 9. [CrossRef]
21. Li, Z.; Wang, D.Z.; Zheng, D.; Yu, L.X. Analytical modeling and analysis of magnetic field and torque for novel axial flux eddy current couplers with PM excitation. *AIP Adv.* **2017**, *7*, 13. [CrossRef]
22. Lubin, T.; Rezzoug, A. Improved 3-D Analytical Model for Axial-Flux Eddy-Current Couplings with Curvature Effects. *IEEE Trans. Magn.* **2017**, *53*, 9. [CrossRef]

23. Huang, X.Z.; Li, L.Y.; Zhou, B.; Zhang, C.M.; Zhang, Z.R. Temperature Calculation for Tubular Linear Motor by the Combination of Thermal Circuit and Temperature Field Method Considering the Linear Motion of Air Gap. *IEEE Trans. Ind. Electron.* **2014**, *61*, 3923–3931. [CrossRef]
24. Lu, Y.P.; Liu, L.; Zhang, D.X. Simulation and Analysis of Thermal Fields of Rotor Multislots for Nonsalient-Pole Motor. *IEEE Trans. Ind. Electron.* **2015**, *62*, 7678–7686. [CrossRef]
25. Vese, I.C.; Marignetti, F.; Radulescu, M.M. Multiphysics Approach to Numerical Modeling of a Permanent-Magnet Tubular Linear Motor. *IEEE Trans. Ind. Electron.* **2010**, *57*, 320–326. [CrossRef]
26. Zhu, X.Y.; Wu, W.Y.; Yang, S.; Xiang, Z.X.; Quan, L. Comparative Design and Analysis of New Type of Flux-Intensifying Interior Permanent Magnet Motors with Different Q-Axis Rotor Flux Barriers. *IEEE Trans. Energy Convers.* **2018**, *33*, 2260–2269. [CrossRef]
27. Le Besnerais, J.; Fasquelle, A.; Hecquet, M.; Pelle, J.; Lanfranchi, V.; Harmand, S.; Brochet, P.; Randria, A. Multiphysics Modeling: Electro-Vibro-Acoustics and Heat Transfer of PWM-Fed Induction Machines. *IEEE Trans. Ind. Electron.* **2010**, *57*, 1279–1287. [CrossRef]
28. Sun, X.K.; Cheng, M. Thermal Analysis and Cooling System Design of Dual Mechanical Port Machine for Wind Power Application. *IEEE Trans. Ind. Electron.* **2013**, *60*, 1724–1733. [CrossRef]
29. Mo, L.H.; Zhang, T.; Lu, Q. Thermal Analysis of a Flux-Switching Permanent-Magnet Double-Rotor Machine with a 3-D Thermal Network Model. *IEEE Trans. Appl. Supercond.* **2019**, *29*, 5. [CrossRef]
30. Valenzuela, M.A.; Ramirez, G. Thermal Models for Online Detection of Pulp Obstructing the Cooling System of TEFC Induction Motors in Pulp Area. *IEEE Trans. Ind. Appl.* **2011**, *47*, 719–729. [CrossRef]

Article

Comparative Analysis of Effectiveness of Resistance and Induction Turnout Heating

Elzbieta Szychta [1] and Leszek Szychta [2,*]

1 Department of Electricity Power Plant, UTP University of Science and Technology in Bydgoszcz, 85-326 Bydgoszcz, Poland; elzbieta.szychta@utp.edu.pl

2 Department of Power Electronics, Electric Machines and Drives, UTP University of Science and Technology in Bydgoszcz, 85-326 Bydgoszcz, Poland

* Correspondence: leszek.szychta@utp.edu.pl; Tel.: +48-608322007

Received: 10 August 2020; Accepted: 22 September 2020; Published: 10 October 2020

Abstract: Turnouts are key parts of rail roads and are exposed to adverse weather conditions such as snowfall, snow drifts, low temperatures, or sleet. Effective protection assures good turnout function and contributes to rail traffic efficiency and safety. Presently, resistance heating (RH) is the most common system of turnout heating in Europe. In this study, we attempted to implement energy-saving induction heating (IH) in order to cut costs of operation and electricity. A turnout heating test stand, including a stock-rail and a switch-rail, was executed in a climatic chamber. Air temperature was constant at the time of heating. Active power received by both the systems was identical for any measurement (450 W). Test results enabled an assessment of switch-rail position and variations of climatic chamber air temperature on growth of turnout temperatures. Effects of heating type on correct lubrication of the slide plate surface were compared. Dynamics of heating variations and their impact on effectiveness of snow or ice removal were defined for both heating systems. Turnout's readiness for switch-rail shifting and lubrication conditions of turnout's moving parts were compared. An in-depth comparative analysis of efficiency of RH and IH turnout heating was undertaken in the conclusion.

Keywords: energy efficiency; induction heating; resistance heating; turnouts; railway; safety of rail traffic; stock-rail; switch-rail

1. Introduction

Turnouts are key parts of rail roads and are exposed to adverse weather conditions such as snowfall, snow drifts, low temperatures, or sleet. A severe winter may block turnouts and disrupt traffic. Effective protection ensures good operation of turnouts and contributes to rail traffic efficiency and safety.

Turnouts are normally comprised of a pair of fixed rails (stock-rail) and a pair of switching rails (switch-rail) that move between the On and Off positions. In the On position, the switch-rail rests against a stock-rail. In the Off position, the switch-rail is separated from the stock-rail so that the latter has no effect on a road wheel. It is important for the stock-rail and the switch-rail to be in good contact in the On position for the purposes of correct turnout operation. A faulty turnout poses the risk of derailing and grave personal and property losses. Although turnouts are provided with sensors that warn of defective operations, they can result in substantial train delays and irritation to passengers. Therefore, it is necessary to periodically heat turnouts in cold climates.

A variety of electric turnout heating systems have been developed, e.g., radiator (e.g., infrared element) heaters, convection (e.g., forced air), electric turnout heating with resistors, or induction heating using eddy currents [1].

Resistance heating (RH) is currently the most common system of turnout heating in Europe (Figure 1). It is installed in more than 18,000 turnouts in Poland with a total installed electric power of approximately 120 MW [2]. RH heats for an average 300 h a season. Application of 330 W per running rail meter provides for sufficiently effective turnout heating during a season. This constitutes a high demand for electricity in a heating year.

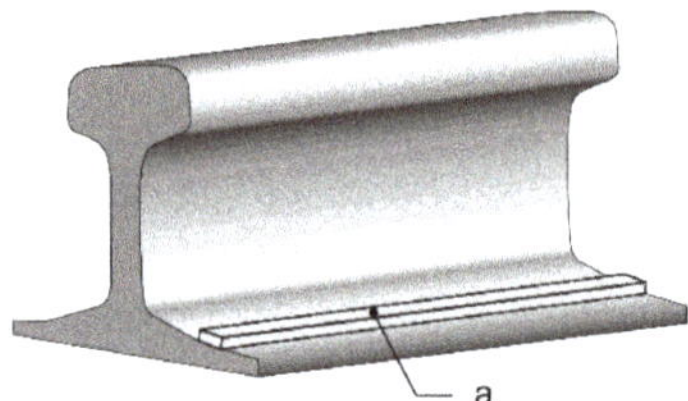

Figure 1. Location of a resistance heater "a" along a rail foot.

2. Problem Description

Resistance turnout heating is one of the most expensive, yet most often used methods. Availability of energy sources has undoubted advantages. Resistance systems of turnout heating, beside low equipment costs, are characterized by high reliability, which is of huge importance to safety and continuity of train traffic.

In this study, to reduce energy consumption of turnout heating, laboratory testing of rail heating with heat insulated heaters in a climatic chamber was undertaken [3,4]. Placement of heat insulation on a rail is shown in Figure 2. Comparative testing in a climatic chamber was conducted of 60 E1 (UIC-60) rail heating with a 330 W/m resistance heater without heat insulation, with heat conducting heat insulation, and with two heat insulations: heat conducting and heat insulating. The chamber was intensively cooled down to −28 °C from the initial rail web temperature of +20 °C (Figure 3) [5]. When the chamber without heat insulation was cooled, the web temperature reached 0 °C after ca. 150 min. The web temperatures of rails wrapped with heat conducting and heat insulating heat insulation were approximately 20 °C greater. Thus, the impact of heat insulation on energy efficiency of heating was substantial. In spite of these efforts, resistance turnout heating generates high operation cost, maintenance, and electricity consumption.

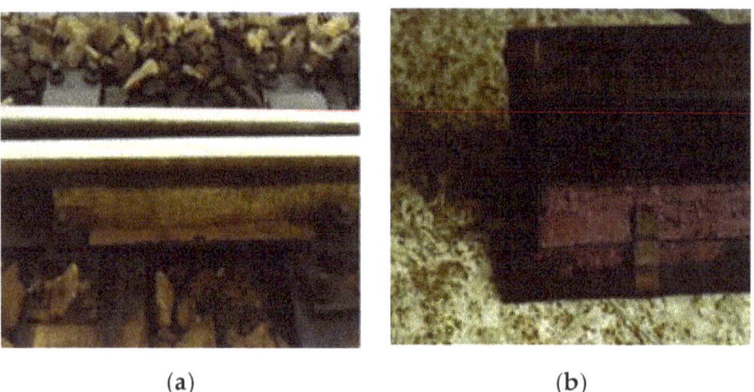

(**a**) (**b**)

Figure 2. Examples of heat insulation mounting: (**a**) heat insulating and (**b**) heat conducting [3,4].

For years, attempts have been made to introduce energy-saving induction heating (IH) of turnouts. Generation of high frequency electricity and its conversion to thermal energy in a ferromagnetic material that is part of a turnout is the essence of induction [6–9]. Induction heating of turnouts can raise their temperature by a dozen Centigrades for effective snow melting.

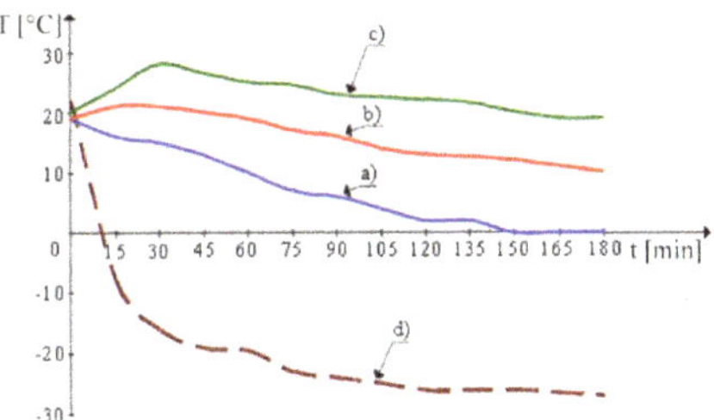

Figure 3. Characteristics of UIC 60 (60 E1) [UIC—*Union internationale des chemins de fer*] rail web heating with a 330 W/m heater: (**a**) without heat insulation, (**b**) with heat conducting heat insulation, (**c**) with heat conducting and insulating heat insulations, and (**d**) temperature in the climatic chamber [5].

In most IH solutions, a half bridge voltage inverter receives the energy from a full wave rectifier connected to single phase mains. Power MOSFETs used in the inverter assure high frequency operation of the induction heating system. The main factors of success that have contributed to the wide acceptance of this technology are as follows [8]:

- Generation of heat sources inside the workpiece to be heated over very short times
- Possibility of concentrating them in specified areas of the workpiece as required by the application
- Rational use of electrical energy for heat generation

IH applications use half bridge inverters to feed heating inductors that generate high frequency electromagnetic fields. An electromagnetic wave penetrates the ferromagnetic material to a depth dependent on the current frequency [6,8,10].

$$\delta = 503\sqrt{\frac{\rho}{\mu f}} \tag{1}$$

where:

ρ: resistivity [Ωm],
μ: relative magnetic permeability of workpiece material,
f: frequency [Hz].

Depth of the electromagnetic wave penetration δ is connected with active power released as heat in the workpiece. There is a dependence between the inductor's external radius r_e and δ, expressed with the factor m for the cylindrical inductor [8]:

$$m = \frac{\sqrt{2}r_e}{\delta} \tag{2}$$

where:
r_e: external radius of the inductor's cylinder.

Assuming constant resistivity and magnetic permeability, slight increments of thermal energy are demonstrated in $m > 4$. Adopting the following rail data [1]: $\rho = 2.7\times 10^{-7}$ [Ωm], $\mu = 100$ and the external radius of the inductor $r_e = 0.001$ [m], heating becomes efficient for the frequency $f > 5.5$ kHz. This calculation should be treated as an estimate, since constant resistivity and magnetic permeability are assumed and the complicated rail design is not addressed. Current frequency of 20 kHz is adopted for purposes of this testing of resistance heating.

Temperature distribution and effectiveness of turnout heating can be tested by means of numerical analysis, experimental testing in a climatic chamber, and on actual facilities. All the methods are important and can supplement one another in different ways, though one cannot replace another.

The numerical method allows for analysis of magnetic field distribution in a rail material and selection of parameters of an effective source of heating. This is very helpful in optimizing the power supply system, shape, and parameters of an inductor, as well as its arrangement in a turnout space.

Experimental testing helps to analyze temperature distribution across turnout elements in specified environment conditions generated in the climatic chamber. This testing addresses heating conditions in 3D space and considers real rail parameters in induction heating, its magnetization characteristic, resistivity, and magnetic permeability dependent on temperature. It should be noted that a standard rail contains 9 different elements and exhibits pearlite structure. Practical verification of heating efficiency is important.

The experimental method of testing in a climatic chamber was adopted for evaluation of induction heating efficiency, whereas we ignored the analytical method [7,11,12]. Efficiency of induction heating was compared to that of the universal resistance heating. Induction heating (IH) uses an innovative technological solution. It consists of installing an inductor in the bottom part of the slide plate, along which the switch-rail travels. The inductor is the source of heat that melts snow or ice on the slide plate's surface and in the space between the switch-rail and stock-rail. This article compared methods of heating, i.e., RH and IH, in terms of permissible variations of ambient temperature at which both the systems work effectively, dynamics of temperature changes across turnout elements, retention of good tribological properties when the switch-rail is moved, and effective ice removal from the rail surface.

3. Review of Turnout Induction Heating Solutions

Several equipment solutions based on induction heating have been proposed for the purpose of effective snow removal from turnouts. One of the first is illustrated in Figure 4 [13]. Two parallel rails R_1 and R_2 are parts of a magnetic circuit. Heating inductors HI_1, HI_2, HI_3 and HI_4 generate a variable magnetic field penetrating the rails and causing them to heat. The inductors comprise coils C_1, C_2, C_3 or C_4 around the cores F_1, F_2, F_3 and F_4 supplied from sources of alternating voltage U_1, U_2, U_3, or U_4. The arrows along the inductors show senses of magnetic fluxes at a given moment in time. Correct operation of the device requires an even number of the inductors. A magnetic flux penetrating the rails is centered between the paired inductors. Rising numbers of the inductors causes their interactions that may increase the magnetic flux and, consequently, temperature in the middle section of a heated rail.

Design of all inductors is identical (Figure 5). The bottom surfaces of R_1 and R_2 are connected with sections of F_A and F_B and insulation I. C encompasses sections of F_A and F_B and is supplied from U. C produces a variable magnetic flux in sections of F_A and F_B. Sections of the magnetic cores are made of a material reducing energy losses due to eddy currents.

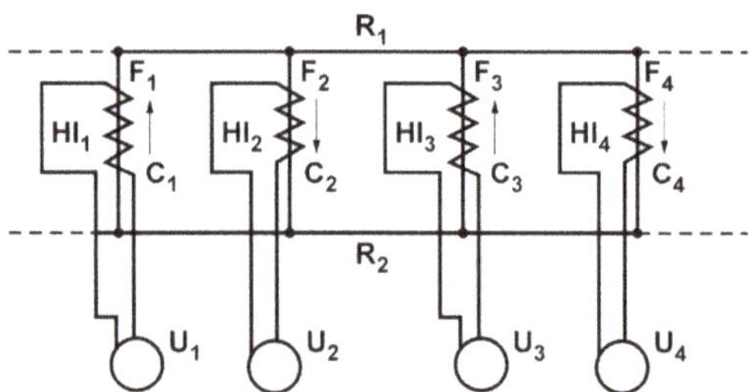

Figure 4. Diagram of an induction heating device circuit; R: rail, HI: inductor, C: coil, F: inductor core, and U: supply voltage [13].

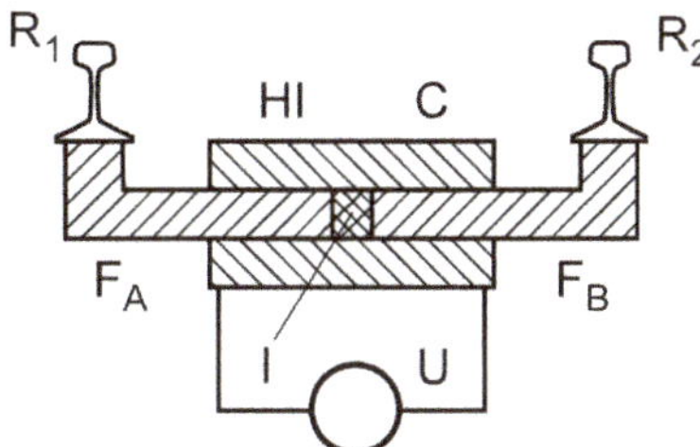

Figure 5. Cross-section of induction heating device; R—rail, HI—inductor, C—coil, F—inductor core, U—supply voltage, I—insulation [13].

The first attempt at turnout heating with 50 Hz variable magnetic field was conducted in Poland in 1978/1979 on five turnouts of Poronin station and 26 turnouts of Tarnów West station. Heating rods in an insulation coating were utilized (Figure 6) [14]. A rod was not in galvanic contact with a stock-rail. The rods were made of copper wrapped in Teflon tape and placed inside a steel guard. A rod was supplied with 3–3.3 V and 50 Hz, while currents across a rod reached 350 A. The operating principle of the method consisted in heating a stock-rail with eddy currents induced therein (Figure 6). Variations of stock-rail head and web temperatures are shown in Figures 7 and 8. Location of a measurement point affected final temperature values. Induction heating proved to have the following advantages over resistance heating: lower electricity consumption, lower operation/maintenance costs, more effective removal of snow/ice as a rail heats more quickly, low safe voltage, lower temperatures of the heater (about 65 °C), longer life of an induction heater, and slower drying of greases.

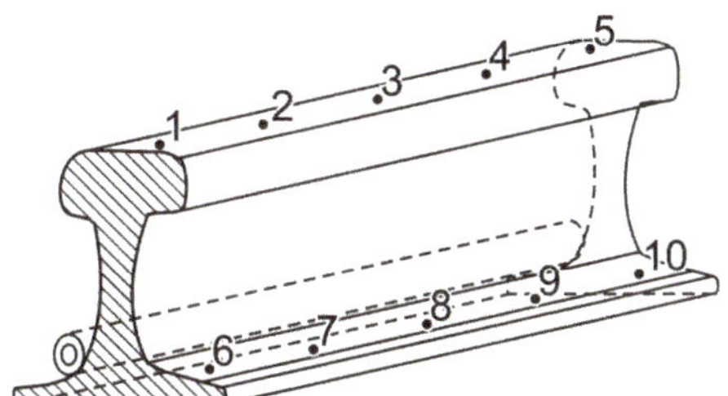

Figure 6. Temperature measurement points on the stock-rail [14].

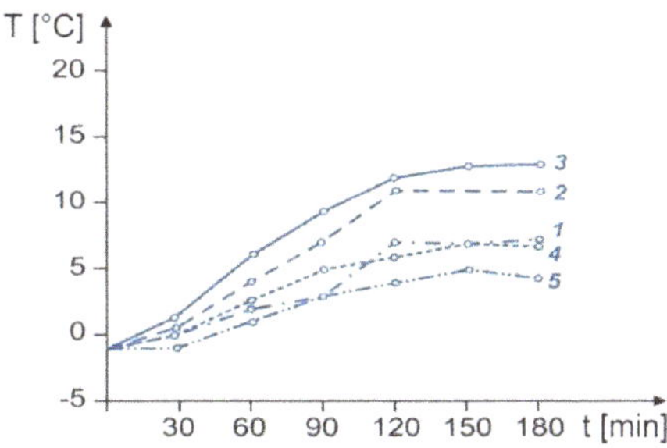

Figure 7. Temperature on the stock-rail head [14].

In spite of its lower energy consumption compared with RH, this method of induction heating was not applied in practice to turnouts. The standard of technology failed to guarantee reliable operation of the system and the solution was abandoned as a result.

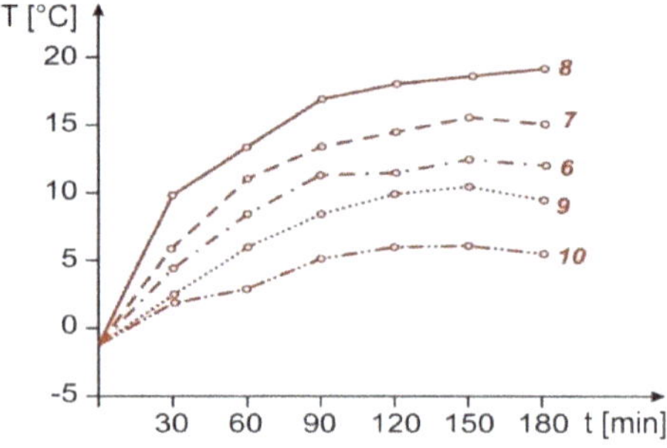

Figure 8. Temperature on the rail web [14].

Development of power electronic equipment has contributed to improved energy efficiency of electricity conversion and miniaturization of inductors used in the process of induction heating. In the age of new technologies, the concept of turnout induction heating has resurfaced. A device made by winding (1) around a substrate plate (2) and causing the latter to induction heat is one such solution (Figure 9) [15]. The area heated expands by winding around the substrate plate. Thermal conduction from the substrate plate to the stock-rail and switch-rail prevents ice formation. Coil does not protrude substantially out of the sleeper and thus does not interfere with maintenance or require dismounting. The solution is effective at snow and ice melting, however its extensive mechanical design makes it impracticable.

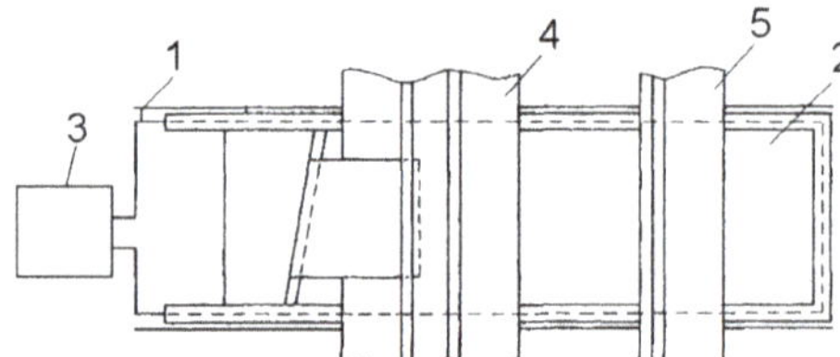

Figure 9. Snow melting system on switch points: 1: winding, 2: substrate plate, 3: power supply and control system, 4: stock-rail, and 5: switch-rail [15].

The system of induction turnout heating depicted in Figure 10 contains an inductor (1) that can be sunk, inbuilt, or incorporated into a sleeper or another turnout element [16]. Protruding parts of the heating system, subject to mechanical damage, are eliminated. The system can operate at high frequencies of the supply voltage (2). An inductor breakdown requires replacement of a rail fragment (3) and more faults may arise when movable switch-rails are heated. Figure 10 shows a solution where the inductor is placed in a duct cut in the rail foot.

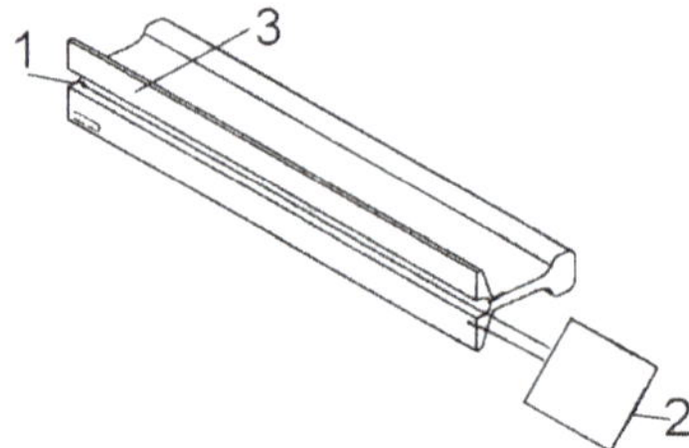

Figure 10. Diagram of rail heating featuring a duct in the rail foot; 1: inductor, 2: power supply and control system, and 3: rail foot [16].

Figure 11 shows a rail heating module consisting of two magnetic field inductors (1), each with two poles (2) oriented towards a rail to be induction heated. Each inductor consists of a magnetic core (3) and windings (4). The poles adhere to the rail web surface. Characteristically, a magnetic flux across one pole has a sense opposite to those of fluxes across the remaining poles. This effect is achieved by modifying polarity of voltage supplied to winding 2 (Figure 11b) [17,18]. The modified polarity improves efficiency of rail heating. The magnetic field of the single pole attracts one of the magnetic fields from one of the other three poles and, as a result, two remaining magnetic fields with the same polarity are obtained. These two magnetic fields counteract one another and the magnetic fields are spread across the rail as a result. By this placement of the coils, uniform heating is achieved.

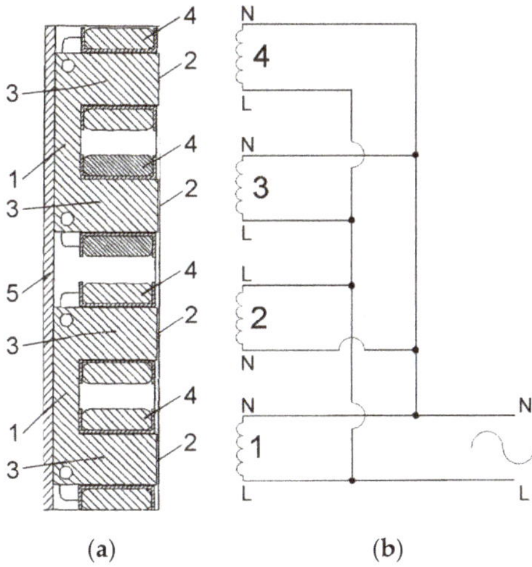

Figure 11. Rail induction heating module: (**a**) module design; 1: inductor, 2: poles, 3: magnetic core, 4: winding, and 5: mounting plate; (**b**) wiring diagram of winding for a single module [17].

Simple design of the induction heater (1), plates of high thermal conductivity, and threaded fixture to the stock-rail web (2) are characteristics of the device shown in Figure 12 [10]. An extensive heating surface, low sensitivity to mechanical damage, and easy access to the apparatus when servicing are advantages of this solution. Unfortunately, it is not clear how the inductor winding is to be mounted. Simplicity of the solution, the possibility of installing the heater between the switch-rail and stock-rail, easy installation without interfering with other electric systems of a rail line, and effective and fast heating of the stock-rail head make this solution particularly interesting.

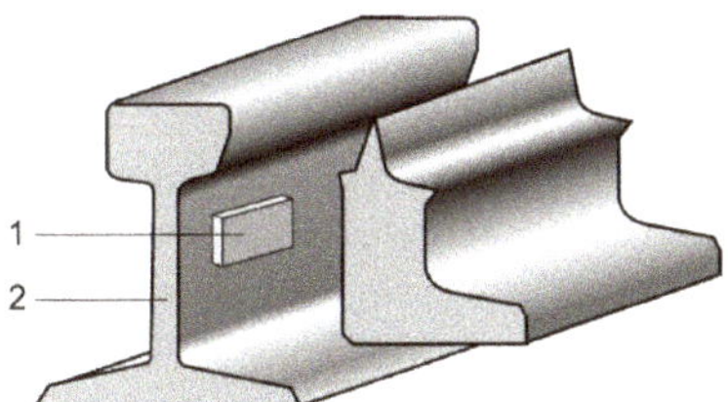

Figure 12. System including a heating plate (1) mounted on the stock-rail web (2) [19].

Another device for melting snow on turnouts by induction heating of the rail web is presented in Figure 13. It consists of a heater (2) and a clamping system (3). The electromagnetic heater (2) adheres to the rail web (1). An inductor producing a variable magnetic field (5) and a ferrite protective plate

(6) are the basic parts of the heater. Placement of a power supply and control system (4) within the heater (2) is characteristic of this solution. Its simple design, high efficiency, fast snow melting, and applicability to all turnout models are among its advantages [20].

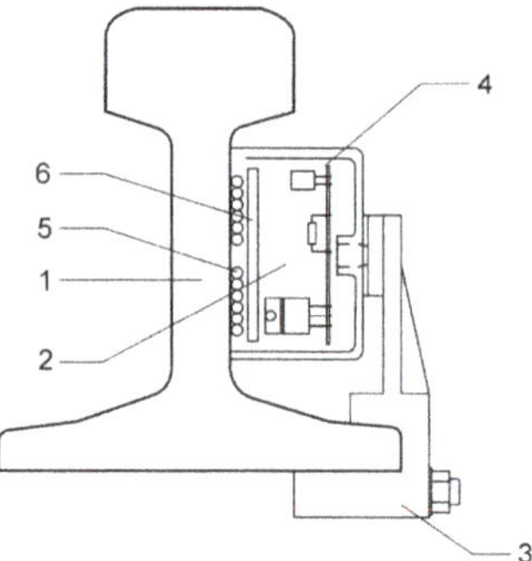

Figure 13. Snow melting device mounted on the stock-rail web; 1: stock-rail web, 2: heater, 3: clamping system, 4: power supply system, 5: inductor, and 6: protective plate [20].

Figure 14 presents a modern solution of turnout induction heating [21,22]. This setup consists of an internal coil that generates a magnetic flux in the main core. The heating plate is made of a material that has poor magnetic properties, causing losses in the magnetic flux. The heat from the heating plate is distributed to the space between the stock-rail and switch-rail. The heating element reaches its operating temperature of about 120–135 °C within five minutes, melting away snow and ice very effectively. Because of this rapid warming, it is enough to turn on the heater only when snowfall has begun. It provides for high energy efficiency saving about 60% costs compared to systems based on resistance heating. This heating method effectively melts snow between the stock-rail and switch-rail. It is not known how efficiently ice is melted from the stock-rail head when the switch-rail is moved away.

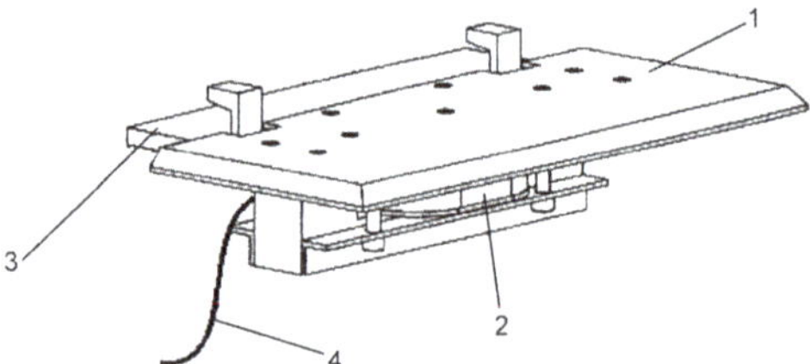

Figure 14. Induction heating system installed under the stock-rail foot; 1: heating plate, 2: inductor, 3: fastening grip, and 4: power supply wire [21,22].

The turnout induction heating equipment illustrated in Figures 9–14 are notable for the variety of novel technological solutions. The authors are not aware of any laboratory test results or practical implementations of these solutions. Absence of such publications is evidence of the scale of the problem of effective turnout de-snowing and the consequent safety of rail traffic. Any results for effects of these systems on rail control systems are not available either.

An original turnout induction heating solution is depicted in Figure 15 [23,24]. Inductor a is installed in the part of the slide plate c along which the switch-rail travels. The inductor winding is supplied from a high-frequency voltage generator c. The solution is simple in its design. Heat released in the process of induction heating increases the slide plate temperature, while temperatures of the stock-rail and switch-rail, which are in direct contact with the slide plate, rise as well. Such a heating system appears energy efficient, since ice or snow should melt rapidly in the space between the stock-rail and switch-rail and on their heads.

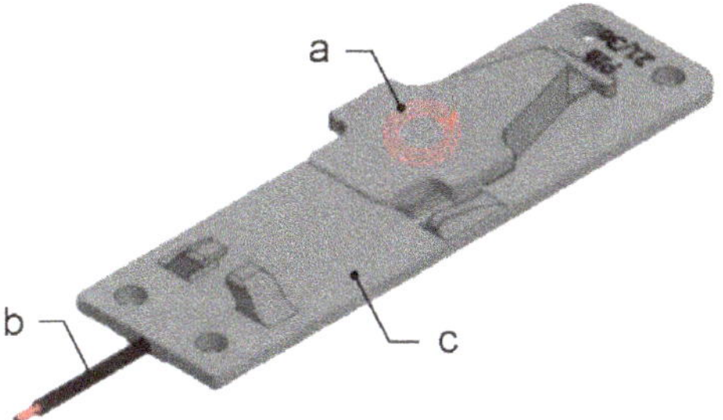

Figure 15. Snow melting device as an inductor for induction turnout heating, placed in a turnout slide plate; (**a**) inductor, (**b**) slide plate, and (**c**) high frequency power supply wire [23,24].

Owing to the original method of inductor attachment and expected high energy efficiency, this article evaluated heating properties of the device compared to those of universally used resistance heating.

4. Data and Methodology of Experimental Turnout Heating

Keeping a switch-rail and stock-rail snow-free for the purpose of safe turnout operation is the key objective of the turnout heating process. Complex design of a turnout, changing position of switch-rail relative to stock-rail, variable weather conditions, and a heating method in place have decisive impact on effectiveness of heating. Energy efficiency of turnout heating should be understood as the process of transforming electricity supplied to a heating device into an effective quantity of thermal energy capable of removing snow and ice from all crucial parts of a turnout over an adequately short period of time. At the same time, energy used to this end should be as low as possible. Safety of rail traffic is the overarching criterion in evaluation of energy efficiency of heating equipment.

Experimental testing of energy efficiency of turnout heating is a long process. The turnout's complex design makes precise analysis of its heating difficult. In real conditions, changeable weather and snow brought by passing trains are some added factors interfering with the analysis. A comparative analysis of energy efficiency of RH and IH systems based on climatic chamber testing is undertaken in this paper.

The comparative analysis of IH (Figure 15) and RH (Figure 1) addresses the following conditions, known from the literature and the authors' own experience [1,3,4,12,14]:

- Turnout heating should aim for high energy efficiency, associated with total snow or ice melting in the area between a switch-rail and stock-rail while consuming a minimum of electricity.
- RH exhibits low efficiency of snow or ice melting on the slide plate from the switch's internal side. This is caused by the fact that only thermal conductivity of the material is taken advantage of. In practice, this may mean insufficient heating of a lubricant on the slide plate plane along which the switch-rail travels. The lubricant maintains good tribological properties at temperatures above −30 °C [25]. However, as the switch-rail travels along an iced slide plate, the lubricant may wear (vanish) faster from the slide plate surface and friction between the switch-rail and slide plate surfaces may increase. Not heating the lubricant but melting ice from the slide plate surface before the switch-rail moves should be the goal.
- IH uniformly heats the entire surface of the slide plate; therefore, lubricant properties remain identical all along the path of the switch-rail's travel.

- IH rapidly melts snow or ice all along the slide plate on the switch-rail's internal and external sides. This improves snow melting above the slide plate surface by thermal radiation in a space delimited by the stock-rail and switch-rail dimensions.
- There is a risk IH will insufficiently melt snow deposits between the slide plates in the space between the stock-rail and switch-rail. This may be particularly dangerous as the switch-rail shifts towards the stock-rail, causing insufficient contact between both the parts.
- As the switch-rail touches the stock-rail in RH, energy efficiency improves considerably since thermal energy losses to the air diminish. Snow and ice in the heated space melt more quickly.
- As the switch-rail touches the stock-rail in RH, temperature of the heating element reaches 200 °C. The rear part of the slide plate between the stock-rail and switch-rail heats rapidly and the lubricant on the slide plate surface dries ([25] gives the lubricant's flash point of 110 °C). The risk of escalating friction between moving parts of the turnout and their freezing to one another emerges.
- The stock-rail can be positioned in relation to a sleeper by means of a plastic washer. This significantly impairs thermal conductivity conditions from the stock-rail to the for both the heating systems.

In general, an in-depth assessment of advantages and disadvantages of both the systems of turnout heating should be attempted, considering experimental results and guided by the overarching criterion of safe rail travel.

In view of the above, the authors have proposed the following assumptions for efficiency assessment of turnout heating:

1. Figures 7 and 8 imply rail temperature may vary considerably when heated. Depending on sensor location, temperature on the rail head ranges between (5–12) °C after 180 min of heating, while the web reaches (5–17) °C. When tested in a climatic chamber, temperature distribution along the rail head or web may vary considerably from temperatures measured at specific locations. Therefore, reading errors of sensors themselves are ignored in analysis of rail heating results and treated as an acceptable part of measurement inaccuracy.
2. Analysis of both the systems' heating efficiency may be liable to error as snow is absent from a turnout in climatic chamber testing. Its melting is the key function of heating systems. To substitute for snowfall, the turnout was iced prior to heating by spraying the switch-rail and stock-rail with water, then the instant of total ice melting was observed.

It should be noted that there is no unambiguous method for comparing both the systems of turnout heating. Taking the above conditions and assumptions into account, the authors have formulated the following criteria of assessing efficiency of turnout heating:

1. For an open turnout, the part of the slide plate between the switch-rail and stock-rail must be free from snow and ice; in addition, snow cannot prevent continuing contact of the switch-rail and stock-rail after the switch-rail changes its position.
2. For a closed turnout—the switch-rail touching the stock-rail—the part of the slide plate between the switch-rail and the medium section of the turnout must be free from snow and ice. The contact between heads of the switch-rail and of the stock-rail cannot be iced, either.
3. Moving parts of the turnout are completely free from snow or ice in a relatively short period of time.
4. Quantity of electricity required to heat a unit distance of rail should be as low as possible.
5. Dynamics of temperature should be maximum possible.

Realizing the foregone conditions is chiefly based on reading temperature variations at designated points of the turnout and visual verification of the moment ice turns into water on selected turnout elements from the time the heating begins. Given temperature values, dynamics of its variations, location of the measurement sensor, and time of the heating, the process of snow melting can be analyzed.

5. Design of a Test Stand

A turnout heating test stand was produced in a climatic chamber. Air temperature was constant at the time of heating. Active power consumed by both the systems was identical for each measurement and equal to 450 W.

- Classic RH, including a 150 cm long bayonet heater, was mounted on the stock-rail foot on the switch-rail's side.
- Induction heating featured three inductors, each mounted on one. Each inductor was supplied from a shared sinusoidal voltage generator with a total power of 450 W (150 W an inductor) [23] were applied (Figure 16).

A 150 cm long switch-rail was laid on the slide plate. For the purposes of testing, the switch-rail was placed in two positions:

- Position a: touching the stock-rail (Figure 17a).
- Position b: 7 cm away from the stock-rail (Figure 17b).

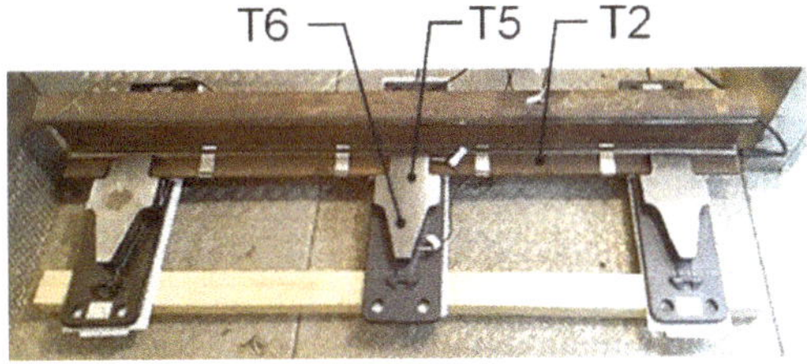

Figure 16. Tested fragment of the turnout stock-rail placed in a climatic chamber subject to resistance heating (RH) and induction heating (IH) [23].

Six PT100 sensors arranged (Figures 16 and 17):

- On the switch-rail foot—T1,
- On the stock-rail foot—T2,
- On the switch-rail head—T3,
- On the stock-rail head—T4,
- In the slide plate part to the side of the stock-rail—T5,
- In the slide plate part to the side of the switch-rail—T6

were used to measure the temperature.

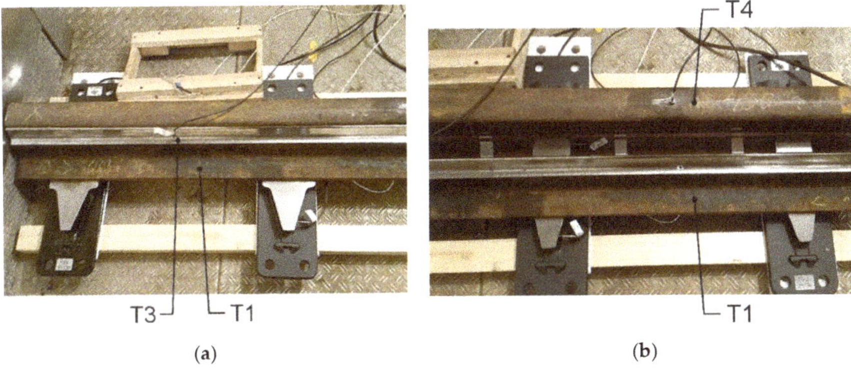

Figure 17. Arrangement of temperature measurement sensors around the turnout for: (**a**) the switch-rail touching the stock-rail; (**b**) the switch-rail away from the stock-rail [23].

6. Results and Discussion of Test Results for Resistance and Induction Turnout Heating

Measurement results for the ambient temperature of −30 °C are illustrated in Figure 18. Two positions of the switch-rail in relation to the stock-rail were assumed: in contact and 7 cm away. For the switch-rail touching the stock-rail in resistance heating, maximum temperature (T1) increment was 15 °C for the switch-rail foot and stock-rail head (Figure 18a). When the switch-rail was shifted away from the switch-rail in resistance heating, temperature of the stock-rail head (T4) was approximately −13 °C while the temperature (T1) of the switch-rail foot reduced dramatically (Figure 18c). Temperature of the stock-rail foot (T2) was identical in both cases and its increment was circa 12 °C (Figure 18a,b).

In the case of induction heating, the temperature (T5) rose by 25 °C for the slide plate to the side of the stock-rail and (T6) by approximately 15 °C for the part of the slide plate towards the switch-rail for both switch-rail positions (Figure 18b,d).

The switch-rail's position has no marked effect on reaching maximum temperatures that are lower 0 °C for both the heating systems. Negative temperatures at each turnout point were noted for both the heating systems. This means they do not operate efficiently at such low temperatures.

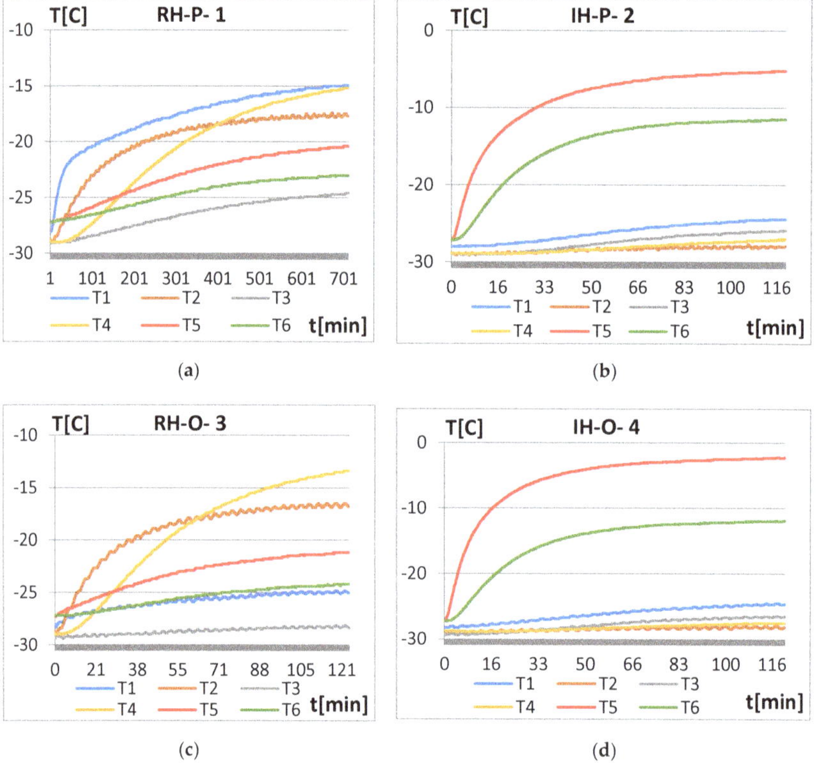

Figure 18. Temperature distribution across a turnout for ambient temperature −30 °C at the points: T1—on the switch-rail foot, T2—on the stock-rail foot, T3—on the switch-rail head, T4—on the stock-rail head, T5—in the slide plate part to the side of the stock-rail, T6—in the slide plate part to the side of the switch-rail; (**a**) resistance heating, the switch-rail touching; (**b**) induction heating, the switch-rail touching; (**c**) resistance heating, the switch-rail away; and (**d**) induction heating, the switch-rail away.

Since the heating is inefficient at −30 °C, the testing was undertaken for −5 °C, the most frequent annual average in temperate climate (Figure 19). Maximum temperatures of selected turnout parts are well in excess of 0 °C, contributing to melting of snow. For resistance heating and the switch-rail in contact with the stock-rail, maximum temperature increment was circa 14–15 °C for the head (T4) and foot of the stock-rail (T2), as well as for the switch-rail foot (T1) (Figure 19a). When the switch-rail was moved away from the stock-rail in resistance heating, temperature (T4) of the stock-rail head rose by ca. 15 °C while temperature (T1) of the switch-rail foot fell dramatically to approximately 0 °C (Figure 19.). Temperature (T2) of the stock-rail foot was identical in both cases and increased by 12 °C (Figure 19a,c).

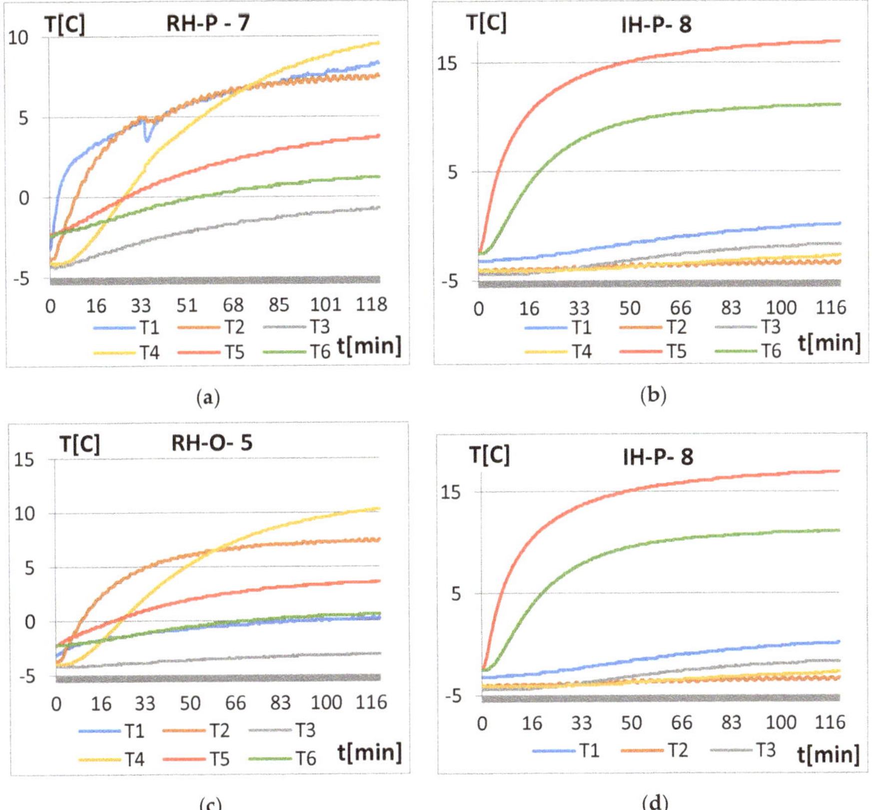

Figure 19. Temperature distribution across a turnout for ambient temperature −5 °C for the slide plate towards the switch-rail (Figure 19b). In the case of induction heating, shifting the at the points: T1—on the switch-rail foot, T2—on the stock-rail foot, T3—on the switch-rail head, T4—on the stock-rail head, T5in the slide plate part to the side of the stock-rail, T6—in the slide plate part to the side of the switch-rail; (**a**) resistance heating, the switch-rail touching; (**b**) induction heating, the switch-rail touching; (**c**) resistance heating, the switch-rail away; and (**d**) induction heating, the switch-rail away.

As far as induction heating is concerned, temperature (T5) grew by 22 °C for the slide plate from the side of the stock-rail and (T6) approximately 16 °C for the slide plate towards the switch-rail (Figure 19b). In the case of induction heating, shifting the switch-rail away caused temperature (T5) of the slide plate from the side of the stock-rail to rise by ca. 25 °C (Figure 19d). Temperatures (T2, T4) of the entire stock-rail are below zero. This was a result of the poor heat conduction between the slide plate and the switch-rail.

It can be said the switch-rail's position and changes of air temperature had no significant impact on increments of turnout temperatures in the foregoing cases. During heat induction, slight rises of stock-rail and switch-rail temperatures were observed, as the turnout parts did not have major roles in the process of snow melting. This indicated thermal energy of induction heating was generated in the slide plate and was for the most part convected in the area of a stock-rail and switch-rail. This was evidence of good snow melting conditions and high heating efficiency. Higher stock-rail temperatures were produced during resistance heating. This caused additional dispersion of thermal energy into the air without influencing snow melting between the stock-rail and switch-rail. Heating efficiency was lower for the same electric power, i.e., 450 W for both the heating systems tested.

Snow melting across the slide plate surface was attempted in order to ensure properly lubricate the slide plate and switch-rail surface. For induction heating, the slide plate temperature was considerably greater than the melting point, whereas the temperature increment was ca. 5 °C in the case of resistance heating when the switch-rail moved away. This value determined the minimum ambient temperature at which the switch-rail may traveled along an ice-free slide plate surface.

There was more effective snow melting in the stock-rail and switch-rail space, as well as more reliable lubrication of moving turnout parts expected in the case of induction heating. In the practice of resistance heating, lubricant was additionally dried in the vicinity of a heating element due to the latter's high temperatures of approximately 200 °C. Tribological parameters of a turnout's moving parts were impaired.

The turnout with a switch-rail shifted toward and away the stock-rail sprayed with water at an ambient temperature of −5 °C. This was done to compare efficiency of both the heating systems. When the switch-rail was away from the stock-rail, the turnout began to be heated after the water had frozen. Ice on the slide plate prevented the switch-rail from moving. During 150 min of resistance heating, an attempt was undertaken to shift the switch-rail, but to no effect. It remained so solidly frozen to the slide plate foot that the upward force could have lifted the entire turnout.

Induction heating was applied in the above conditions of a water sprayed turnout. Water drops began appearing on the slide plate feet after five minutes and it was possible to move the switch-rail. This means resistance forces to turnout drive reduce quickly and quality of lubrication of the turnout's moving parts is improved compared to resistance heating.

Similar testing was conducted for the switch-rail touching the stock-rail. It was only after 90 min of resistance heating that the switch-rail could be shifted away from the stock-rail due to some ice on the slide plate inside the turnout. On the other hand, the slide plate, stock-rail head, and switch-rail head remained iced together after 30 min of induction heating. The switch-rail could be shifted away, nonetheless. This may suggest greater start-up force in the initial phase of moving the switch-rail away from the stock-rail.

In general, the switch-rail was more ready to shift and lubrication conditions of moving turnout parts were better for the induction heating of an iced turnout with the switch-rail moved away from and moved to the stock-rail than for resistance heating.

Dynamics of temperature variations of selected turnout elements are an important criterion when evaluating turnout heating. This parameter is closely correlated with a system of turnout heating control. Greater dynamics also influence effectiveness of heating, expressed as a high capacity for heat emissions to melt snow.

Based on the temperature distribution across the stock-rail web by means of turnout automatic controls, it was assumed that the heating element temperature in constant operation should vary within (+2–+6) °C (Figure 20). In Polish weather conditions, the minimum ambient temperature of −10 °C can be assumed at which a heating system is capable of maintaining a turnout snow-free. Snow is rare at lower temperatures. For the minimum ambient temperature T_{am} = −10 °C and assuming a maximum temperature of a heated turnout part equal to +6 °C, a maximum temperature increment ΔT_{max} = 16 °C required for the purpose of comparative analysis was adopted.

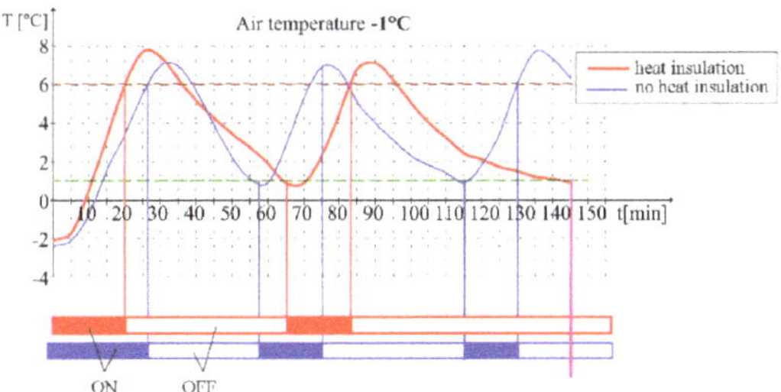

Figure 20. Temperature of stock-rail web for the control system in RH at the ambient temperature $T = -1$ °C; (**a**) with heat insulations and (**b**) without heat insulation [3].

Given these assumptions, the dynamics of temperature increment are defined as:

$$\Delta\tau = \frac{\Delta T}{\Delta t} \tag{3}$$

where: $\Delta T = T - T_{am}$—temperature increment over the initial value of T_{am},

T: measured temperature,
T_{am}: ambient temperature,
Δt: time interval of temperature increment [min].

Dynamics of temperature increment for the stock-rail web of a turnout installed in the field was 5/20 [C/min] = 0.25 [C/min] (Figure 20). In turnout testing in a climatic chamber (Figure 18a), the foot temperature increment for the RH system was 7/23 [C/min] = 0.3 [C/min]. Both the results were comparable, which validated the correctness of adopting the testing methodology with a climatic chamber.

Analysis of temperature increment across one of temperature measurement points for both the RH and IH systems was undertaken to evaluate energy efficiency of turnout heating. Temperature increment ΔT of this point had a considerable impact on effectiveness of snow melting across moving parts of the turnout and in the space between the stock-rail and the switch-rail. Temperature increment ΔT defined the difference between a current T and initial temperature T_{am}= −5 °C (Figure 21) and T_{am} = −15 °C (Figure 22).

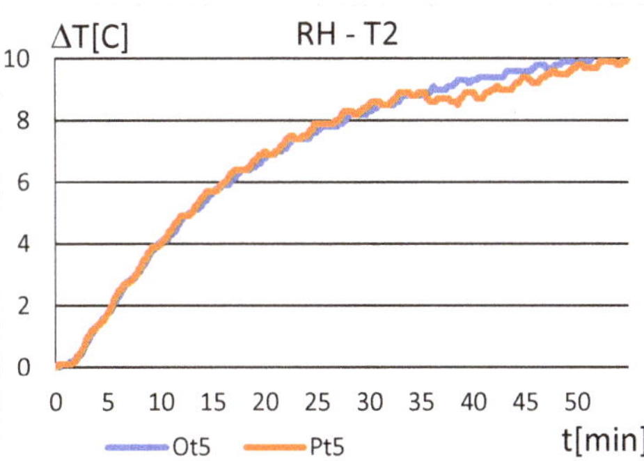

Figure 21. Increment temperature of the stock-rail web (T2) in resistance heating RH for: Ot5—the switch-rail moved away from the stock-rail and the ambient temperature −5 °C; Pt5—switch-rail in contact with the stock-rail and the ambient temperature −5 °C.

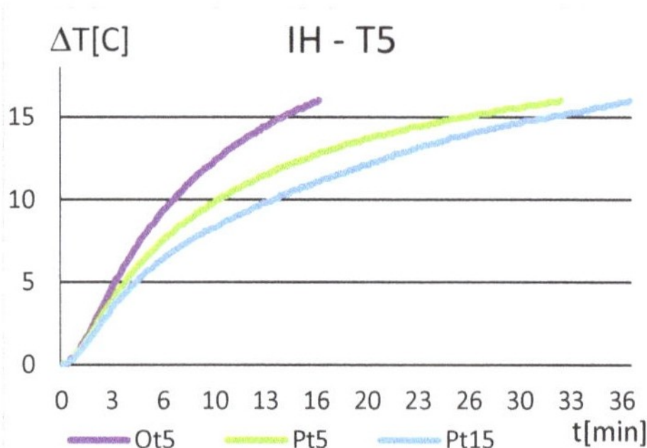

Figure 22. Slide plate increment temperature from the side of stock-rail (T5) in induction heating IH for: Ot5—the switch-rail moved away from the stock-rail and the ambient temperature −5 °C; Pt5—switch-rail in contact with the stock-rail and the ambient temperature −5 °C; Pt15—switch-rail in contact with the stock-rail and the ambient temperature −15 °C.

Temperature variations (T2) of the stock-rail foot for the switch-rail shifted away from (Ot) and touched (Pt) the stock-rail. Thus, they were selected for resistance heating RH (Figure 21). The temperature distribution (T2) was limited to the time interval of 50 min, when effective heating of the stock-rail foot took place. The temperature rose by about 10 °C. The ambient temperature at the time of testing was T_{am} = −5 °C. For T_{am} < −15 °C. The RH system was inefficient, attaining maximum temperatures below zero.

Dynamics of temperature increment in the adopted time interval were maximum—i.e., approximately 0.4—for an interval of around 10 min (Figure 23). Temperature increment (T2) and its dynamics were independent from the position of the switch-rail in relation to the stock-rail.

Temperature increments (T5) of the slide plate towards the stock-rail for the switch-rail shifted away from (Ot) and touching (Pt) the stock-rail were selected for the induction heating IH (Figure 22). The testing was conducted for the ambient temperatures T_{am} = −5 °C and T_{am} = −15 °C. The IH system was assumed to be inefficient for lower ambient temperatures.

The temperature distribution (T5) was limited to the time interval of 35 min. The temperature increments gained their maximum required value then: ΔT_{max} = 16 °C (Figure 22). The time interval needed to reach ΔT_{max} depended on the switch-rail's position (Ot or Pt) and ambient temperature T_{am}. When the switch-rail was in contact with the stock-rail, the heating time approximately doubled. This was not a major limitation since snow melting in the space between the switch-rail and the stock-rail was not required in the circumstances. The 10 °C lower ambient temperature contributed to extension of the heating time by several minutes.

Dynamics of temperature increment in the IH were maximum in the range $\Delta\tau\in$(1–1.4) for approximately 5 min (Figure 24). Shifting the switch-rail away from the stock-rail or increasing the ambient temperature contribute to greater dynamics of the temperature increment.

In general, the maximum dynamics of temperature increment in induction heating were approximately 3–4 times higher than in resistance heating, thus becoming maximum twice as soon. As a result, the slide plate was faster to reach its temperature set by the IH control system than in the RH, thereby melting snow more efficiently.

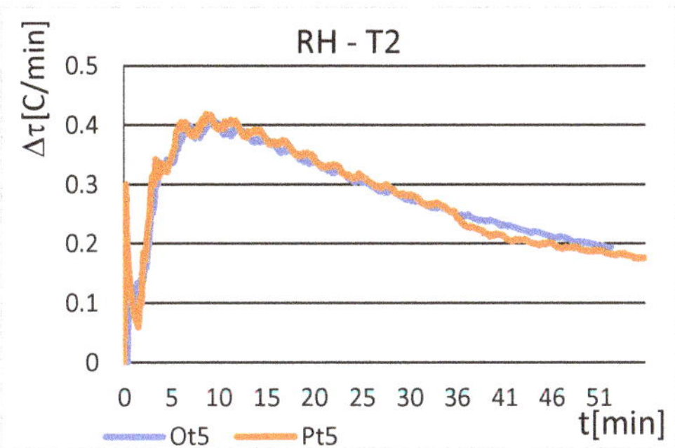

Figure 23. Dynamics of temperature increment across the stock-rail web (T2) in resistance heating RH for: Ot5—the switch-rail moved away from the stock-rail and the ambient temperature −5 °C; Pt5—switch-rail in contact with the stock-rail and the ambient temperature −5 °C.

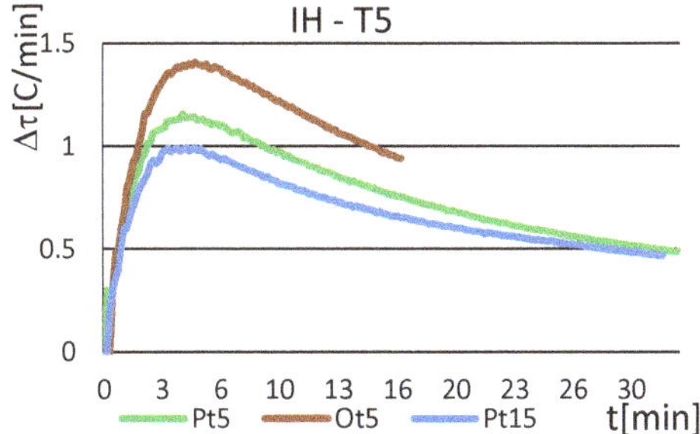

Figure 24. Dynamics of temperature increment across the slide plate from the side of stock-rail (T5) in induction heating IH for: Pt5—switch-rail in contact with the stock-rail and the ambient temperature −5 °C; Ot5—the switch-rail moved away from the stock-rail and the ambient temperature −5 °C; Pt15—switch-rail in contact with the stock-rail and the ambient temperature −15 °C.

Results of the climatic chamber testing also required analysis with regard to their uncertainty. The uncertainty of the measurements were related to a higher number of possible sources of impact, i.e., imperfect realization of a measured quantity, incomplete knowledge of impact of external conditions on the measurement procedure, finite resolution capabilities of measurement equipment, simplifying approximations, and assumptions adopted for the testing.

Figures 6 and 7 indicate that temperatures across the stock-rail head (T4) varied at the same instant when the rail was heated. Temperature variations between points 5 and 3 may reach ca. 10 °C. For the purpose of climatic chamber testing, the sensor (T4) was placed 0.5 m from one rail's end and 1 m from the other end. Rail head temperatures in the remaining positions are unknown. A rail was considerably longer in real turnouts, which may have added effect on temperature.

The imperfect temperature measurements made the results display slight oscillations (Figures 18 and 19). This could be explained with variable conditions in the climatic chamber. The temperature varied a little. Figure 25 shows a correlation between temperature in the climatic chamber (T_{am}) and temperature of the switch-rail foot (T1) in the time interval of 500 s. As T_{am} grows, temperature (T1) increased, and vice versa. Low-amplitude oscillations appeared in a broader time interval.

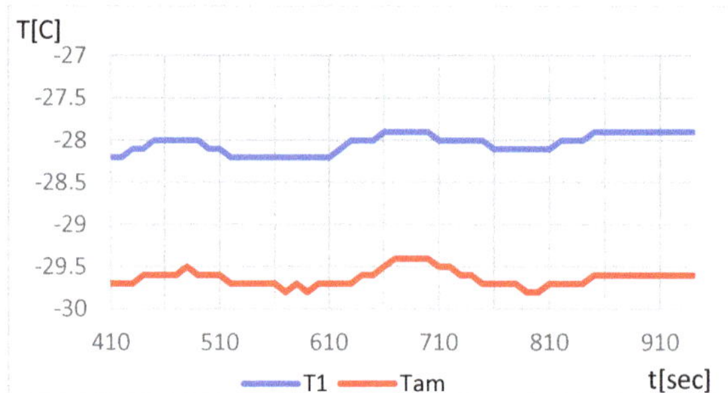

Figure 25. Temperature variations: on the switch-rail foot (T1) and in the climatic chamber (T_{am}).

Temperature changes on the switch-rail foot (T1) exhibited a sudden drop at 33 min (Figure 19a). This was most likely caused by shifting of measurement wires and loss of precise contact.

Oscillations of a significant amplitude were observed in the initial phase of the measurements. According to Equation (3), dynamics of temperature variations defined how quickly temperature increased from the start of the heating process. As the measurement accuracy of 0.1 °C was adopted, the temperature change ΔT (3) was constant, 0.1 °C, in the time interval of 110 s. As the time of heating increased, temperature dynamic variations diminished. Therefore, temperature variability in the climatic chamber had some impact on the process of oscillations (Figure 25).

7. Conclusions

Experimental testing with an identical electric power supplied to both systems of turnout heating, suggesting the following conclusions:

- In resistance heating, active energy supplied to a heater is converted into thermal energy emitted by the heater directly into the space between the switch-rail and the stock-rail. In addition, the stock-rail is heated, which contributes to energy dispersion, given the absence of any external heat insulation, and to reduced energy efficiency.
- During induction heating, active energy supplied to an inductor is converted into thermal energy in the slide plate, from where it penetrates directly into the space between the switch-rail and stock-rail, contributing to snow melting. Ineffective energy dispersion is minimum.
- Lubrication conditions of moving turnout parts are better in induction heating.
- Dynamics of temperature increment in induction heating are greater, which improves responses of a workpiece to temperatures set by the control system and contributes to more efficient melting of snow.
- Maximum temperature increment during induction heating assures continuous operation of the heating system at lower ambient temperatures than for resistance heating.

It should be said in summary that these results point to a greater heating efficiency of induction than of resistance heating. An ultimate assessment and comparison of both the system's heating efficiency is only possible in field conditions, since sources of thermal energy are positioned elsewhere, losses of the energy are emitted to the air in different ways, and energy sources for snow melting are diverse in the RH and IH heating systems under discussion.

Author Contributions: Conceptualization, E.S. and L.S.; methodology, L.S. and E.S.; validation, E.S., formal analysis, L.S. and E.S.; resources, L.S.; data curation, L.S., writing—original draft preparation, L.S. and E.S.; visualization, E.S. All authors have read and agreed to the published version of the manuscript.

Funding: This research received no external funding.

Conflicts of Interest: The authors declare no conflict of interest.

References

1. Szychta, E.; Szychta, L.; Luft, M.; Kiraga, K. Application of 3D simulation methods of the process of induction heating of rail turnouts. In *Infrastructure Design, Signalling and Security in Railway*; Perpinya, X., Ed.; InTech: Rijeka, Croatia, 2012; ISBN 978-953-51-0448-3.
2. Białoń, A.; Mikulski, J. Wpływ typu ogrzewania rozjazdów na zużycie energii elektrycznej. *Electr. Rev.* **2009**, *9*, 37–39.
3. Brodowski, D.; Andrulonis, J. Ogrzewanie rozjazdów kolejowych. *Probl. Kolejnictwa* **2002**, *135*, 45–87.
4. Brodowski, D.; Andrulonis, J. Efektywność ogrzewania rozjazdów kolejowych. *Probl. Kolejnictwa* **2000**, *122*, 25–34.
5. Flis, M.; Wołoszyn, M. Energy Efficiency Analysis of Railway Turnout Heating with a Simplified Snow Model Using Classical and Contactless Heating Method. In Proceedings of the International Interdisciplinary PhD Workshop, IIPhDW 2018, Świnoujście, Poland, 9–12 May 2018.
6. Burca, A.; Trip, N.D.; Leuca, T.; Dudrik, J. Consideration on the Analysis of an Induction Heating System. *IEEE Trans. Ind.* **2014**, *5*, 2504–2508.
7. Jintang, Y.; Haiduan, Q.; Gongfa, G.; Xiaoliang, Z. Numerical Simulation of the Quenching Process of U71Mn Rail Head. In Proceedings of the International Conferece on Measuring Technology and Mechatronics Automation, Changsha, China, 13–14 March 2010.
8. Lupi, S. *Fundamentals of Electroheat*; Springer International Publishing: Cham, Switzerland, 2017.
9. Rudnev, V.; Loveless, D.; Cook, R.L. *Handbook of Induction Heating*; CRC Press: Boca Raton, FL, USA, 2017.
10. Phadungthin, R.; Haema, J. Application Study on Induction Heating Using Half Bridge LLC Resonant Inverter. In Proceedings of the 12th IEEE Conference on Industrial Electronics and Applications (ICIEA), Siem Reap, Cambodia, 18–20 June 2017.
11. Komeza, K.; Lefik, M.; Szychta, E.; Szychta, L. Analysis of Resistance Heating Replacement by Induction Heating for Railway Tracks. In Proceedings of the XVII International Symposium on Electromagnetic Fields in Mechatronics, Electrical and Electronics Engineering, Poznan, Poland, 10–12 September 2015.
12. Wołoszyn, M.; Jakubuk, K.; Flis, M. Analysis of resistive and inductive heating of railway turnouts. *Electr. Rev.* **2016**, *1*, 54–57. [CrossRef]
13. Obata, Y. Device for Deicing Rails. U.S. Patent 3,648,017, 7 March 1972.
14. Horoszko, E. Indukcyjne ogrzewanie rozjazdów kolejowych. *Elektrowarme Int.* **1978**, *6*, 17–23.
15. Kondo, S.; Uetani, Y.; Hama, Y.; Iguchi, J.; Tasaki, H.; Fujisawa, A.; Higashide, S. Line Switch Part Snow Melting Device. U.S. Patent 6664521 B1, 16 December 2003.
16. Reichle, D.L. Impedance Heating for Railroad Track Switch. U.S. Patent 6727470 B2, 27 April 2004.
17. Lennart, A. Rail Heating Device. EP Patent 1 582 627 A1, 5 October 2005.
18. Lennart, A. Magnetic Heating Device. U.S. Patent 7,315,011 B2, 1 January 2008.
19. Chuanbin, S. Fast Snow Melting Device of Railroad Switch. CN Patent 101629402A, 10 January 2010.
20. Xibao, S. Railway Turnout Snow Melting Device Adopting Electromagnetic Heating. CN Patent 201896278 U, 13 July 2011.
21. Falldin, A. Railway Track Heating Device. U.S. Patent 10619310 B2, 14 April 2020.
22. Indheater. Innovative and Reliable Switch Heater. Rail&Industry. Available online: http://www.yumpu.com (accessed on 13 February 2020).
23. Cholewa, A.; Kreft, O.; Buda, T. Inteligentny system indukcyjnego ogrzewania rozjazdów—TRACK TEC. In Proceedings of the Konferencja Naukowo-Techniczna Nowoczesne Technologie w Realizacji Projektów Inwestycyjnych Transportu Kolejowego, Jurata, Poland, 5–7 April 2017.
24. Cholewa, A. Rail Switch Heating Device. European Patent Application EP 3 141 657 B1, 2 May 2018.
25. Milczarek, D.; Naduk, E. Wymagania dotyczące środków do smarowania rozjazdów kolejowych. *Prace Inst. Kolejnictwa* **2016**, *149*, 22–26.

Article

Entropy Generation in a Dissipative Nanofluid Flow under the Influence of Magnetic Dissipation and Transpiration

Dianchen Lu [1], Muhammad Idrees Afridi [2], Usman Allauddin [3], Umer Farooq [1] and Muhammad Qasim [1,*]

[1] Department of Mathematics, Faculty of Science, Jiangsu University, Zhenjiang 212013, China; dclu@ujs.edu.cn (D.L.); umer_farooq@comsats.edu.pk (U.F.)

[2] Department of Computing, Abasyn University, Islamabad 45710, Pakistan; Muhammad.idrees@abasynisb.edu.pk

[3] Department of Mechanical Engineering, NED University of Engineering & Technology, Karachi 75270, Pakistan; usman.allauddin@neduet.edu.pk

* Correspondence: mqasim@ujs.edu.cn

Received: 18 June 2020; Accepted: 25 August 2020; Published: 21 October 2020

Abstract: The present study explores the entropy generation, flow, and heat transfer characteristics of a dissipative nanofluid in the presence of transpiration effects at the boundary. The non-isothermal boundary conditions are taken into consideration to guarantee self-similar solutions. The electrically conducting nanofluid flow is influenced by a magnetic field of constant strength. The ultrafine particles (nanoparticles of Fe_3O_4/CuO) are dispersed in the technological fluid water (H_2O). Both the base fluid and the nanofluid have the same bulk velocity and are assumed to be in thermal equilibrium. Tiwari and Dass's idea is used for the mathematical modeling of the problem. Furthermore, the ultrafine particles are supposed to be spherical, and Maxwell Garnett's model is used for the effective thermal conductivity of the nanofluid. Closed-form solutions are derived for boundary layer momentum and energy equations. These solutions are then utilized to access the entropy generation and the irreversibility parameter. The relative importance of different sources of entropy generation in the boundary layer is discussed through various graphs. The effects of space free physical parameters such as mass suction parameter (S), viscous dissipation parameter (Ec), magnetic heating parameter (M), and solid volume fraction (ϕ) of the ultrafine particles on the velocity, Bejan number, temperature, and entropy generation are elaborated through various graphs. It is found that the parabolic wall temperature facilitates similarity transformations so that self-similar equations can be achieved in the presence of viscous dissipation. It is observed that the entropy generation number is an increasing function of the Eckert number and solid volume fraction. The entropy production rate in the $Fe_3O_4 - H_2O$ nanofluid is higher than that in the $CuO - H_2O$ nanofluid under the same circumstances.

Keywords: nanofluid; heat transfer; entropy generation; viscous dissipation; magnetic heating

1. Introduction

The Navier-Stokes equations, which are second-order nonlinear partial differential equations, govern the viscous fluid–fluid flow. The exact solution of the complete Navier–Stokes equations has not yet been computed. However, closed-form solutions can be established in certain physical circumstances under reasonable suppositions [1–5]. Exact solutions are important since such solutions can be utilized to validate asymptotic analytical and numerical solutions. Crane [6] found the closed-form solution of the simplified Navier-Stokes equations under the boundary layer approximations to analyze the flow over a stretched surface. Some researchers determined the closed-form solutions of boundary layer flow after the pioneering work of Crane with various physical conditions [7–11].

It is essential to examine heat transfer issues in industrial engineering. Recently, heat transfer analysis has been limited to the first law of thermodynamics, which only concerns energy conservation during the interactions of the systems and surroundings. It deals solely with the amount of energy regardless of its quality. Moreover, the first law does not distinguish between heat and work. It assumes that work and heat are fully interchangeable, but work is high-quality energy and can be fully converted into heat, while heat is low-quality energy and cannot be fully converted into work. Heat is an unorganized form of energy. The law of entropy shows that the entropy increase in the cold object is higher than the decrease of entropy in the hot object. This means that the final state is more random in the thermodynamic system. This analysis suggests that the heat transfer phenomenon decreases energy quality or increases the system entropy. To investigate this energy quality reduction, Bejan [12,13] proposed a method called entropy minimization that is based on the law of entropy. The law of entropy (second law of thermodynamics) is used to maintain energy quality [14–20]. In addition to heat transfer, frictional heating and magnetic dissipation also generate entropy in fluid flow problems [21–25].

Conventional working fluids such as kerosene, gasoline, water, engine oil, and fluid mixtures have exceptionally poor thermal conductivity, as demonstrated by the vast number of industries dealing with these conventional working fluids. However, due to their inefficiency in thermal conductivity, they face several problems. The use of nanoscale elements in base fluids is one of the most important techniques used to resolve this deficiency. Such a mixture of nanometer-sized particles and a working fluid is called a nanofluid. In comparison to base liquids, nanofluids possess high thermal conductivity [26–32]. Many researchers firmly agree on the remarkable characteristics of nanofluids. Over the past two decades, this new type of fluid has attracted the attention of many researchers. Nanofluid studies have a variety of important applications, such as product provision for cancer, cooling systems, nuclear power plant cooling, and computer equipment cooling. Hsiao [33] conducted stagnation nanofluid energy conversion analysis for the conjugate problem of conduction–convection and heat source/sink. Ma et al. [34] explored the gravitational convection term of heat management in a shell and tube heat exchanger filled with a $Fe_3O_4 - H_2O$ nanoliquid by utilizing a lattice Boltzmann scheme. Wakif et al. [35] reported the impacts of thermal radiation and surface roughness on the complex dynamics of water transporting alumina and copper oxide nanoparticles. Hsiao [36] reported nanofluid flow for conjugating mixed convection and radiation with interactive physical characteristics. In a channel with active heaters and coolers, a numerical simulation was introduced by Ma et al. [37] to examine the impacts of magnetic field on heat transfer in a $MgO - Ag - H_2O$ nanoliquid. Prasad et al. [38] examined the upper-convected Maxwell three-dimensional rotational flow with a convective boundary condition and zero mass flux for the concentration of nanoparticles. Frictional heating is the conversion of fluid kinetic energy to heat due to the frictional forces between all the neighboring fluid layers. Frictional heating is the main factor in the study of heat transfer in boundary layer flows. Since large velocity gradients exist within the boundary layer, the viscous dissipation effects cannot be neglected. When there is a viscous dissipation, a term for viscous dissipation is incorporated into the energy equation [39–46].

In this research, the exact solutions of transformed nonlinear dimensionless momentum and energy equations that occur in the magnetohydrodynamic (MHD) boundary layer flow of nanofluid are obtained. The goal of the work, apart from providing a benchmark solution for numerical simulation, is the parametric analysis of entropy generation. The work also describes how boundary conditions facilitate similarity transformations to get self-similar equations. The literature review reveals that nonsimilar problems are treated as self-similar problems. Furthermore, the entropy generation analysis exists in literature, but the analysis is limited to the low temperature difference between the boundary and bulk fluid. The present work is free from such a constraint and is valid for both low and high temperature differences. In addition, the terms for frictional heating and magnetic dissipation are added to the energy equation and the expression for entropy generation. To the best of our knowledge, no one has reported the exact solutions for nanofluid flow induced by a linearly stretching surface with a parabolic temperature profile at the boundary. Obtained exact solutions are used for calculating entropy generation and the Bejan number. Visual representations are used to investigate the effects of physical parameters on the nanofluid flow, thermal field, entropy generation profile, and Bejan number.

2. Statement of the Problem and Governing Equations

Consider the electrically conducting and dissipative nanofluid flow over a stretching surface as shown in Figure 1. The nanofluid is supposed to be a mixture of base fluid (water) and nanoparticles Fe_3O_4/CuO. The Cartesian coordinate system (X, Y) is chosen in such a way that the $X-$axis is taken along the solid boundary and the $Y-$axis is normal to it. Let $U_w(X) = U_o X$ be the velocity of the stretching boundary and $T_w(X) = T_b + C_o X^2$ be the temperature variation at the surface of the stretching boundary; here, T_b and the subscript w represent the bulk fluid temperature and the condition at the solid boundary, while U_o and C_o represent the dimensional constants. The imposed magnetic field is constant and of strength B_o. The generalized Ohm's law in the absence of an electrical field is $\vec{j} = \sigma_{nf}\left(\vec{q} \times \vec{B}_o\right)$, where σ_{nf} and $\vec{q}\left(\vec{U}, \vec{V}\right)$ show the electrical conductivity of nanofluid and bulk velocity field of the nanofluid, respectively. The magnetic force $\bar{j} \times B_o$ and magnetic dissipation $\frac{\vec{j}.\vec{j}}{\sigma_{nf}}$ are simplified to $-\sigma_{nf} B_o^2 U$ and $\sigma_{nf} B_o^2 U^2$, respectively.

The equations governing the incompressible nanofluid flow for the present problem are

$$\frac{\partial U}{\partial X} + \frac{\partial V}{\partial Y} = 0, \tag{1}$$

$$U\frac{\partial U}{\partial X} + V\frac{\partial U}{\partial Y} = \nu_{nf}\frac{\partial^2 U}{\partial Y^2} - \frac{\sigma_{nf} B_o^2 U}{\rho_{nf}}, \tag{2}$$

$$\left(U\frac{\partial T}{\partial X} + V\frac{\partial T}{\partial Y}\right) = \left(\frac{1}{\rho C_p}\right)_{nf}\left(k_{nf}\frac{\partial^2 T}{\partial Y^2} + \mu_{nf}\left(\frac{\partial U}{\partial Y}\right)^2 + \sigma_{nf} B_o^2 U^2\right) \tag{3}$$

The imposed boundary conditions are as follows:

$$\left.\begin{array}{c} U(X,0) = U_w(X) = U_o X,\ V(X,0) = V_w,\ T(X,0) = T_w(X) = T_b + C_o X^2 \\ U(X, Y \to \infty) \to 0,\ T(X, Y \to \infty) \to T_b \end{array}\right\} \tag{4}$$

The governing self-similar equations are obtained from Equations (2) and (3) by using the following dimensionless variables:

$$\eta = Y\sqrt{\frac{U_o}{\nu_{bf}}},\ U = U_o X f'(\eta),\ V = -\sqrt{U_o \nu_{bf}} f(\eta),\ \theta(\eta) = \frac{T - T_b}{T(X,0) - T_b} \tag{5}$$

Equations (2) and (3) under the transformation in Equation (5) become

$$\frac{G_1}{G_o} f''' + f f'' - f'^2 - \frac{G_3}{G_o} M^2 f' = 0, \tag{6}$$

$$\frac{G_5}{G_4}\theta'' + \frac{G_1}{G_4}\mathrm{EcPr} f''^2 + \mathrm{Pr} f \theta' + \frac{G_3}{G_4}\mathrm{Ec}M^2\mathrm{Pr} f'^2 - 2\mathrm{Pr}\theta f' = 0 \tag{7}$$

The imposed boundary conditions are transformed to

$$f(0) = -\frac{V_w}{\sqrt{U_o \nu_{bf}}} = S,\ f'(0) = 1,\ f'(\eta \to \infty) = 0 \tag{8}$$

$$\theta(0) = 1,\ \theta(\eta \to \infty) = 0 \tag{9}$$

where $G_o = (1-\phi) + \phi\left(\frac{\rho_s}{\rho_{bf}}\right)$, $G_1 = (1-\phi)^{-2.5}$, $G_3 = \frac{\sigma_{nf}}{\sigma_{bf}}$, $G_4 = 1 - \phi + \phi\left(\frac{(\rho C_p)_s}{(\rho C_p)_{bf}}\right)$, $G_5 = \frac{k_{nf}}{k_{bf}}$, and $Ec = \frac{U_w^2}{(C_p)_{bf}\,(T(X,0)-T_b)}$ (Eckert number), and the subscripts bf and s are used for base fluid and nanoparticles, respectively. $\Pr = \frac{v_{bf}}{\alpha_{bf}}$ (Prandtl number); α_{bf} indicates base fluid thermal diffusivity; $M^2 = \frac{\sigma_{bf} B_0^2}{\rho_{bf} U_0}$; $S = -\frac{V_w}{\sqrt{U_o v_{bf}}}$ and shows the dimensionless mass-transfer parameter; and v_{nf}, σ_{nf}, ρ_{nf}, k_{nf}, and $(\rho C_p)_{nf}$ are defined in Table 1. The thermophysical properties of CuO, Fe_3O_4, and working fluid (H_2O) are shown in Table 2.

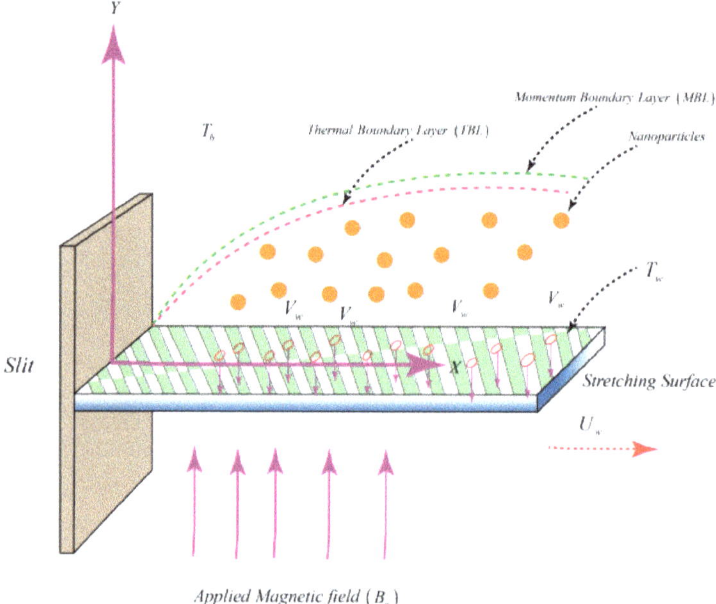

Figure 1. Physical flow model and coordinate system.

Table 1. Effective thermophysical properties of nanofluid [47–52].

Thermophysical Property of Nanofluid	Symbol	Defined
Thermal conductivity	k_{nf}	$k_{nf} = \frac{(k_s+2k_{bf})-2\phi(k_{bf}-k_s)}{(k_s+2k_{bf})+\phi(k_{bf}-k_s)} k_{bf}$ here, ϕ represents sold volume fraction of nanoparticles.
Viscosity	μ_{nf}	$\mu_{nf} = \frac{\mu_{bf}}{(1-\phi)^{2.5}}$
Electric conductivity	σ_{nf}	$\sigma_{nf} = 1 + \frac{3\left(\frac{\sigma_s}{\sigma_{bf}}-1\right)\phi}{\left(\frac{\sigma_s}{\sigma_{bf}}+2\right)-\left(\frac{\sigma_s}{\sigma_{bf}}-1\right)\phi} \sigma_{bf}$
Heat capacitance	$(\rho C_p)_{nf}$	$(\rho C_p)_{nf} = (1-\phi)(\rho C_p)_{bf} + \phi(\rho c_p)_s$
Density	ρ_{nf}	$\rho_{nf} = (1-\phi)\rho_{bf} + \phi\rho_s$

Table 2. Thermophysical properties of CuO, Fe_3O_4, and working fluid (H_2O).

Physical Properties	H_2O	CuO	Fe_3O_4
C_p (J/kgK)	4179	531.8	670
k (W/mK)	0.613	76.5	6.0
$\rho\left(\text{kg/m}^3\right)$	997.1	6320	5200
$\sigma\left(\text{S}\times\text{m}^{-1}\right)$	5180	2.7×10^{-8}	25,000
Pr (-)	6.8	-	-

3. Solution Methodology

3.1. Closed-Form Solution of Momentum Balance Equation

The closed-form exact solution of Equation (6) with associated boundary conditions of Equation (8) is supposed as follows:

$$f(\eta) = C_1 + C_2e^{-\beta\eta},\ \beta > 0 \tag{10}$$

Using the first two boundary conditions defined in Equation (8), the computed arbitrary constants C_1 and C_2 are

$$C_1 = S + \frac{1}{\beta},\ C_2 = -\frac{1}{\beta} \tag{11}$$

Putting Equation (11) into Equation (10), we get

$$f(\eta) = S + \frac{1}{\beta}\left(1 - e^{-\beta\eta}\right) \tag{12}$$

The above closed-form solution trivially satisfies the far-field boundary condition as defined in Equation (8) for $\beta > 0$. To find β, we insert Equation (12) into Equation (6) and get

$$\frac{G_1}{G_o}\beta^2 - S\beta - 1 - \frac{G_3}{G_o}M^2 = 0 \tag{13}$$

By solving the above equation, we have

$$\beta = G_o\left(\frac{S + \sqrt{S^2 + 4\frac{G_1}{G_o}\left(1 + \frac{G_3}{G_o}M^2\right)}}{2G_1}\right) > 0. \tag{14}$$

The closed-form solution of the boundary value problem (Equations (6) and (7)) is given by

$$f(\eta) = S + \frac{2G_1}{G_o\left(S + \sqrt{S^2 + 4\frac{G_1}{G_o}\left(1 + \frac{G_3}{G_o}M^2\right)}\right)}\left(1 - e^{-G_o\left(\frac{S+\sqrt{S^2+4\frac{G_1}{G_o}\left(1+\frac{G_3}{G_o}M^2\right)}}{2G_1}\right)\eta}\right) \tag{15}$$

3.2. Solution of Energy Balance Equation via Laplace Transform

Equation (7) is decoupled from Equation (6) by substituting Equation (12) into Equation (7) as follows:

$$\frac{G_5}{G_4}\theta'' + \frac{G_1}{G_4}\text{EcPr}\beta^2e^{-2\beta\eta} + \text{Pr}\left(S + \frac{1}{\beta}\left(1 - e^{-\beta\eta}\right)\right)\theta' + \frac{G_3}{G_4}\text{Ec}M^2\text{Pr}e^{-2\beta\eta} - 2\text{Pr}\theta e^{-\beta\eta} = 0 \tag{16}$$

To get rid of exponential coefficients, we define a new variable, ξ, as follows:

$$\xi = \frac{\Pr}{\beta^2}e^{-\beta\eta} \tag{17}$$

By utilizing the above transformation, Equation (7) and the related boundary conditions take the following form:

$$\xi\frac{d^2\theta}{d\xi^2} + \frac{d\theta}{d\xi}\left(K + \frac{\xi}{G}\right) + \xi L - 2\frac{\theta}{G} = 0, \tag{18}$$

$$\theta\left(\frac{\Pr}{\beta^2}\right) = 1,\ \theta(0) = 0 \tag{19}$$

with

$$K = 1 - \frac{\Pr(1+\beta S)}{G\beta^2},\ L = \frac{Ec\beta^2}{G\Pr}\left(\frac{G_1}{G_4}\beta^2 + \frac{G_3}{G_4}M^2\right) \text{ and } G = \frac{G_5}{G_4}. \tag{20}$$

By employing Laplace transform on Equation (18) and then using Equation (19), we obtain

$$\frac{d\Theta(\zeta)}{d\zeta} + \Theta(\zeta)\left[\frac{\zeta(2-K) + \frac{3}{G}}{\zeta\left(\zeta + \frac{1}{G}\right)}\right] = \frac{L}{\zeta^3\left(\zeta + \frac{1}{G}\right)} \tag{21}$$

where $\Theta(\zeta)$ is the Laplace transform of the function $\theta(\xi)$. Equation (21) is a Leibnitz first-type linear equation with integrating factor

$$e^{\int\frac{\zeta(2-K)+\frac{3}{G}}{\zeta\left(\zeta+\frac{1}{G}\right)}d\zeta} = \frac{\zeta^3}{(G\zeta+1)^{1+K}}. \tag{22}$$

Solving Equation (21) by utilizing Equation (22), we have

$$\Theta(\zeta) = \frac{L}{\zeta^3(-K-1)} + c\frac{(G\zeta+1)^{K+1}}{\zeta^3} \tag{23}$$

By taking Laplace inverse of Equation (23), we get

$$\theta(\xi) = \frac{L\xi^2}{2(-K-1)} + \frac{c}{2G^{-K-1}\Gamma(-K-1)}\left(\xi^2 * \xi^{-2-K}e^{\left(\frac{-\xi}{G}\right)}\right) \tag{24}$$

Here, an asterisk (*) indicates convolution and Γ shows a gamma function. The convolution of two functions, $F(\xi)$ and $G(\xi)$, is defined as follows:

$$F(\xi) * H(\xi) = \int_0^{\xi} F(\xi-\varepsilon)H(\varepsilon)d\varepsilon \tag{25}$$

By taking $F(\xi) = \xi^2$ and $H(\xi) = e^{\frac{-\xi}{G}}\xi^{-K-2}$, Equation (24) takes the following form:

$$\theta(\xi) = \frac{L\xi^2}{2(-K-1)} + \frac{c}{2G^{-K-1}\Gamma(-K-1)}\int_0^{\xi}(\xi-\varepsilon)^2 e^{\frac{-\varepsilon}{G}}\varepsilon^{-K-2}d\varepsilon. \tag{26}$$

By employing the transformation $\varepsilon = \xi u$, the above equation takes the following form:

$$\theta(\xi) = -\frac{L\xi^2}{2(K+1)} + \frac{c\xi^{1-K}}{2G^{-K-1}\Gamma(-K-1)} \int_0^1 (1-u)^2 e^{\frac{-u\xi}{G}} u^{-K-2} du. \tag{27}$$

By utilizing the integral form of Kummer's confluent hypergeometric function, i.e., $M_{1,1}\left(-K-1; -K+2; \frac{-\xi}{G}\right) = \frac{\Gamma(2-K)}{2\Gamma(-1-K)} \int_0^1 (1-u)^2 e^{\frac{-u\xi}{G}} u^{-K-2} d$, Equation (27) becomes

$$\theta(\xi) = -\frac{L\xi^2}{2(K+1)} + \frac{cG^{K+1}\xi^{1-K}}{\Gamma(2-K)} M_{1,1}\left(-K-1; -K+2; \frac{-\xi}{G}\right). \tag{28}$$

The boundary condition at the surface of the stretching surface $\theta(0) = 0$ is satisfied identically. However, the constant of integration c is obtained by using the far-field boundary condition $\theta\left(\xi = \frac{\Pr}{\beta^2}\right) = 1$ and is given by

$$c = \frac{\Gamma(2-m)\left(\frac{2(K+1)+L\left(\frac{\Pr}{\beta^2}\right)^2}{2(K+1)}\right)}{G^{K+1}\left(\frac{\Pr}{\beta^2}\right)^{1-K} M_{1,1}\left(-1-K; 2-K; -\frac{\Pr}{G\beta^2}\right)}. \tag{29}$$

Finally, by inserting Equation (29) into Equation (28) and using the transformation $\xi = \frac{\Pr}{\beta^2} e^{-\beta\eta}$, we obtain the exact solution of the energy equation:

$$\theta(\eta) = -\frac{1}{2}\frac{L}{(K+1)}\left(\frac{\Pr e^{-\beta\eta}}{\beta^2}\right)^2 + \frac{\left(\frac{\Pr e^{-\beta\eta}}{\beta^2}\right)^{1-K} M_{1,1}\left(-1-K; 2-K; -\frac{\Pr e^{-\beta\eta}}{G\beta^2}\right)\left(1 + \frac{L}{2(1+K)}\left(\frac{\Pr}{\beta^2}\right)^2\right)}{\left(\frac{\Pr}{\beta^2}\right)^{1-K} M_{1,1}\left(-K-1; 2-K; -\frac{\Pr}{G\beta^2}\right)}. \tag{30}$$

4. Analysis of Entropy Generation

The rate of entropy generation in the presence of heat dissipation phenomenon with magnetic heating is given by

$$\dot{E}'''_{Gen} = \frac{k_{nf}}{T^2}\left(\frac{\partial T}{\partial Y}\right)^2 + \frac{\mu_{nf}}{T}\left(\frac{\partial U}{\partial Y}\right)^2 + \frac{\sigma_{nf} B_0^2 U^2}{T}, \tag{31}$$

Using Equation (6), Equation (31) becomes

$$\frac{\dot{E}'''_{Gen}}{\left(\dot{E}'''_{Gen}\right)_0} = Ns = \underbrace{G_5\frac{\theta'^2}{(\theta+\Lambda)^2}}_{N_H} + \underbrace{\frac{G_1 \Pr Ec f''^2}{(\theta+\Lambda)}}_{N_F} + \underbrace{G_3\frac{\Pr M^2 Ec f'^2}{(\theta+\Lambda)}}_{N_M}. \tag{32}$$

Here, $\left(\dot{E}'''_{Gen}\right)_0 = \frac{k_{bf} U_o}{\nu_{bf}}$ indicates characteristic entropy generation; Ns indicates entropy production rate in dimensionless form; $\Lambda = \frac{T_b}{T_w - T_b}$ shows the temperature parameter; and N_H, N_F, and N_M represent the dimensionless form of entropy generation due to heat transfer, viscous dissipation, and magnetic heating, respectively.

By utilizing the obtained exact solutions, the three sources of entropy generation stated above take the following forms:

$$N_H = \frac{1}{(\Pr)_{eff}} \frac{\left[\left(-\frac{1}{2}\frac{L}{(K+1)}\left(\frac{\Pr e^{-\beta\eta}}{\beta^2}\right)^2 + \frac{\left(\frac{\Pr e^{-\beta\eta}}{\beta^2}\right)^{1-K} M_{1,1}\left(-1-K\,;2-K\,;-\frac{\Pr e^{-\beta\eta}}{G\beta^2}\right)\left(1+\frac{L}{2(1+K)}\left(\frac{\Pr}{\beta^2}\right)^2\right)}{\left(\frac{\Pr}{\beta^2}\right)^{1-K} M_{1,1}\left(-K-1\,;2-K\,;-\frac{\Pr}{G\beta^2}\right)}\right)'\right]^2}{\left(\Lambda-\frac{1}{2}\frac{L}{(K+1)}\left(\frac{\Pr e^{-\beta\eta}}{\beta^2}\right)^2 + \frac{\left(\frac{\Pr e^{-\beta\eta}}{\beta^2}\right)^{1-K} M_{1,1}\left(-1-K\,;2-K\,;-\frac{\Pr e^{-\beta\eta}}{G\beta^2}\right)\left(1+\frac{L}{2(1+K)}\left(\frac{\Pr}{\beta^2}\right)^2\right)}{\left(\frac{\Pr}{\beta^2}\right)^{1-K} M_{1,1}\left(-K-1\,;2-K\,;-\frac{\Pr}{G\beta^2}\right)}\right)^2}. \tag{33}$$

$$N_F = \frac{Ec\Pr\left(\frac{-S+\sqrt{S^2+4\frac{G_1}{G_o}\left(1+\frac{G_3}{G_o}M^2\right)}}{2G_1} e^{-\eta\left(\frac{S+\sqrt{S^2+4\frac{G_1}{G_o}(1+\frac{G_3}{G_o}M^2)G_o}}{2G_1}\right)}\right)^2 G^3{}_o}{\left(\Lambda-\frac{1}{2}\frac{L}{(K+1)}\left(\frac{\Pr e^{-\beta\eta}}{\beta^2}\right)^2 + \frac{\left(\frac{\Pr e^{-\beta\eta}}{\beta^2}\right)^{1-K} M_{1,1}\left(-1-K\,;2-K\,;-\frac{\Pr e^{-\beta\eta}}{G\beta^2}\right)\left(1+\frac{L}{2(1+K)}\left(\frac{\Pr}{\beta^2}\right)^2\right)}{\left(\frac{\Pr}{\beta^2}\right)^{1-K} M_{1,1}\left(-K-1\,;2-K\,;-\frac{\Pr}{G\beta^2}\right)}\right)}. \tag{34}$$

and

$$N_H = \frac{M^2 Ec\Pr\left(e^{-\eta\left(\frac{S+\sqrt{S^2+4\frac{G_1}{G_o}(1+\frac{G_3}{G_o}M^2)G_o}}{2G_1}\right)}\right)^2}{\left(\Lambda-\frac{1}{2}\frac{L}{(K+1)}\left(\frac{\Pr e^{-\beta\eta}}{\beta^2}\right)^2 + \frac{\left(\frac{\Pr e^{-\beta\eta}}{\beta^2}\right)^{1-K} M_{1,1}\left(-1-K\,;2-K\,;-\frac{\Pr e^{-\beta\eta}}{G\beta^2}\right)\left(1+\frac{L}{2(1+K)}\left(\frac{\Pr}{\beta^2}\right)^2\right)}{\left(\frac{\Pr}{\beta^2}\right)^{1-K} M_{1,1}\left(-K-1\,;2-K\,;-\frac{\Pr}{G\beta^2}\right)}\right)}. \tag{35}$$

4.1. Bejan Number

To compare the spatial distribution of entropy generation in a flow field due to various sources, an irreversibility ratio parameter known as Bejan number (Be) is defined as given below

$$Be = \frac{\frac{k_{nf}}{T^2}\left(\frac{\partial T}{\partial Y}\right)^2 \Rightarrow (Entropy\ generation\ due\ to\ heat\ transfer)}{\left(\frac{k_{nf}}{T^2}\left(\frac{\partial T}{\partial Y}\right)^2 + \frac{\mu_{nf}}{T}\left(\frac{\partial U}{\partial Y}\right)^2 + \frac{\sigma_{nf}B_0^2U^2}{T}\right) \Rightarrow (Total\ entropy\ generation)} \tag{36}$$

After the utilization of similarity variables, Equation (36) takes the following form:

$$Be = \frac{G_5\frac{\theta'^2}{(\theta+\Lambda)^2} \Rightarrow N_H}{\left(G_5\frac{\theta'^2}{(\theta+\Lambda)^2} + \frac{G_1\Pr Ecf''^2}{(\theta+\Lambda)} + G_3\frac{\Pr M^2 Ecf'^2}{(\theta+\Lambda)}\right) \Rightarrow (N_H+N_F+N_M)} \tag{37}$$

5. Results and Discussion

The nondimensional complicated differential equations (momentum and energy equations) are solved by taking into consideration the exponential form solution and the Laplace transform. The exact expressions are obtained for entropy generation via heat transfer, magnetic heating, and frictional heating. The dimensionless entropy production (Ns), velocity $f'(\eta)$, and temperature $\theta(\eta)$ are plotted against η by taking various values of relevant parameters. The Bejan number (Be) profile is also plotted against the similarity variable η by considering different values of the relevant embedded parameters. All the figures are plotted by taking water as a base fluid. Nanoparticles of Fe_3O_4/CuO are dispersed in H_2O.

Figure 2a demonstrates the impact of mass suction (S) on the velocity of $Fe_3O_4 - H_2O$ and $CuO - H_2O$ nanoliquids. The decrement in motion is seen for both $Fe_3O_4 - H_2O$ and $CuO - H_2O$ nanoliquids with increasing (S). For a fixed value of (S), the velocity of the $CuO - H_2O$ nanoliquid is higher than the velocity of the $Fe_3O_4 - H_2O$ nanoliquid. Furthermore, the velocity of both nanoliquids satisfies the boundary condition at $\eta \to \infty$ asymptotically. Figure 2b demonstrates the influence of the magnetic parameter $\left(M^2\right)$ on $f'(\eta)$. It is seen that $f'(\eta)$ reduces as M^2 increases. It is a well-known fact that the Lorentz force acts as a decelerating force for fluid flow and varies directly as M^2 increases. Due to this fact, $f'(\eta)$ varies inversely with M^2. Furthermore, the velocity of the $Fe_3O_4 - H_2O$ nanoliquid is lower than the velocity of the $CuO - H_2O$ nanoliquid, and this is because of the low density of $Fe_3O_4 - H_2O$ compared to $CuO - H_2O$. Figure 3a shows the variation of temperature $\theta(\eta)$ with S by taking $M = 1$, $\phi = 0.1$, $Ec = 0.5$, and $\Pr = 6.8$. The temperature drop is observed with increasing values of S. The width of the thermal boundary layer (TBL) of the $Fe_3O_4 - H_2O$ nanoliquid is greater than that of the $CuO - H_2O$ nanoliquid. Furthermore, the difference in TBL thickness reduces as S increases. The effects of M^2 on temperature $\theta(\eta)$ are presented in Figure 3b. It is seen that $\theta(\eta)$ is augmented as M^2 increases. The rising behavior of temperature is because of magnetic heating. The effective thermal conductivity of nanoliquids is directly related to the solid volume fraction of nanoparticles (ϕ), and this augments the temperature of nanoliquids, as shown in Figure 3c. Furthermore, the width of TBL is smaller for base fluid H_2O and larger for $Fe_3O_4 - H_2O$. This is due to the low thermal conductivity of water and the high effective thermal conductivity of the $Fe_3O_4 - H_2O$ nanoliquid. Figure 3d reveals the influence of the Eckert number (Ec) on $\theta(\eta)$. It is found that increasing Ec leads to a rising temperature. The dissipation function implies that frictional heating varies directly with velocity gradients, and the velocity gradients are high in the vicinity of stretching surface. Due to this fact, the temperature shoots up suddenly, resulting in a higher Eckert number in the vicinity of the stretching plate, as shown in Figure 3d.

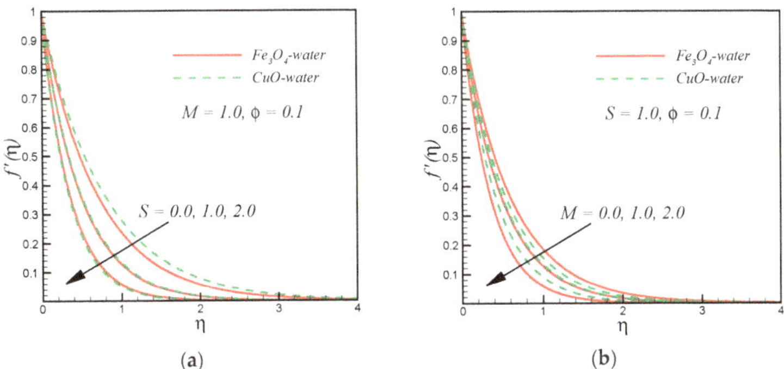

(a) (b)

Figure 2. Variation of the velocity profile $f'(\eta)$ with (**a**) S and (**b**) M.

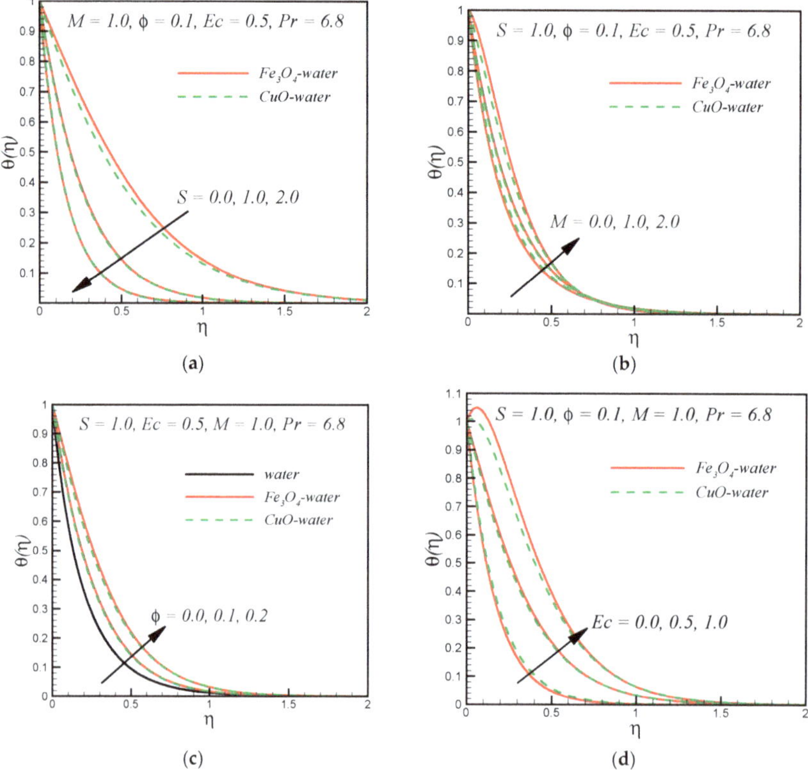

Figure 3. Variation of the temperature profile $\theta(\eta)$ with (**a**) S, (**b**) M, (**c**) ϕ,and (**d**) Ec.

Figure 4a portrays the effects of the Eckert number (Ec) on the entropy generation number (Ns). As seen from the plot, Ns is directly related to the Eckert number. This happens since frictional heating increases with the increasing Eckert number. The entropy generation in the $Fe_3O_4 - H_2O$ nanoliquid than that in the $CuO - H_2O$ nanoliquid. Furthermore, the surface of the solid boundary is the region where maximum entropy is generated. The features of mass suction (S) on Ns are revealed in Figure 4b. As S increases, entropy generation rises at the solid wall and its vicinity, but the opposite trend is observed to start at a certain distance away from the boundary. Furthermore, entropy generation is higher in the $Fe_3O_4 - H_2O$ nanoliquid at the solid boundary and its neighborhood as compared to the $CuO - H_2O$ nanoliquid, but the trend becomes the opposite at a certain distance from the boundary. The nature of entropy generation (Ns) with disparate values of the solid volume fraction of nanoparticles (ϕ) is shown in Figure 4c. From this plot, it can be seen that Ns increases as ϕ increases. This increase in Ns is due to the boost of heat transfer with increasing ϕ. It is well known that the magnetic force is nonconservative. The entropy generation is directly related to the nonconservative forces, and this fact is depicted in Figure 4d. The variations of Ns with temperature difference function (Λ) are presented in Figure 4e. The Ns decreases with increasing values of Λ. Figure 5a shows that the Bejan number (Be) has a maximum value at the surface of the stretching boundary for a nonzero suction parameter (S). In the case of an impermeable stretching boundary, the entropy generation in the $Fe_3O_4 - H_2O$ nanoliquid is due to dissipative forces (viscous and magnetic) near and on the boundary, which are high in comparison to those of the $CuO - H_2O$ nanoliquid. An opposite trend is observed to start at a certain vertical distance from the stretching surface. In the case of $S > 0$, the entropy generation on the stretching surface and inside the boundary layer due to magnetic and viscous heating is more dominant in the $Fe_3O_4 - H_2O$ nanoliquid as compared to the $CuO - H_2O$ nanoliquid. It is noticed from Figure 5b that *Be* is directly related to the solid volume fraction (ϕ) in the region away from the stretching boundary. In the vicinity of an elastic boundary, the opposite trend is observed. From Figure 5c, it can be seen that the Bejan number diminishes as Λ increases. Furthermore, the entropy generation by nonconservative forces (viscous and magnetic) is higher in the $Fe_3O_4 - H_2O$ nanoliquid than in the $CuO - H_2O$ nanoliquid.

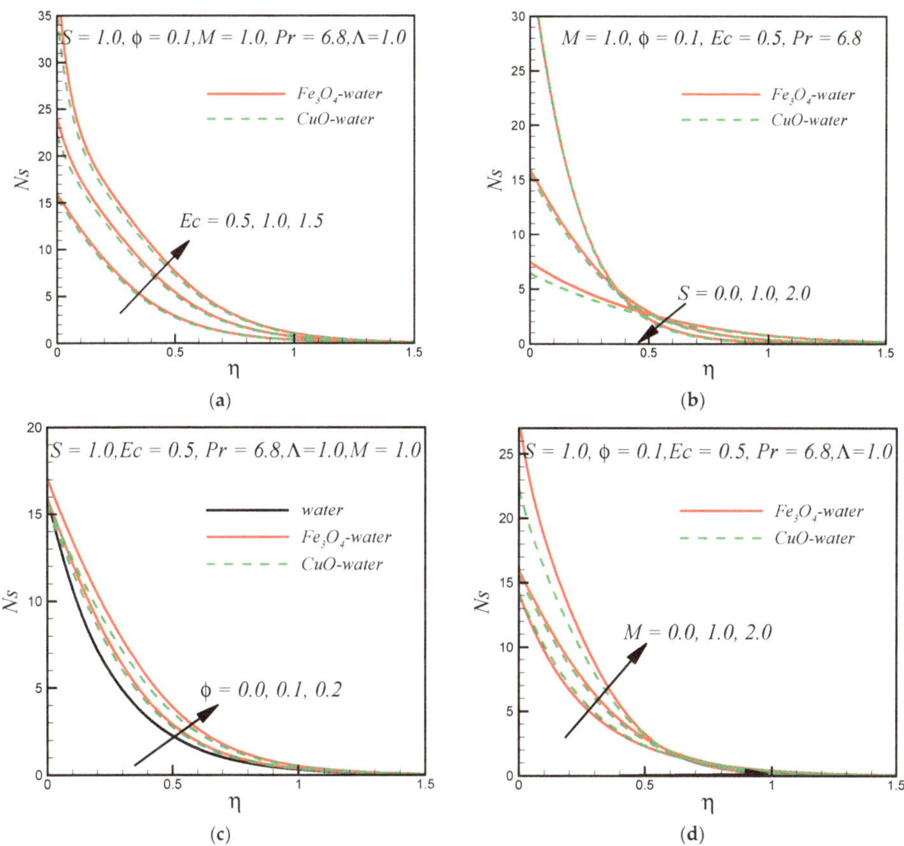

Figure 4. *Cont.*

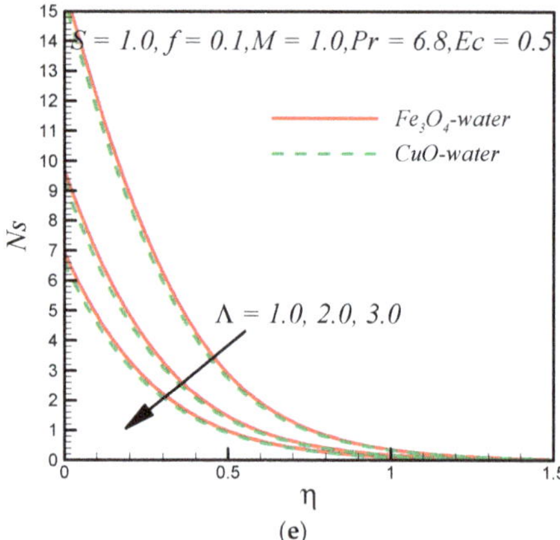

Figure 4. Variation of entropy generation profile $Ns(\eta)$ with (**a**) Ec, (**b**) S, (**c**) ϕ, (**d**) M, and (**e**) Λ.

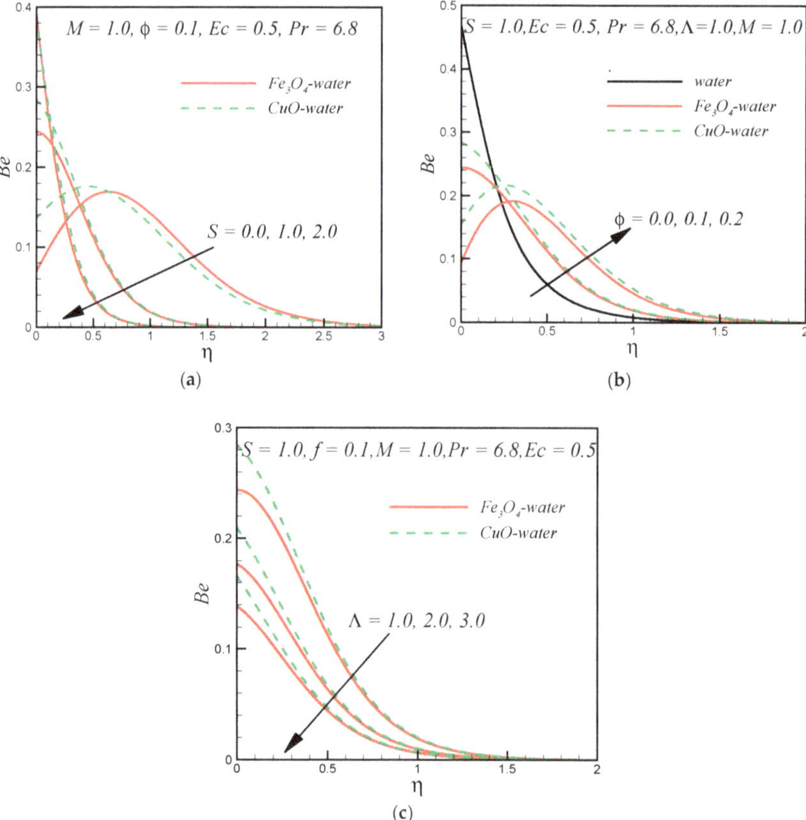

Figure 5. Variation of Bejan number $Be(\eta)$ with (**a**) S, (**b**) M and (**c**) ϕ.

6. Concluding Remarks

In this study, we investigated flow, heat transfer, and entropy production in a dissipative nanofluid flow under the influence of a magnetic field. The following findings can be drawn from the exact results:

- The decrement in motion is seen for both $Fe_3O_4 - H_2O$ and $CuO - H_2O$ nanofluids with increasing S and M^2.
- The velocity of the $CuO - H_2O$ nanofluid is higher than that of the $Fe_3O_4 - H_2O$ nanofluid.
- The temperature $\theta(\eta)$ is observed to decrease with increasing values of S.
- The temperature $\theta(\eta)$ increases as M^2, ϕ, and Ec increase.
- The thermal boundary layer (TBL) width of the $Fe_3O_4 - H_2O$ nanoliquid is greater than that of the $CuO - H_2O$ nanoliquid.
- The entropy generation number (Ns) is directly related to the Eckert number (Ec) and solid volume fraction (ϕ).
- Entropy generation (Ns) by nonconservative forces is higher in the $Fe_3O_4 - H_2O$ nanoliquid than in the $CuO - H_2O$ nanoliquid.

Author Contributions: M.I.A. and M.Q. formulated the problem. D.L. and U.A. solved the problem. U.F. and M.I.A. computed the results. All the authors equally contributed in writing and proofreading the paper. All authors have read and agreed to the published version of the manuscript.

Funding: This work was supported by the China Post-doctoral science foundation, China (Grant No. 2018M632237).

Acknowledgments: The authors would like to acknowledge the Department of Mathematics, Faculty of Science, Jiangsu University, Zhenjiang China for technical support.

Conflicts of Interest: The authors declare no conflict of interest.

Nomenclature

C_o	(KL^{-2})	Dimensional constant
Be	(Dimensionless)	Bejan number
B_0	$(MT^{-2}I^{-1})$	The applied magnetic field. ("I" shows electric current)
$(C_p)_{bf}$	$(L^2T^{-2}K^{-1})$	Specific heat at a constant pressure of a base fluid
$(C_p)_{nf}$	$(L^2T^{-2}K^{-1})$	Specific heat at a constant pressure of nanofluid
Ec	(Dimensionless)	Eckert number
$f(\eta)$	(Dimensionless)	Velocity normal to the solid surface
$f'(\eta)$	(Dimensionless)	Velocity along the solid surface
$\vec{j}$	$L^{-2}I$	Current density
k_{nf}	$(MLT^{-3}K^{-1})$	Thermal conductivity of nanofluid
k_{bf}	$(MLT^{-3}K^{-1})$	Thermal conductivity of the base fluid
k_s	$(MLT^{-3}K^{-1})$	Thermal conductivity of nanoparticle
M^2	(Dimensionless)	Magnetic parameter
N_H	(Dimensionless)	Entropy generation due to heat transfer
N_F	(Dimensionless)	Entropy generation due to viscous dissipation
N_M	(Dimensionless)	Entropy generation due to the magnetic field
N_s	(Dimensionless)	Entropy generation number
Pr	(Dimensionless)	Prandtl number
S	(Dimensionless)	Mass transfer parameter
$\dot{E}'''_{Gen}$	$(ML^{-1}K^{-1}T^{-3})$	Rate of volumetric entropy generation
$(\dot{E}'''_{Gen})_o$	$(ML^{-1}K^{-1}T^{-3})$	Characteristic entropy generation
T	(K)	The temperature inside the boundary layer

$T_w(x)$	(K)	The temperature at the solid boundary
T_b	(K)	The temperature of fluid outside the thermal boundary layer
$U_w(x)$	$\left(LT^{-1}\right)$	The velocity of a stretching sheet
U	$\left(LT^{-1}\right)$	Velocity component along the surface of the solid body
U_o	(T^{-1})	Constant
V	$\left(LT^{-1}\right)$	Velocity component normal to the surface of the solid body
V_w	$\left(LT^{-1}\right)$	Normal velocity component at the boundary
X, Y	(L)	Cartesian coordinates

Greek Symbols

η	(Dimensionless)	Similarity variable
μ_{bf}	$\left(ML^{-1}T^{-1}\right)$	Dynamic viscosity of a base fluid
μ_{nf}	$\left(ML^{-1}T^{-1}\right)$	Dynamic viscosity of nanofluid
ν_{nf}	(L^2T^{-1})	Kinematic viscosity of nanofluid
ρ_{nf}	$\left(ML^{-3}\right)$	Nanofluid density
ρ_{bf}	$\left(ML^{-3}\right)$	The density of a base fluid
ρ_s	$\left(ML^{-3}\right)$	Density of nanoparticles
σ_{nf}	$\left(M^{-1}L^{-3}T^3I^2\right)$	Electric conductivity
σ_{bf}	$\left(M^{-1}L^{-3}T^3I^2\right)$	The electric conductivity of a base fluid
σ_s	$\left(M^{-1}L^{-3}T^3I^2\right)$	The electric conductivity of nanoparticle
$\theta(\eta)$	(Dimensionless)	Temperature
ϕ	(Dimensionless)	The solid volume fraction of nanoparticles
Λ	(Dimensionless)	Temperature difference parameter

References

1. Wang, C.Y. Exact solutions of the steady-state Navier-Stockes equations. *Annu. Rev. Fluid Mech.* **1991**, *23*, 159–177. [CrossRef]
2. Hui, W.H. Exact solutions of the unsteady two-dimensional Navier-Stokes equations. *J. Appl. Math. Phys.* **1987**, *38*, 689–702. [CrossRef]
3. Polyanin, A.D. Exact solutions to the Navier-Stokes equations with generalized separation of variables. *Dokl. Phys.* **2001**, *46*, 726–731. [CrossRef]
4. Al-Mdallal, Q.M. A new family of exact solutions to the unsteady Navier–Stokes equations using canonical transformation with complex coefficients. *Appl. Math. Comput.* **2008**, *196*, 303–308. [CrossRef]
5. Daly, E.; Basser, H.; Rudman, M. Exact solutions of the Navier-Stokes equations generalized for flow in porous media. *Eur. Phys. J. Plus* **2018**, *133*, 173. [CrossRef]
6. Crane, L.J. Flow past a stretching plate. *Zeitschrift für Angewandte Mathematik und Physik* **1970**, *4*, 645–647. [CrossRef]
7. Fang, T.; Zhang, J.; Yao, S. Slip MHD viscous flow over a stretching sheet—An exact solution. *Commun. Nonlinear Sci. Numer. Simul.* **2009**, *14*, 3731–3737. [CrossRef]
8. Fang, T.; Zhang, J. Closed-form exact solutions of MHD viscous flow over a shrinking sheet. *Commun. Nonlinear Sci. Numer. Simul.* **2009**, *14*, 2853–2857. [CrossRef]
9. Liu, I.-C. Exact Solutions for a Fluid-Saturated Porous Medium with Heat and Mass Transfer. *J. Mech.* **2011**, *21*, 57–62. [CrossRef]
10. Turkyilmazoglu, M. Mixed convection flow of magnetohydrodynamic micropolar fluid due to a porous heated/cooled deformable plate: Exact solutions. *Int. J. Heat Mass Transf.* **2017**, *106*, 127–134. [CrossRef]
11. Khan, Z.H.; Qasim, M.; Ishfaq, N.; Khan, W.A. Dual Solutions of MHD Boundary Layer Flow of a Micropolar Fluid with Weak Concentration over a Stretching/Shrinking Sheet. *Commun. Theor. Phys.* **2017**, *67*, 449. [CrossRef]
12. Bejan, A. A Study of Entropy Generation in Fundamental Convective Heat Transfer. *J. Heat Transf.* **1979**, *101*, 718–725. [CrossRef]

13. Bejan, A. The thermodynamic design of heat and mass transfer processes and devices. *Int. J. Heat Fluid Flow* **1987**, *8*, 258–276. [CrossRef]
14. Butt, A.S.; Ali, A.; Mehmood, A. Entropy analysis in MHD nanofluid flow near a convectively heated stretching surface. *Int. J. Exergy* **2016**, *20*, 318–342. [CrossRef]
15. Das, S.; Chakraborty, S.; Jana, R.N.; Makinde, O.D. Entropy analysis of unsteady magneto-nanofluid flow past accelerating stretching sheet with convective boundary condition. *Appl. Math. Mech.* **2015**, *36*, 1593–1610. [CrossRef]
16. Hakeem, A.K.A.; Govindaraju, M.; Ganga, B.; Kayalvizhi, M. Second law analysis for radiative MHD slip flow of a nanofluid over a stretching sheet with non-uniform heat source effect. *Sci. Iran.* **2016**, *23*, 1524–1538. [CrossRef]
17. Rashidi, M.M.; Freidoonimehr, N. Analysis of Entropy Generation in MHD Stagnation-Point Flow in Porous Media with Heat Transfer. *Int. J. Comput. Methods Eng. Sci. Mech.* **2014**, *15*, 345–355. [CrossRef]
18. Makinde, O.D. Entropy analysis for MHD boundary layer flow and heat transfer over a flat plate with a convective surface boundary condition. *Int. J. Exergy* **2012**, *10*, 142. [CrossRef]
19. Butt, A.S.; Ali, A.; Mehmood, A. Numerical investigation of magnetic field effects on entropy generation in viscous flow over a stretching cylinder embedded in a porous medium. *Energy* **2016**, *99*, 237–249. [CrossRef]
20. Ding, H.; Li, Y.; Lakzian, E.; Wen, C.; Wang, C. Entropy generation and exergy destruction in condensing steam flow through turbine blade with surface roughness. *Energy Convers. Manag.* **2019**, *196*, 1089–1104. [CrossRef]
21. Al-Odat, M.Q.; Damseh, R.A.; Al-Nimr, M.A. Effect of Magnetic Field on Entropy Generation Due to Laminar Forced Convection Past a Horizontal Flat Plate. *Entropy* **2004**, *4*, 293–303. [CrossRef]
22. Rashidi, M.M.; Mohammadi, F.; Abbasbandy, S.; Alhuthali, M.S. Entropy Generation Analysis for Stagnation Point Flow in a Porous Medium over a Permeable Stretching Surface. *J. Appl. Fluid Mech.* **2015**, *8*, 753–765. [CrossRef]
23. Das, S.; Jana, R.N.; Makinde, O.D. Entropy generation in hydromagnetic and thermal boundary layer flow due to radial stretching sheet with Newtonian heating. *J. Heat Mass Transf. Res.* **2015**, *2*, 51–61.
24. Ajibade, A.O.; Jha, B.K.; Omame, A. Entropy generation under the effect of suction/injection. *Appl. Math. Model.* **2011**, *35*, 4630–4646. [CrossRef]
25. Govindaraju, M.; Ganga, B.; Hakeem, A.K.A. Second law analysis on radiative slip flow of nanofluid over a stretching sheet in the presence of lorentz force and heat generation/absorption. *Front. Heat Mass Transf.* **2017**, *8*, 1–8. [CrossRef]
26. Milanese, M.; Iacobazzi, F.; Colangelo, G.; Risi, A. An investigation of layering phenomenon at the liquid–solid interface in Cu and CuO based nanofluids. *Int. J. Heat Mass Transf.* **2016**, *103*, 564–571. [CrossRef]
27. Iacobazzi, F.; Milanese, M.; Colangelo, G.; Lomascolo, M.; Risi, A. An explanation of the Al_2O_3 nanofluid thermal conductivity based on the phonon theory of liquid. *Energy* **2016**, *116*, 786–794. [CrossRef]
28. Colangelo, G.; Milanese, M.; De Risi, A. Numerical simulation of thermal efficiency of an innovative Al_2O_3 nanofluid solar thermal collector: Influence of nanoparticles concentration. *Therm. Sci.* **2017**, *21*, 2769–2779. [CrossRef]
29. Colangelo, G.; Favale, E.; Milanese, M.; de Risi, A.; Laforgia, D. Cooling of electronic devices: Nanofluids contribution. *Appl. Therm. Eng.* **2017**, *127*, 421–435. [CrossRef]
30. Colangelo, G.; Favale, E.; Miglietta, P.; Milanese, M.; de Risi, M. Thermal conductivity, viscosity and stability of Al_2O_3-diathermic oil nanofluids for solar energy systems. *Energy* **2016**, *95*, 124–136. [CrossRef]
31. Colangelo, G.; Favale, E.; Milanese, M.; Starace, G.; De Risi, A. Experimental Measurements of Al_2O_3 and CuO Nanofluids Interaction with Microwaves. *J. Energy Eng.* **2017**, *143*, 04016045. [CrossRef]
32. Iacobazzi, F.; Milanese, M.; Colangelo, G.; de Risi, A. A critical analysis of clustering phenomenon in Al_2O_3 nanofluids. *J. Therm. Anal. Calorim.* **2018**, *135*, 371–377. [CrossRef]
33. Hsiao, K.-L. Stagnation electrical MHD nanofluid mixed convection with slip boundary on a stretching sheet. *Appl. Therm. Eng.* **2016**, *98*, 850–861. [CrossRef]
34. Ma, Y.; Mohebbi, R.; Rashidi, M.; Yang, Z.; Sheremet, M. Nanoliquid thermal convection in I-shaped multiple-pipe heat exchanger under magnetic field influence. *Phys. A Stat. Mech. Its Appl.* **2020**, *550*, 124028. [CrossRef]

35. Wakif, A.; Chamkha, A.; Thumma, T.; Animasaun, I.L.; Sehaqui, R. Thermal radiation and surface roughness effects on the thermo-magneto-hydrodynamic stability of alumina–copper oxide hybrid nanofluids utilizing the generalized Buongiorno's nanofluid model. *J. Therm. Anal. Calorim.* **2020**, 1–20. [CrossRef]
36. Hsiao, K.-L. Nanofluid flow with multimedia physical features for conjugate mixed convection and radiation. *Comput. Fluids* **2014**, *104*, 1–8. [CrossRef]
37. Ma, Y.; Mohebbi, R.; Rashidi, M.M.; Yang, Z. MHD convective heat transfer of Ag-MgO/water hybrid nanofluid in a channel with active heaters and coolers. *Int. J. Heat Mass Transf.* **2019**, *137*, 714–726. [CrossRef]
38. Prasad, K.V.; Vaidya, H.; Makinde, O.D.; Vajravelu, K.; Wakif, A.; Basha, H. Comprehensive examination of the three-dimensional rotating flow of a UCM nanoliquid over an exponentially stretchable convective surface utilizing the optimal homotopy analysis method. *Front. Heat Mass Transf.* **2020**, *14*. [CrossRef]
39. Gebhart, B. Effects of viscous dissipation in natural convection. *J. Fluid Mech.* **1962**, *14*, 225–232. [CrossRef]
40. Desale, S.; Pradhan, V.H. Numerical Solution of Boundary Layer Flow Equation with Viscous Dissipation Effect Along a Flat Plate with Variable Temperature. *Procedia Eng.* **2015**, *127*, 846–853. [CrossRef]
41. Vajravelu, K.; Hadjinicolaou, A. Heat transfer in a viscous fluid over a stretching sheet with viscous dissipation and internal heat generation. *Int. Commun. Heat Mass Transf.* **1993**, *20*, 417–430. [CrossRef]
42. Alsabery, A.; Saleh, H.; Hashim, I. Effects of Viscous Dissipation and Radiation on MHD Natural Convection in Oblique Porous Cavity with Constant Heat Flux. *Adv. Appl. Math. Mech.* **2017**, *9*, 463–484. [CrossRef]
43. Mohamed, M.K.A.; Sarif, N.M.; Noar, N.A.Z.M.; Salleh, M.Z.; Ishak, A. Viscous dissipation effect on the mixed convection boundary layer flow towards solid sphere. *Trans. Sci. Technol.* **2016**, *3*, 59–67.
44. Jamaludin, A.; Nazar, R.; Khan, I. *AIP Conference Proceedings*; Boundary Layer Flow and Heat Transfer in a Viscous Fluid Over a Stretching Sheet with Viscous Dissipation, Internal Heat Generation and Prescribed Heat Flux; AIP Publishing LLC: New York, NY, USA, 2017; Volume 1870, p. 040029. [CrossRef]
45. Makinde, O.D. Effects of viscous dissipation and Newtonian heating on boundary-layer flow of nanofluids over a flat plate. *Int. J. Numer. Methods Heat Fluid Flow* **2013**, *23*, 1291–1303. [CrossRef]
46. Motsumi, T.G.; Makinde, O.D. Effects of thermal radiation and viscous dissipation on boundary layer flow of nanofluids over a permeable moving flat plate. *Phys. Scr.* **2012**, *86*, 1–8. [CrossRef]
47. Minkowycz, W.J.; Sparrow, E.M.; Abraham, J.P. *Nanoparticle Heat Transfer and Fluid Flow*; CRC Press: Boca Raton, FL, USA, 2012.
48. Bianco, V.; Manca, O.; Nardini, S.; Vafai, K. *Heat Transfer Enhancement of Nanofluids*; CRC Press: Boca Raton, FL, USA, 2015.
49. Kleinstreuer, C.; Feng, Y. Experimental and Theoretical Studies of Nanofluid Thermal Conductivity Enhancement: A Review. *Nanoscale Res. Lett.* **2011**, *6*, 229. [CrossRef]
50. Bashirnezhad, K.; Bazri, S.; Safaei, M.R.; Goodarzi, M.; Dahari, M.; Mahian, O.; Dalkılıça, A.S.; Wongwises, S. Viscosity of nanofluids: A review of recent experimental studies. *Int. Commun. Heat Mass Transf.* **2016**, *73*, 114–123. [CrossRef]
51. Sheikholeslami, M. *Nanofluid Heat and Mass Transfer in Engineering Problems*; Intech Open: Rijeka, Croatia, 2017.
52. Sajid, M.U.; Ali, H.M. Recent advances in application of nanofluids in heat transfer devices: A critical review. *Renew. Sustain. Energy Rev.* **2019**, *103*, 556–592. [CrossRef]

Article

Effects of Welding Time and Electrical Power on Thermal Characteristics of Welding Spatter for Fire Risk Analysis

Yeon Je Shin and Woo Jun You *

Department of Architecture & Fire Safety, Dongyang University, Yeongju 36040, Korea; jeje5842@naver.com
* Correspondence: wjyou@dyu.ac.kr; Tel.: +82-54-630-1297

Received: 10 November 2020; Accepted: 4 December 2020; Published: 9 December 2020

Abstract: To predict the fire risk of spatter generated during shielded metal arc welding, the thermal characteristics of welding spatter were analyzed according to different welding times and electrical powers supplied to the electrode. An experimental apparatus for controlling the contact angle between the electrode and base metal as well as the feed rate was prepared. Moreover, the correlations among the volume, maximum diameter, scattering velocity, maximum number, and maximum temperature of the welding spatter were derived using welding power from 984–2067 W and welding times of 30 s, 50 s, and 70 s. It was found that the volume, maximum diameter, and maximum number of welding spatters increased proportionally as the welding time and electrical power increased, but the scattering velocity decreased as the particle diameter increased regardless of the welding time and electrical power. When the measured maximum temperature of the welding spatter was compared with an empirical formula, the accuracy of the results was confirmed to be within ±7% of the experimental constant $C = 112.414 \times P_e^{-0.5045}$. Results of this study indicate quantitatively predicting the thermal characteristics of welding spatter is possible for minimizing the risk of fire spread when the electrode type and welding power is known.

Keywords: shielded metal arc welding; welding spatter; electrode; electrical power; welding time

---

1. Introduction

Fire risks in construction sites may occur when flammable gases, liquids, or substances reach their ignition points owing to the scattered welding spatter [1–6]. Shielded metal arc welding (SMAW), a method of joining metals by generating an arc and heating the weld metal zone by applying electrical power between the base metal and electrode, has been widely used in industrial sites since the method of SMAW is applied to almost all repairing of cast iron in air or steel under water [7–10]. However, it involves the risk of fire spreading to nearby combustibles caused by high temperatures because the scattered welding spatters are larger comparable to those of gas metal arc welding (GMAW), which uses plasma [11,12]. Especially, the fire hazards from the SMAW at building construction sites can occur when welding spatters make contact with the inward of a pipe or other enclosed space filled with flammable vapor or liquid [12–19]. In addition, the polarity of electrode can cause changes in welding spatter diameter and number [7,19–23]. From the viewpoint of fire technology, analyzing the thermal characteristics of welding spatter is one of the widely used methods to predict fire spread, where related research has already been conducted.

Hagiwara et al. [24,25] conducted an experimental study on the particle size distribution of welding spatter according to the electrical power supplied to the electrode. They found that 90% of the particles had diameters of less than 1 mm and analyzed the fire spread phenomenon in combustibles, such as benzene, acetone, and urethane foam. This study, however, appears to have limitations in quantitatively analyzing the thermal characteristics of welding spatter crucial to fire risks, which are caused by the electrical power and depend on the particle size.

Hagimoto et al. [19] calculated the particle size according to the electrode diameter when the same electrical power was supplied to the electrode and found that approximately 80% of the particles were scattered to a distance of 0.5 m, 15% to 0.5–1.0 m, and 5% to more than 1 m. They reported that fire can spread to combustibles (urethane foam etc.) when large particles of diameters 0.9–3.0 mm are scattered to a distance of more than 3.5 m.

Brandi et al. [26] analyzed the correlation between the material properties of the electrode core and fire risks using standard mineral dressing techniques. They found that the porosity and density of the welding spatter varied according to the electrical power and stressed the importance of the electrode physical properties for satisfying the ignition requirements of combustibles.

Results from previous studies show that the conditions of fire spread to combustibles during welding vary due to the varying thermal characteristics of welding spatter depending on the electrical power [18,26,27]. Therefore, Shin and You [27] calculated the particle size distribution and mean particle size of welding spatter by assuming a steady-state maximum temperature of the welding spatter for igniting combustibles and proposed an equation for predicting the mean particle temperature based on the energy conservation relationship. According to them, predicting the maximum temperature of the welding spatter is possible when the electrical power, total volume, mean size, and scattering velocity of the particles are known. As these parameters (except the electrical power) vary depending on the electrical power, it is necessary to analyze the relationships among the main factors according to the experimental conditions. This is necessary for the quantitative analysis of the risk of fire spread due to scattered particles. Therefore, in this study, we propose a method for predicting fire risks by quantitatively deriving the thermal characteristics of welding spatter according to the welding time and electrical power.

2. Material and Methods

2.1. Theoretical Approach

Figure 1 shows the schematic of the total volume of the welding spatter during welding. The total mass of the welding spatter ($\Delta m_{p,total}$) can be calculated according to Equation (1) after measuring the mass melted on the base metal ($\Delta m_{b,p}$) and is dependent on the welding time (Δt) and electrical power (P_e). The core inside the electrode is made of steel (ρ_{iron} = 7860 kg/m^3) and the coating outside the electrode contains sodium silicate ($\rho_{\text{Sodium Silicate}}$ = 2400 kg/m^3). However, it is possible to calculate the volume of a single particle (ΔV_i) using the relationship $\Delta V_i = \Delta m_i/\rho_i$ only when the mixed ratios of materials are given for each welding spatter.

$$\Delta m_{p,total} = \Delta m_{el} - \Delta m_{b,p} \tag{1}$$

where Δm_{el}, $\Delta m_{b,p}$, and $\Delta m_{p,total}$ are the masses of the electrode and solidified weld metal attached to the base metal and total mass of scattered particles, respectively. In a previous study, the mean particle temperature was predicted by assuming the steady-state condition of the initial temperature of welding spatter as the maximum value for the ignition combustibles, as shown in Equation (2) [27].

$$T_{P,S} = T_\infty + \frac{P_e - \sigma\varepsilon A_{b,s}\left(T_{s,b}^4 - T_{sur}^4\right)}{NhA_{P,S}} \tag{2}$$

where $T_{p,s}$, T_∞, T_{sur}, P_e, σ, ε, $A_{b,s}$, $T_{s,b}$, N, h, and $A_{p,s}$ are the mean particle temperature, surrounding temperature, surface temperature, electrical power (P_e), Stefan–Boltzmann constant, emissivity, surface area and surface temperature of base metal, average number of particles, convective heat transfer coefficient, and surface area of particle, respectively. The mean particle size, $d_{p,m}$, and convective heat transfer coefficient, h, can be obtained using Equations (3) and (4), respectively [14,27,28].

$$N = \frac{6V_{p,total}}{\pi d_{p,m}^3} \tag{3}$$

$$Nu_D = \frac{hd_{p,m}}{k} = 2 + 0.6Re_D^{0.5}Pr^{1/3} \tag{4}$$

where $V_{p,total}$, $d_{p,m}$, Nu_D, k, Re, and Pr are the total volume of particles, mean diameter of particles, Nusselt number, thermal conductivity, Reynolds number ($Re_D = \rho u_{p,m}d_{p,m}/\mu$), and Prandtl number ($Pr = C_p\mu/k$)), respectively. Therefore, the temperature distribution prediction shown in Equation (2) is possible when $V_{p,total}$, $d_{p,m}$, and $u_{p,m}$ can be calculated using Equations (3) and (4). As the total volume, mean diameter, and scattering velocity of particles vary according to the welding time and electrical power, the functional relationship given by Equation (5) must be also determined [14,27].

$$N, d_m, h \sim f(P_e, \Delta t) \tag{5}$$

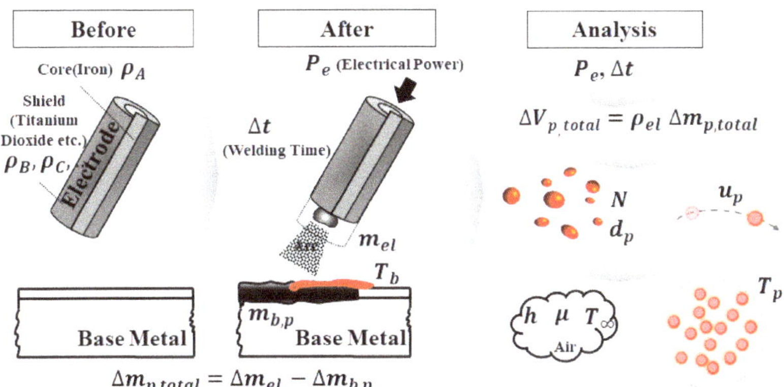

Figure 1. Schematic of welding spatter in shielded metal arc welding (SMAW) and assumptions for energy balance between particles and base metal.

2.2. Experimental Apparatus

Figure 2 shows the semi-automated SMAW experimental apparatus, which was constructed by maintaining perpendicularity between the electrode and base metal constant (welding angle θ = 90°) and specifying the maximum feed rate of the welding torch as 7 mm/s. This made it possible to analyze the size and scattering velocity of the particles. As shown in the figure, welding spatters were scattered under different welding times (Δt) and electrical power (P_e), and the scattering velocity and mean temperature of the welding spatters were measured using a high-speed camera (model: phantom Miro M/R/LC310, USA) and a thermal imaging camera (model: Fluke Tix501). The scattering velocity and mean temperature of the welding spatter were measured using a high-speed camera (model: phantom LC310) and a thermal imaging camera (model: Fluke Tix501). The particle size distribution was determined using Image J software after collecting all welding spatter in a 50 × 46 × 64 cm^3 acrylic box. Table 1 lists the specifications of the experimental apparatus and the experimental conditions for the average values of three times results are denoted in Table 2.

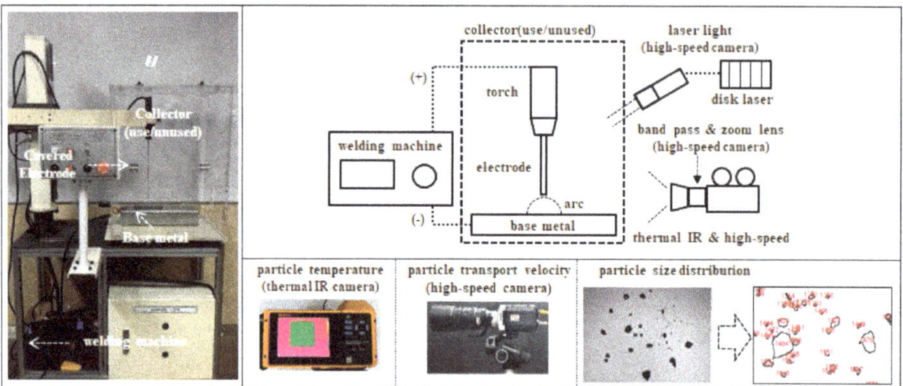

Figure 2. Schematic and images of experimental apparatus for welding spatter analysis.

Table 1. Specifications of experiment apparatus.

Equipment	Specification
Welding machine	Output current (20~220) A, Rated input voltage 220 V, Electric power (0~2.5) kW Rated duty cycle 60%, Model: Rolwal MMA-200E
Thermal imaging camera	Infrared resolution 640 × 480, Temp. measurement range −20 to 650 °C, Accuracy ±2 °C or 2%, Frame rate 60 Hz, Model: Fluke Tix 501
High-speed camera	Resolution 640 × 480, Sampling rate 10,000 fps, Model: phantom Miro M/R/LC310 Lens: Nikon 105 mm, 2× converter Band-pass filter: Φ 50 mm, 810 nm/12 nm, CN code 90022000
Electronic energy meter	230 AC, 60 Hz, 16 A/3680 W (Model: KEM2500)
Precision balance	Max. load weight 320 g, Accuracy 0.1 mg, Model: PX224KR
Electrode	High titanium oxide type electrode (AWS E-6013) Core: Iron (65–75%), Coating: Titanium dioxide (10–15%), Feldspar (5–10%), Mn (1–5%), Sodium silicate (1–5%), Limestone (1–5%), Mica (1–5%)

Table 2. Experimental conditions to study the effects of thermal characteristics of welding spatter on the welding time and electrical power.

Test Number	Welding Time, Δt (s)	Welding Current (A)	Welding Voltage (V)	Electrical Power, P_e (W)
Case #1	30			
Case #2	50	80	12	984
Case #3	70			
Case #4	30			
Case #5	50	100	13	1337.3
Case #6	70			
Case #7	30			
Case #8	50	130	14	1802.0
Case #9	70			
Case #10	30			
Case #11	50	150	17	2067.5
Case #12	70			

Welding polarity: DC-, Contact angle: 90°, Arc length: 5 mm, Base metal: Mild steel (SS400); Electrode diameter, d_{el}: 4.0 mm, Material properties of electrode given in Table 1.

3. Results and Discussion

3.1. Volume of Welding Spatter

Figure 3 shows the variation of the measured reduction rate (u_{rate}) of the electrode length according to the electrical power (P_e) in the case of welding times (Δt) of 30 s, 50 s, and 70 s. It is seen that as P_e increased, u_{rate} also increased proportionally as the mass of the electrode welded to the base metal ($\Delta m_{b,p}$) increased. When P_e was constant, a constant value of u_{rate} was calculated, which was consistent with that obtained using Equation (6) within ±4% for the average values u_{rate} regardless of Δt.

$$u_{rate} = a_1 + b_1 \times P_e \tag{6}$$

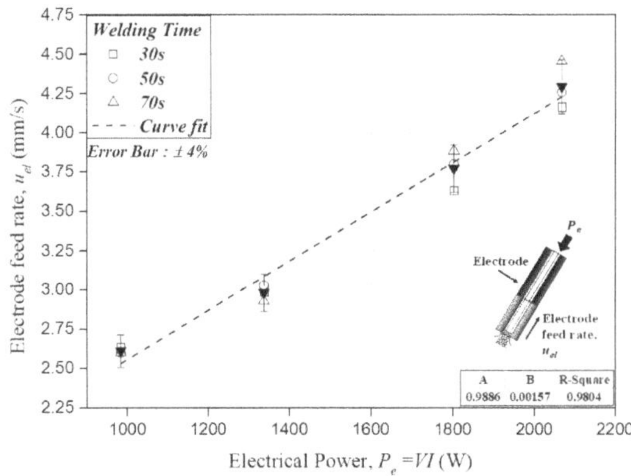

Figure 3. Electrode feed rate depending on the electrical power for welding time = 30 s, 50 s, and 70 s.

Here, a_1 and b_1 are experimental constants. It is estimated that a_1 = 0.989 mm/s and $b_1 = 0.157 \times 10^{-2}$ W-mm/s are obtained according to the base metal and electrode specifications listed in Table 1, and the electrical power ranges between 984 and 2067 W.

Figure 4 shows the variation in the measured mass loss of the electrode (Δm_{el}) and the mass welded to the base metal ($\Delta m_{b,p}$) according to the electrical power (P_e) for the welding times (Δt) of 30 s, 50 s, and 70 s. In this figure, the symbols enclosed in brackets represent Δm_{el}, which increased proportionally to u_{rate}. When Δt increased keeping P_e constant, Δm_{el} increased in proportion to u_{rate}. Therefore, when Equation (6) and the electrode density measured using a load cell (ρ = 4726 kg/m^3) are applied, Δm_{el} is given by Equation (7) expressed by the dotted line, which agrees with the measured value within an error range of approximately ±5%.

$$\Delta m_{el} = u_{rate} \times \Delta t \times \left(\frac{\pi}{4} d_{el}^2\right) \times \rho_{el} \tag{7}$$

where d_{el} is the electrode diameter is used as the reference value of 4.0 mm. Notably in the figure, the measured value of $\Delta m_{b,p}$ increased in proportion to the magnitudes of Δt and P_e, as shown by the closed symbol value. In addition, 88.6% Δm_{el} was found to be welded to the base metal on average. This result indicates that approximately 11.4% Δm_{el} was responsible for generating welding spatter when P_e was supplied to the electrode. The energy transmitted to the electrode can be simplified through the assumption shown in Equation (8).

$$\sigma \varepsilon A_{b,s}\left(T_{s,b}^4 - T_{sur}^4\right) \equiv 0.886 P_e \tag{8}$$

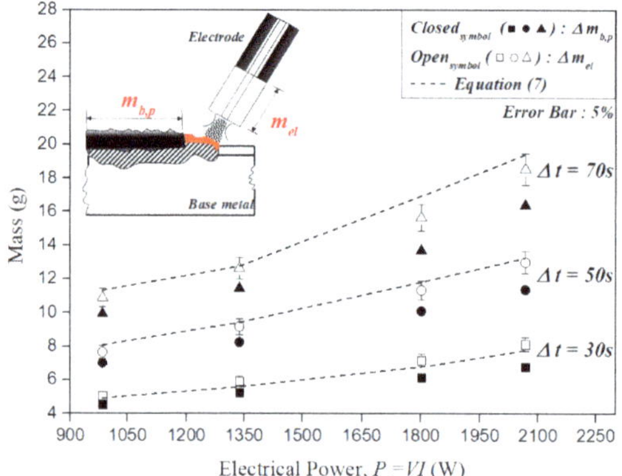

Figure 4. Experimental variation of Δm_{el} and $\Delta m_{b,p}$ according to P_e for Δt = 30 s, 50 s, and 70 s.

Figure 5a shows the variation of the total mass of the particles scattered from the electrode ($\Delta m_{p,total}$) using Equation (1), mass reduction of the electrode (Δm_{el}), and mass welded to the base metal ($\Delta m_{b,p}$) for the welding times (Δt) of 30 s, 50 s, and 70 s according to the electrical power. The result of curve-fitting the calculated values of $\Delta m_{p,total}$ with increasing electrical power (P_e) under the same Δt values is shown in Equation (9).

$$m_{p,total} = a_2 \cdot P_e^{b_2} \tag{9}$$

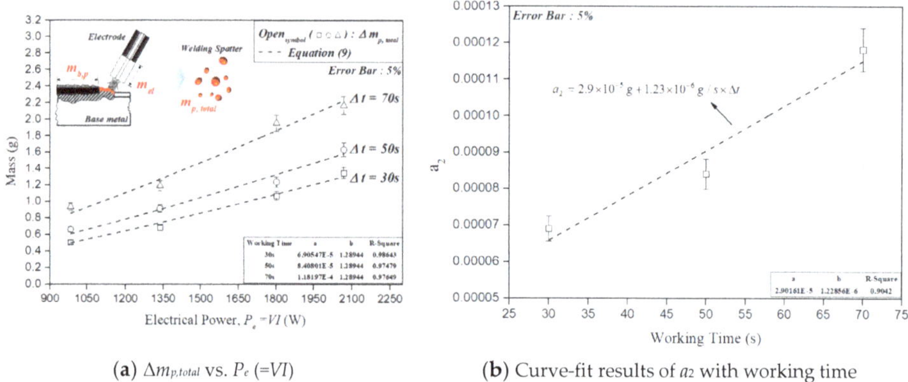

(**a**) $\Delta m_{p,total}$ vs. P_e (=VI) (**b**) Curve-fit results of a_2 with working time

Figure 5. Effects of electrical power on the total mass of welding particles at welding times of 30 s, 50 s, and 70 s.

It was found that a_2 is related to Δt as shown in Figure 5b, and this tendency is shown in Equation (10) when b_2 is constant at 1.28944.

$$a_2 = 2.9 \times 10^{-5}\text{g} + 1.23 \times 10^{-6}\text{g/s} \times \Delta t \tag{10}$$

Figure 6 shows the density values (ρ_i) of a single scattered welding particle measured by calculating the mass ($m_{p,i}$) and volume ($\Delta V_{p,i}$) of the particle using diameters ($d_{p,i}$) of 1.736, 2.023, 2.294, and 2.352 mm. Because the electrode contains various metal components, as shown in Table 1, ρ_i may vary depending on the material composition inside the shield and core [26,29]. In particular, the mass proportions of metals that constitute each particle must be determined to obtain the total volume of scattered welding spatter ($\Delta V_{p,total}$), but limitations exist in analyzing the density when measuring each mixed component for at least 1000 small particles with a diameter of 0.1 mm or less. Therefore, we attempted to analyze the thermal characteristics of welding spatter by assuming $\rho_{p,total} \equiv \rho_{el}$ (4726 kg/m^3) and calculating $\Delta V_{p,total}$ as shown in Equation (11).

$$\Delta V_{p,total} = \frac{\Delta m_{p,total}}{\rho_{el}} = \frac{(2.9 \times 10^{-5} + 1.23 \times 10^{-6} \times \Delta t) \times P_e^{\,b_2}}{\rho_{el}} \quad (11)$$

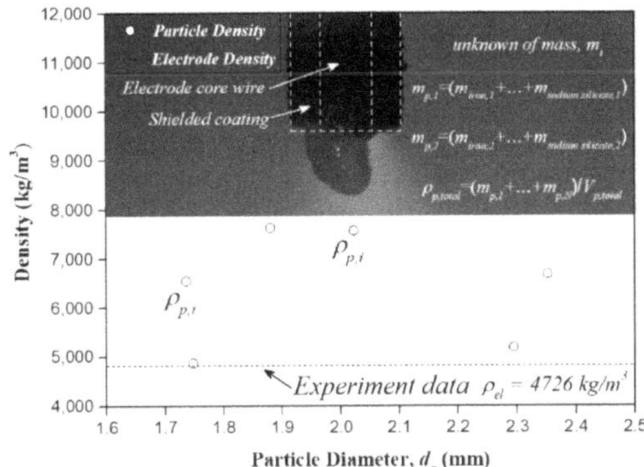

Figure 6. Measured value of one particle density using d_p = 1.73–2.35 mm.

3.2. Diameter and Number of Welding Spatters

Figure 7 shows the fraction (N_i/N_{total}) of the number of particles, which is the ratio of particles with diameter (N_i) to the total number of scattered particles (N_{total}), according to P_e at Δt = 30 s, 50 s, and 70 s. As mentioned before, the particle size distribution was determined using Image J software after collecting welding spatter in a 50 × 46 × 64 cm^3 acrylic space, and it was analyzed by excluding the diameters of 0.3 mm or less due to the resolution. It was found that the mean particle diameter ($d_{p,m}$) was approximately 0.3 mm regardless of Δt and P_e, which is similar to the results of previous studies ($d_{p,m}$ < 0.5 mm) [9,11]. Therefore, the mean number of particles (N) can be calculated using Equation (4). The main purpose of this study was to predict fire risks according to the thermal characteristics of scattered particles; however, the mean number of particles (N) was expressed using the maximum number of particles (N_{max}) to analyze the thermal characteristics according to the maximum particle diameter ($d_{p,max}$) generated during welding.

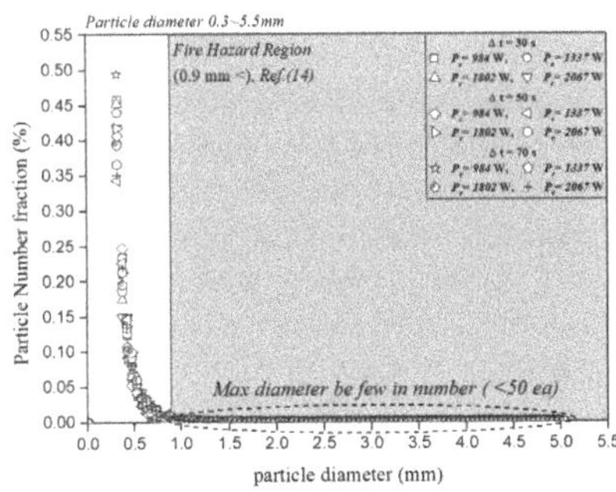

(**a**) Particle number fraction vs. particle diameter

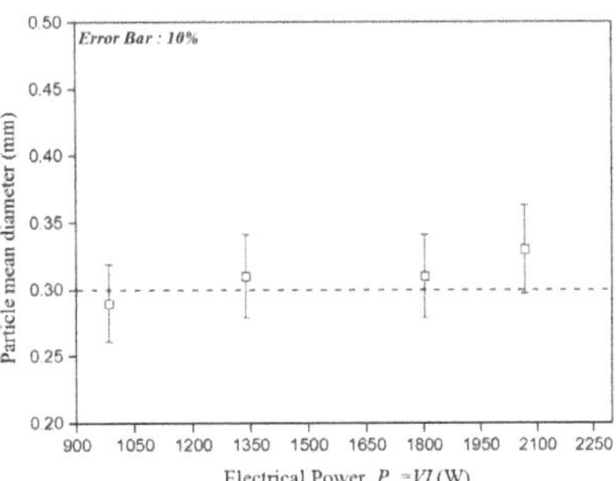

(**b**) Particle mean diameter vs. electrical power

Figure 7. Results of the (**a**) fraction of particle number, and (**b**) particle mean diameter when P_e = 984, 1337, 1802, and 2067 W and Δt = 30 s, 50 s, and 70 s.

As the maximum particle size, $d_{p,max}$, is still undetermined, it is necessary to analyze $d_{p,max}$ according to Δt and P_e to solve Equation (12).

$$N_{max} = \frac{6V_{total}}{\pi d_{p,max}^3},\ N = N_{max}\left(d_{p,max}/d_{p,m}\right)^3 \quad (12)$$

Figure 8 shows the results of analyzing the maximum diameter of scattered particles ($d_{p,max}$) according to the electrical power (P_e) at Δt = 30 s, 50 s, and 70 s. Each measured value represents the average of the maximum diameter obtained in three repeated experiments. Apparently, the size of the particles scattered from the electrode increased as Δt increased under a constant P_e because the temperature around the weld zone of the base metal increased. In addition, under the same Δt, the size of scattered particles increased in proportion to the melted mass of the electrode as P_e increased as shown in Equation (13).

$$d_{p,max} = a_3 \times P_e^{\,b_3} \quad (13)$$

where a_3 and b_3 are experimental constants. When $b_3 = 0.4825$, a_3 can be calculated by Equation (14) and plotted in Figure 8b.

$$a_3 = 2.194 \times 10^{-2} + 9.38 \times 10^{-4} \times \Delta t \quad (14)$$

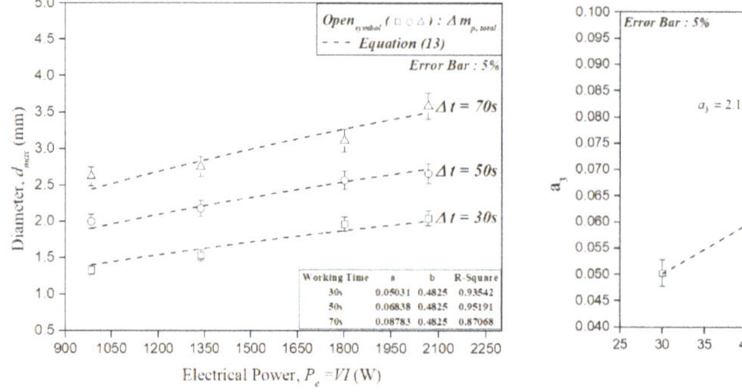

(**a**) Maximum particle diameter vs. electrical power

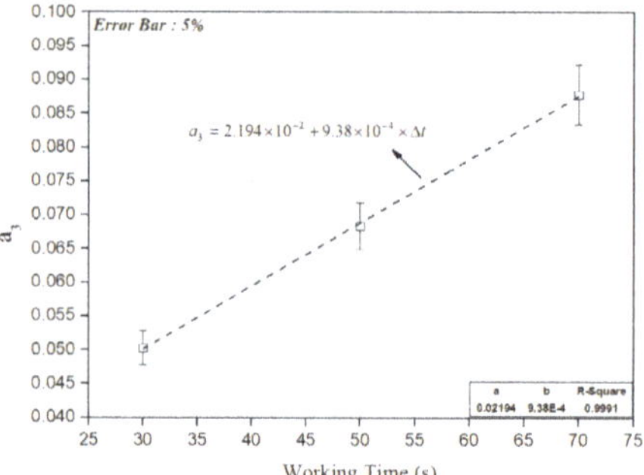

(**b**) The experiment coefficient of a_3 vs. Welding time

Figure 8. Effects of electrical power on the distribution of the particles and the max diameter of the particles at welding time = 30 s, 50 s, and 70 s.

3.3. *Velocity of Welding Spatter*

Figure 9a shows the results of measuring the mean particle velocity according to the particle diameter under an electrical power of 984 W using a high-speed camera (model: phantom LC310) and particle tracking velocimetry (PTV). It is observed that the scattering velocity showed a tendency to decrease as the particle diameter increased under the experimental conditions of the welding time (Δt) and electrical power (P_e), as shown in Table 2. The correlation between the maximum diameter of scattered particles ($d_{p,max}$) and the scattering velocity ($u_{p,max}$) was analyzed, as shown in Figure 9b. $u_{p,max}$ decreased as $d_{p,max}$ increased with a difference of less than ±10% depending on the values of Δt and P_e, and the relationship shown in Equation (15) was found.

$$u_{p,max} = 1.3 \times d_{p,max}^{-1.004} \quad (15)$$

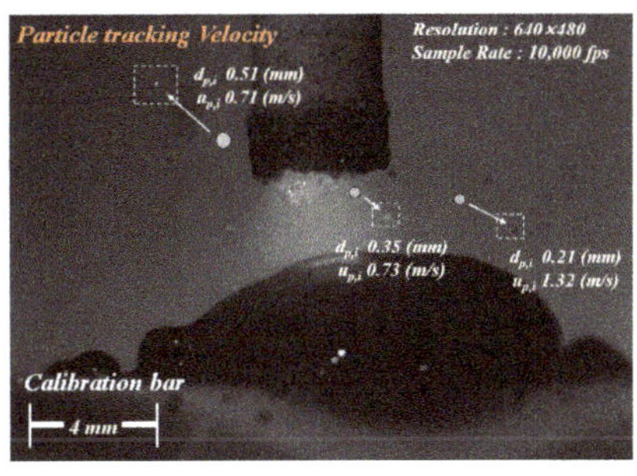

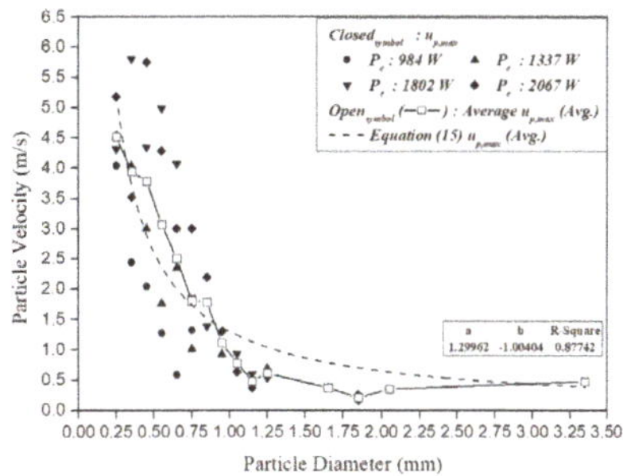

(**a**) Velocity contour of welding particle (**b**) Particle velocity vs. particle diameter

Figure 9. Results of the particle velocity according to the particle diameter.

3.4. Thermal Characteristics of Welding Spatter

Using the results of the total volume of particles ($V_{p,total}$), maximum number of particles (N_{max}), particle diameter ($d_{p,max}$), and scattering velocity ($u_{p,max}$) according to the Δt and P_e obtained in Section 3.1 to determine the convective heat transfer coefficient shown in Equation (4), Equation (16) can be formed.

$$h_{max} = \left(2 + 0.6Re_{d,max}^{0.5}Pr^{1/3}\right)k(T_{ref})/d_{p,max}, \quad (16)$$

where $Re_{d,max}$ is the Reynolds number ($Re_{d,max} = \rho_p u_{p,max} d_{p,max}/\mu$) considering the maximum particle diameter ($d_{p,max}$). In particular, the thermal conductivity (k), specific heat (Cp), viscosity (μ), and density (ρ) of air vary depending on the reference temperature ($T_{ref} = (T_{p,s} + T_\infty)/2$), as shown in Figure 10a. In this study, these can be calculated using Equations (17)–(20) based on the data given by this study [30].

$$k(T_{rep}) = -7.16\times10^{-3} + 1.72\times10^{-4}\times T - 2.45\times10^{-7}\times T^2 + 2.29\times10^{-10}\times T^3 - 9.81\times10^{-14}\times T^4 + 1.63\times10^{-17}\times T^5 \quad (17)$$

$$c_p(T_{rep}) = 1.05 - 2.89\times10^{-4}\times T + 6.83\times10^{-7}\times T^2 - 3.4\times10^{-10}\times T^3 + 2.25\times10^{-14}\times T^4 + 1.55\times10^{-17}\times T^5 \quad (18)$$

$$\mu(T_{rep}) = 4.26\times10^{-6} + 4.93\times10^{-8}\times T + -1.33\times10^{-11}\times T^2 + 2.36\times10^{-15}\times T^3 \quad (19)$$

$$\rho(T_{rep}) = 374.074\times T^{-1.0114} \quad (20)$$

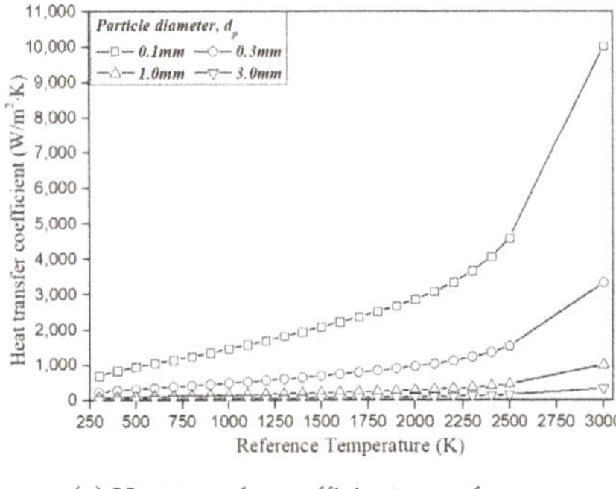

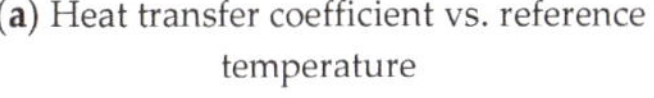

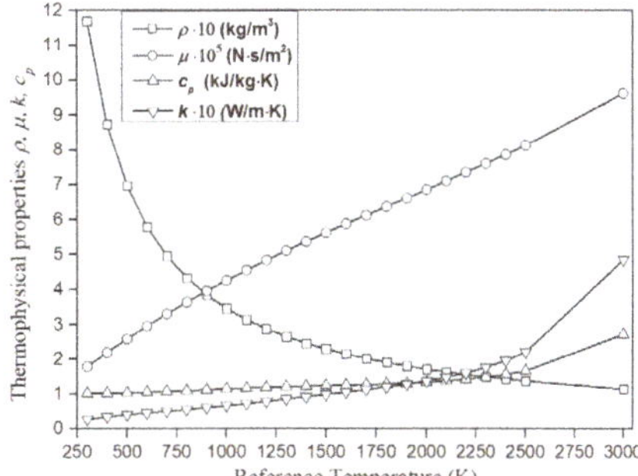

(**a**) Heat transfer coefficient vs. reference temperature (**b**) Thermophysical properties (ρ, μ, k, c_p) vs. reference temperature

Figure 10. Effects of the heat transfer coefficient of the maximum particle dimeter and velocity according to thermophysical properties (ρ, μ, k, c_p).

Figure 10a shows the results of analyzing the convective heat transfer coefficient, h, according to T_{ref} when $d_{p,max}$ = 0.1, 0.3, 1.0, and 3.0 mm. h decreased as $d_{p,max}$ increased while the scattering velocity ($u_{p,max}$) decreased to 13.12, 4.35, 1.30, and 0.43 m/s at different $d_{p,max}$ values obtained by Equation (15). Therefore, Equation (2), for calculating the mean particle temperature considering the welding time and electrical power, can be expressed as in Equation (21).

$$T_{P,s} = T_{\infty} + \frac{0.13P_e}{N_{max}h_{max}A_{P,max}r_{ratio}}, \tag{21}$$

where N_{max}, h_{max}, $A_{p,max}$, and r_{ratio} are the total number of particles, heat transfer coefficient, surface area of a particle, and a constant calculated by replacing $d_{p,m}$ with $d_{p,max}$, respectively, when welding particles are at their maximum size. In particular, the mean number of particles (N) used in Equation (3) and the mean surface area ($A_{\mathrm{p,m}}$) are related as $N = N_{max}(d\_{p,m}/d\_{p,max})^{-3}$ and $A_{p,m} = A_{p,max}(d\_{p,m}/d\_{p,max})^2$, whereas the convective heat transfer coefficient (h) for determining the temperature of the welding spatter is given by $h = h_{max} \times (d\_{p,m}/d\_{p,max})^{-0.5} \times (u_{p,m}/u_{p,max})^{0.5}$. Therefore, r_{ratio} is related as,

$$r_{ratio} \sim \left(\frac{u_{p,m}}{u_{p,max}}\right)^{0.5}\left(\frac{d_{p,m}}{d_{p,max}}\right)^{-1.5}, \tag{22}$$

where, $u_{p,m}/u_{p,max}$ represents the ratio of the mean velocity to the maximum velocity. As it varies depending on the diameter, it can be expressed as Equation (23).

$$\left(\frac{u_{p,m}}{u_{p,max}}\right)^{0.5} = C\left(\frac{d_{p,m}}{d_{p,max}}\right)^{0.5}, \tag{23}$$

where C is the experimental constant which may vary depending on the velocity difference. Because k, Cp, μ, and ρ used to obtain the convective heat transfer coefficient are functions of $T_{p,s}$ as shown in Equations (17)–(20), it is necessary to solve Equation (13) for the maximum particle size, Equation (15) for the maximum velocity, Equation (16) for the convective heat transfer coefficient, and Equation (23) to perform iterative calculations for predicting the maximum temperature of the welding spatter.

Figure 11a shows the results of maximum temperature ($T_{p,max}$) measured by capturing the welding spatter images accumulated on the acrylic collection plate at 60 fps for 70 s using a thermal imaging camera (Model: Fluck Ti520) at Δt = 70 s and P_e = 1337 W. As shown in the figure, the approximate maximum and mean particles were 432 °C and 347 °C, respectively.

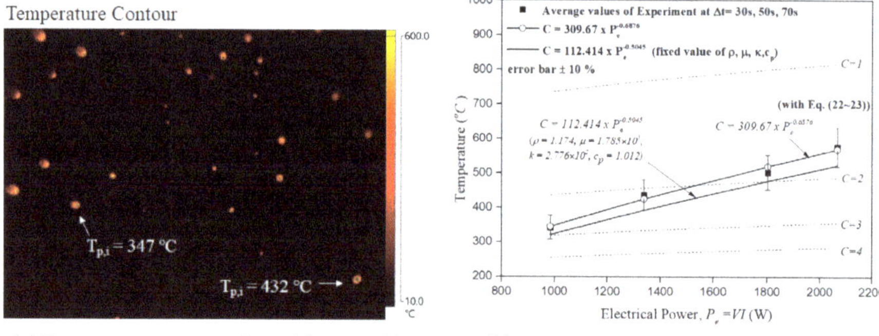

(**a**) Temperature contour of a welding particle (**b**) Particle temperature vs. supply power

Figure 11. The results of temperature contour and the comparison of the prediction with experiment values of C.

Figure 11b shows the results calculated using the maximum particle temperature measurements and Equation (21) under the experimental conditions of Δt and P_e shown in Table 2. The experimental values agreed with the mean values with a difference of up to ±10% depending on Δt, and the temperature tended to increase as P_e increased. It should be noted that the maximum difference due to the time change was small (within ±2C) as shown in Equation (21) when C was constant, but the increasing tendency of the temperature in proportion to P_e was found to be consistent with the experimental values. However, when the values of C were 1, 2, 3, and 4, the slope at which the maximum particle temperature increased over the increase in P_e was smaller than the experimental value. This appears to be due to different values of C when the difference between the mean and maximum scattering velocities of the welding spatter increased along with P_e. At each P_e, the value of C can be obtained using Equations (24) and (25).

$$C_1 = 309.67 \times P_e^{-0.6876} \tag{24}$$

$$C_2 = 112.414 \times P_e^{-0.5045} \tag{25}$$

C_1 is a constant calculated by performing iterative calculations using Equations (17)–(20) for predicting the maximum particle temperature, whereas C_2 is calculated using the values of the density, thermal conductivity, viscosity coefficient, and specific heat at room temperature (298 K). Therefore, the maximum particle temperature can be predicted within an error range of approximately 5% using the equation to solve C_2.

Figure 12 shows the results of calculating the maximum temperature and diameter of the welding particles when the welding time (Δt) ranged from 30–70 s and P_e from 984–2067.5 W. "Fire hazard region" means the possibility of fire spreading to combustible materials such as polyurethan foam as mentioned in Ref [14,15,27], and "No ignition region" means the minimized conditions of fire spread. Based on previous studies, it can be confirmed that the maximum welding time and electrical power are 10 s and 1150 W, respectively, when the minimum particle size of welding spatter for the risk of fire spread is 0.9 mm, and the minimum temperature is 350 °C [19]. Therefore, the results of this study indicate that it is possible to calculate the electrical power for minimizing the risk of fire due to welding spatter when the electrode type and welding time are known.

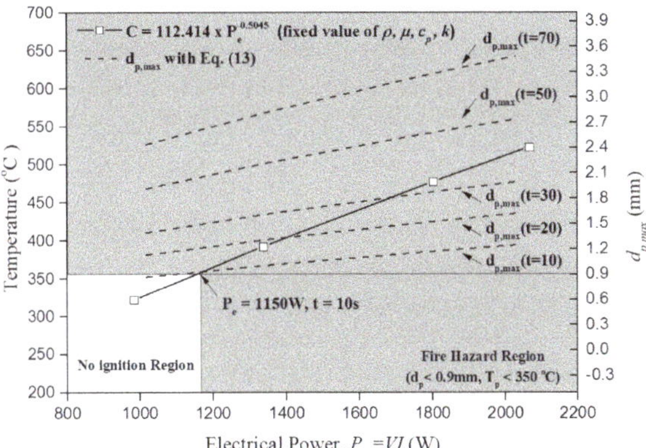

Figure 12. Predicted results of the maximum diameter and temperature of the welding particle depending on the electrical power and welding time.

4. Summary

In this study, the volume, maximum diameter, scattering velocity, and maximum number of welding spatter for shielded metal arc welding (SMAW) were analyzed according to the electrical power and welding time. When the electrical power was varied for welding times of 30 s, 50 s, and 70 s, the following results were derived.

First, when the mass of the electrode and scattered particles was calculated, an empirical formula was derived, which showed an increase in the mass of scattered particles when the electrical power increased at a constant welding time. In particular, the mass of scattered welding spatter represented approximately 11.45% of the total mass of the consumed electrode on average. The densities of the scattered particles were found to vary between 4876–7572 kg/m^3 depending on the volume fraction of the core and coating composition of the electrode as referred by manufacturer.

Second, it was found that the mean diameter of welding spatter was approximately 0.3 mm, which was constant regardless of the welding time and electrical power. The maximum particle size, which has an important impact on fire risks, however, showed a tendency to increase in proportion to the welding time and electrical power. An empirical formula considering the maximum particle size was also derived to predict the temperature of the scattered welding spatter.

Third, the scattering velocity differed with differences of up to ±91% according to the welding time and electrical power. This appears to be due to the fact that materials with significantly different densities were mixed, which affected the momentum of the welding spatter while they were generated from the electrode. However, the scattering velocity decreased as the particle diameter increased.

Fourth, empirical formulas for the volume, maximum diameter, and scattering velocity of the welding spatter according to the welding time and electrical power were derived and compared with the maximum temperature measurements during the welding process. Results showed a good agreement between the compared values within an error range of approximately 10%. After verifying this accuracy, the case in which the minimum temperature of welding spatter was 350 °C or higher and the particle size was 0.9 mm was analyzed. It was found that fire risks can be minimized when a maximum welding time of 10 s and maximum electrical power of 1150 W are used. It should be noted that the maximum temperature of the welding spatter increased in proportion to the electrical power regardless of the welding time. The results of this study are expected to be used as important data for quantitatively presenting measures to minimize the fire risk of welding spatter.

Author Contributions: Conceptualization, Y.J.S. and W.J.Y.; Data curation, W.J.Y.; Formal analysis, W.J.Y.; Funding acquisition, W.J.Y.; Investigation, Y.J.S.; Methodology, W.J.Y.; Project administration, W.J.Y.; Software, Y.J.S.; Supervision, W.J.Y.; Validation, Y.J.S. and W.J.Y.; Visualization, Y.J.S.; Writing—original draft, Y.J.S.; Writing—review & editing, W.J.Y. All authors have read and agreed to the published version of the manuscript.

Funding: This research was funded by Korea Agency for Infrastructure Technology Advancement, grant number (19CTAP-C151893-01).

Conflicts of Interest: The authors declare no conflict of interest. The funders had no role in the design of the study; in the collection, analyses, or interpretation of data; in the writing of the manuscript, or in the decision to publish the results.

References

1. Messler, R.W., Jr. *Principles of Welding: Processes, Physics, Chemistry, and Metallurgy*; Wiley: London, UK, 2008.
2. DuPont, J.N.; Lippold, J.C.; Kiser, S.D. *Welding Metallurgy and Weldability of Nickel-Base Alloys*; Wiley: London, UK, 2009.
3. Weman, K. *Welding Processes Handbook*, 2nd ed.; Cambridge Woodhead Publishing: Cambridge, UK, 2003.
4. Babrauskas, V. *Ignition Handbook: Principles and Applications to Fire Safety Engineering, Fire Investigation, Risk Management, and Forensic Science*; Fire Science Publishers: Issaquah, WA, USA, 2003.
5. National Fire Protection Association (NFPA). *NFPA Standard 51B: Standard for Fire Prevention During Welding, Cutting, and Other Hot Work, Technical Report*; National Fire Protection Association: Quincy, MA, USA, 2009.
6. Schonherr, W. Fire risks with welding torches and manual arc welding-globules, spatter and how far they can be thrown. *Schweiss. Schneid.* **1982**, *34*, E74–E77.
7. Tomków, J.; Fydrych, D.; Wilk, K. Effect of Electrode Waterproof Coating on Quality of Underwater Wet Welded Joints. *Materials* **2020**, *13*, 2947. [CrossRef] [PubMed]

8. Srisuwan, N.; Kumsri, N.; Yingsamphancharoen, T.; Kaewvilai, A. Hardfacing welded ASTM A572-based, high-strength low-alloy steel: Welding, characterization, and surface properties related to the wear resistance. *Metals* **2019**, *9*, 244. [CrossRef]
9. Omajene, J.E.; Martikainen, J.; Kah, P.; Pirinen, M. Fundamental difficulties associated with underwater wet welding. *Int. J. Eng. Res. Appl.* **2014**, *4*, 26–31.
10. Łabanowski, J.; Fydrych, D.; Rogalski, G.; Samson, K. Underwater welding of duplex stainless steel. *Solid State Phenom.* **2012**, *183*, 101–106. [CrossRef]
11. Janusas, G.; Jutas, A.; Palevicius, A.; Zizys, D.; Barila, A. Static and vibrational analysis of the GMAW and SMAW joints quality. *J. Vibroeng.* **2012**, *14*, 1220–1226.
12. Urban, J.L. Spot Ignition of Natural Fuels by Hot Metal Particles. Ph.D. Theses, University of California Berkeley, Berkeley, CA, USA, 2017.
13. Mikkelsen, K. An Experimental Investigation of Ignition Propensity of Hot Work Processes in the Nuclear Industry. Master's Thesis, University of Waterloo, Waterloo, ON, Canada, 2014.
14. Song, J.; Wang, S.; Chen, H. Safety distance for preventing hot particle ignition of building insulation materials. *Theor. Appl. Mech. Lett.* **2014**, *4*, 034005. [CrossRef]
15. Urban, J.L. Spot Fire Ignition of Natural Fuels by Hot Aluminum Particles. *Fire Technol.* **2017**, *54*, 797–808. [CrossRef]
16. Liu, Y.; Urban, J.L.; Xu, C.; Fernandez-Pello, C. Temperature and Motion Tracking of Metal Spark Sprays. *Fire Technol.* **2019**, *55*, 2143–2169. [CrossRef]
17. Kim, Y.S.; Eagar, T.W. Analysis of metal transfer in gas metal arc welding. *Weld. J.* **1993**, *72*, 269-s.
18. Tanaka, T. On the inflammability of combustible materials by welding spatter. *Rep. Natl. Res. Inst. Police Sci.* **1977**, *30*, 51–58.
19. Hagimoto, N.; Kinoshita, K. Scattering and Igniting Properties of Sparks Generated in an Arc Welding. In *6th Indo Pacific Congress on Legal Medicine and Forensic Sciences*; Indo Pacific Association of Law, Medeicine and Science: Kobe, Japan, 1998; pp. 863–866.
20. Shigeta, M.; Peansukmanee, S.; Kunawong, N. Qualitative and quantitative analyses of arc characteristics in SMAW. *Weld. World* **2016**, *60*, 355–361. [CrossRef]
21. Pistorius, P.G.; Liu, S. Changes in Metal Transfer Behavior during Shielded Metal Arc Welding. *Weld. J.* **1997**, *76*, 305–315.
22. Narasimhan, P.N.; Mehrotra, S.; Raja, A.R.; Vashista, M.; Yusufzai, M.Z.K. Development of hybrid welding processes incorporating GMAW and SMAW. *Mater. Today Proc.* **2019**, *18*, 2924–2932. [CrossRef]
23. Xu, X.; Liu, S.; Bang, K.S. Comparison of metal transfer behavior in electrodes for shielded metal arc welding. *Int. J. Korean Weld. Soc.* **2004**, *4*, 15–22.
24. Hagiwara, T.; Yamano, K.; Nishida, Y. Ignition risk to combustibles by welding spatter. *J. Jpn. Assn. Fire Sci. Eng.* **1982**, *32*, 8–12.
25. Kinoshita, K.; Hagimoto, Y. Temperature measurement of falling spatters of arc welding. In Proceedings of the 22nd Annual Meeting Japan Society for Safety Engineering, Japan Society for Safety Engineering, Tokyo, Japan; 1989; pp. 145–148.
26. Brandi, B.; Taniguchi, C.; Liu, S. Analysis of metal transfer in shielded metal arc welding. *Suppl. Weld. J.* **1991**, *70*, 261s–270s.
27. Shin, Y.J.; You, W.J. Analysis of the thermal characteristics of welding spatters in SMAW using simplified model in fire technology. *Energies* **2020**, *13*, 2266. [CrossRef]
28. Ranz, W.E.; Marshall, W.R. Evaporation from drop, Part 1. *Chem. Eng. Prog.* **1952**, *48*, 141–146.
29. Marques, E.S.V.; Silva, F.J.G.; Paiva, O.C.; Pereira, A.B. Improving the mechanical strength of ductile cast iron welded joints using different heat treatments. *Materials* **2019**, *226*, 2263. [CrossRef] [PubMed]
30. Incropera, F.P.; Lavine, A.S.; Bergman, T.L.; DeWitt, D.P. *Principles of Heat and Mass Transfer*; Wiley: Hoboken, NJ, USA, 2013.

Publisher's Note: MDPI stays neutral with regard to jurisdictional claims in published maps and institutional affiliations.

MDPI
St. Alban-Anlage 66
4052 Basel
Switzerland
Tel. +41 61 683 77 34
Fax +41 61 302 89 18
www.mdpi.com

Energies Editorial Office
E-mail: energies@mdpi.com
www.mdpi.com/journal/energies

www.ingramcontent.com/pod-product-compliance
Lightning Source LLC
LaVergne TN
LVHW070119170726
843515LV00007B/1716

* 9 7 8 3 0 3 6 5 0 7 5 0 7 *